Coastal Geomorphology

Coastal Geomorphology
An Introduction

by

Eric Bird

Principal Fellow, Department of Geography,
University of Melbourne, Australia

JOHN WILEY & SONS, LTD
Chichester • New York • Weinheim • Brisbane • Singapore • Toronto

Other Wiley Editorial Offices

John Wiley & Sons, Inc., 605 Third Avenue,
New York, NY 10158-0012, USA

WILEY-VCH Verlag GmbH, Pappelallee 3,
D-69469 Weinheim, Germany

Jacaranda Wiley Ltd, 33 Park Road, Milton,
Queensland 4064, Australia

John Wiley & Sons (Asia) Pte Ltd, 2 Clementi Loop #02-01,
Jin Xing Distripark, Singapore 129809

John Wiley & Sons (Canada) Ltd, 22 Worcester Road,
Rexdale, Ontario M9W 1L1, Canada

Library of Congress Cataloging-in-Publication Data

Bird, E.C.F. (Eric Charles Frederick), 1930–
    Coastal geomorphology : an introduction / by Eric Bird.
       p.   cm.
    Includes bibliographical references (p.).
    ISBN 0-471-89976-3 (alk. paper) – ISBN 0-471-89977-1 (alk. paper)
    1. Coasts.   I. Title.

  GB451.2.B55   2000
  551.45′7 – dc21   00-032095

British Library Cataloguing in Publication Data

A catalogue record for this book is available from the British Library

ISBN 0-471-89976-3 (Hardback)   0-471-89977-1 (Paperback)

Typeset in 9/11 pt Times by C.K.M. Typesetting, Salisbury, Wiltshire.
Printed and bound in Great Britain by Bookcraft (Bath) Ltd, Midsomer Norton.
This book is printed on acid-free paper responsibly manufactured from sustainable forestry, in which at least two trees are planted for each one used for paper production.

Contents

Preface

This introduction to the study of coastal geomorphology provides a background for people interested in how coastal features, such as cliffs, beaches, spits and deltas have developed, and how they are changing. It is intended for people coming newly to the subject, for students, and for ecologists, engineers, planners and developers concerned with the coast.

Coastal geomorphology is a broad subject that has developed rapidly, and now generates about 400 publications each year. It has become difficult to produce an introductory textbook, for topics covered in chapters in previous textbooks have subsequently been dealt with at book length, as in the Wiley *Coastal Morphology and Research* series. A comprehensive treatise on coastal geomorphology would now require a massive volume that would certainly be too expensive for students. This book provides a concise introduction that draws attention to unsolved problems and matters on which there are differences of opinion, and gives references to more detailed research work. The coverage is necessarily selective, and somewhat personal, drawing upon my studies of coasts in various parts of the world over the past four decades.

The book discusses the shaping of coastal landforms and examines the changes that are taking place in response to coastal processes. It demonstrates the dynamic nature of coastal landforms and provides a background for analytical planning and management decisions in coastal areas subject to continuing change. One of the problems in producing an introductory textbook on coastal geomorphology is the need to be selective in quoting examples of coastal features and process relationships, bearing in mind that most readers come from Britain, Europe, North America or Australasia, and are likely to be more interested in local and accessible examples. Reference can be made to *The World's Coastline* (edited by Bird and Schwartz, Van Nostrand Reinhold, New York, 1985) for examples from various other coasts. Place names in England are identified by county, in the United States and Australia by state, and elsewhere by country.

The book begins with an introduction to concepts and terminology, and the factors that have affected coastal evolution and coastline changes (Chapter 1). This is followed by a discussion of waves, tides, currents and other nearshore processes (Chapter 2), and a study of the effects of land and sea level changes, notably the Late Quaternary marine transgression, which has played a major part in shaping modern coastlines (Chapter 3). Cliffs and rocky shores are discussed in Chapter 4, and beaches in Chapter 5, the greater length of these chapters reflecting the dominance of these topics in recent coastal studies.

Spits, barriers and related features are considered in Chapter 6, and the formation of coastal dunes in Chapter 7. The intertidal zone, including salt marshes, mangroves and seagrass terraces, is examined in Chapter 8, and the features of river estuaries and coastal lagoons in Chapter 9. Chapter 10 deals with deltas produced by deposition at river-mouths, and Chapter 11 with the various kinds of reefs built by corals, algae and other organisms on the shore and in coastal waters. Changes likely to occur on coastlines in response to the predicted world-wide rise in sea level, resulting from global warming by the enhanced greenhouse effect, are discussed in Chapter 12. It was difficult to decide where to place a discussion of the various attempts that have been made to classify coastal landforms, and so this has been included as an appendix. A list of references provides a guide to more detailed information, including many pre-1990 publications that remain relevant.

Acknowledgements
I am grateful to Juliet Bird for assistance in processing the text, and to Chandra Jayasuriya and Blaise Vinot for help with maps and diagrams. I would also like to thank the staff of the Earth Sciences Library in the University of Melbourne for assistance in the course of exploring the coastal literature.

Eric Bird
Black Rock, March 1999

List of Figures

List of Tables

1

Introduction

COASTAL GEOMORPHOLOGY

Coastal geomorphology deals with the evolution of coastal landforms (such as cliffs, rocky shores, beaches, dunes, estuaries, lagoons and deltas), the processes at work on them and the changes taking place. Coastal geology is concerned with the rock formations, structures and sediments that are found in coastal regions, and provides a background for coastal geomorphology.

Apart from incidental comments by classical Greek and Roman observers, and by Leonardo da Vinci, the first systematic attempts to explain coastal landforms were by nineteenth century scientists such as Charles Lyell and Charles Darwin, and the pioneer American geomorphologist William Morris Davis. While a great deal of work was done in the twentieth century on various parts of the world's coastline, particularly in Europe and North America, it is only in the past few decades that coastal research has become widespread, and there is still plenty of opportunity for original contributions.

TERMINOLOGY

The coast consists of a number of zones (Figure 1.1). The shore is the zone between the water's edge at low

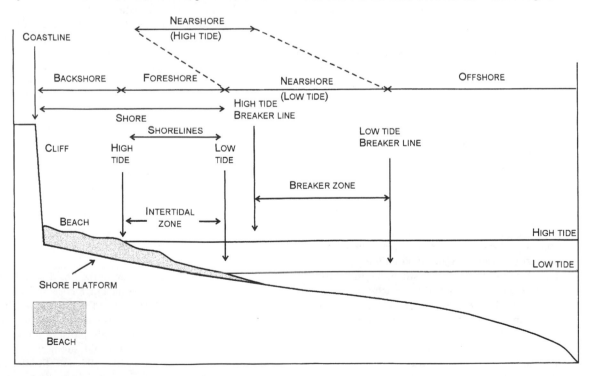

Figure 1.1 Coastal terminology.

tide and the upper limit of effective wave action, usually extending to the cliff base. It includes the foreshore, exposed at low tide and submerged at high tide, and the backshore, extending above normal high tide level, but inundated by exceptionally high tides or by large waves during storms. The shoreline is strictly the water's edge, migrating to and fro as the tide rises and falls.

The nearshore zone, comprising the surf zone (with broken waves) and the swash zone (covered as each wave runs up the foreshore), also migrates to and fro as the tides rise and fall. The breaker zone (where waves are disrupted) is bordered seaward by the offshore zone, extending to an arbitrary limit in deep water. The terms offshore, onshore and longshore are also used to describe directions of flow of wind, water, or sediment.

A beach is an accumulation of loose sediment, such as sand, gravel or boulders, sometimes confined to the backshore but often extending across the foreshore as well. Shingle is beach gravel, especially where the stones are well-rounded.

The coast is a zone of varying width, including the shore and extending to the landward limit of penetration of marine influence: the crest of a cliff, the head of a tidal estuary, or the solid ground that lies behind coastal dunes, lagoons, and swamps. The term coastline indicates the land margin at normal high tide (behind the backshore zone), and may be a cliff or the seaward margin of dunes or dry land. In American literature the term shoreline (or seaboard) is often used as a synonym for coastline, while the coast is elaborated to the coastal zone. The preference here is to acknowledge that the shoreline moves to and fro as tides rise and fall, so that there is a low-tide shoreline, a mid-tide shoreline, and a high-tide shoreline.

ANCIENT COASTLINES

Coastlines have existed since oceans first formed on the earth's surface, about 4000 million years ago, but it is difficult to find early coastlines because most of the evidence has been removed by erosion or concealed by deposition. Deposits associated with coastlines that existed in Mesozoic and Tertiary times can be found in the stratigraphy of southern Britain. On the Haldon Hills, east of Dartmoor in Devon, there are pebbly sands with corals and mollusc shells which represent a Lower Cretaceous beach, deposited about 110 million years ago. Other fragments of ancient coastlines have been preserved far inland. In the quarry on Kank Hill, near Kutna Hora, about 70 kilometres east of Prague in the Czech Republic, it is possible to stand on the Upper Cretaceous shore, where the beach that marks the limits of the Cenomanian sea, formed about 95 million years ago, rests on an irregular wave-worn surface of Pre-Cambrian gneiss (Ager 1980).

Evidence of former coastlines becomes clearer in the most recent of the geological periods, the Quaternary, which comprises the Pleistocene (which began about 2.3 million years ago) and the succeeding Holocene (the last 10 000 years). The Quaternary period was one of major global climate and sea level fluctuations, and Quaternary coastlines can be found above and below present sea level (Chapter 3). There are Late Pleistocene beaches and shore platforms standing above the present sea level on many coasts, notably in south-west England and around Scotland (p. 33), while submerged Pleistocene coastlines, including cliffs, shore platforms and beaches, have been detected on the sea floor off the coasts of California and Japan (p. 37). Coastal plains built forward by deposition, as in the south-eastern United States, may include stranded remnants of coastline features of Pleistocene and Holocene ages, containing evidence of past conditions that have generally been lost on receding cliffed coasts.

During cold climate phases of the Quaternary, when glaciers and ice sheets became extensive, global sea level was much lower than it is now, and when the climate of the Ice Age gave place to milder conditions there was a major world-wide sea level rise. Existing coastal landforms have been largely shaped within the past 6000 years, when the sea has stood at or close to its present level, with global climate much as it is now. Some coasts have older (relict) features, inherited from earlier environments when the sea stood higher or lower, or when the climate was warmer or colder, wetter or drier, or stormier or calmer than it is now.

COASTLINE MORPHOLOGY

Maps and charts show that few of the world's coastlines are straight. Exceptions are the north coast of Madura in Indonesia, probably related to a major fault line, and the 800 kilometre east coast of Madagascar, also fault-guided but fringed by sandy barriers shaped by Indian Ocean swell (p. 8). Coastal submergence has produced embayed coasts with valley-mouth inlets, as on the Atlantic coasts of the United States and western Britain. Many beach-fringed coasts have gently curved concave outlines (e.g. the Texas coast and Encounter Bay in South Australia), as have some cliffed coasts cut in soft rock formations (e.g. Lyme Bay in Dorset and the Nullarbor coast in Australia). Other coasts are more intricate, with headlands and embayments (e.g. South China), branching inlets and ramifying peninsulas (e.g. Sulawesi in Indonesia and the Kimberley coast in northern Australia), numerous islands (e.g. Burma and south-west Finland), or protruding deltas.

The mathematician Mandelbrot (1967) saw coastlines as analogues of fractal curves, which maintain the same general pattern regardless of how much they are magnified. Similar coastline features occur on a variety of scales. Beach cusps, for example, retain their shape as their dimensions increase or decrease in relation to incident wave heights (p. 125), but a particular beach cusp is not subdivided into smaller, nested beach cusps, and the beaches on which they occur are not as a rule cuspate on a larger scale. It is true that coastal promontories and embayments occur on various scales, from continental down to a particular headland and cove, but their pattern is not maintained hierarchically as the scale changes.

The total length of the world's coastline is probably close to a million kilometres, including the coasts of the very many small islands. Measurements of coastline length are necessary for statements about the proportions of various types of coastline around the world (pp. 47, 297) or the lateral extent of erosion and accretion on beaches (p. 147). Such measurements can be made by counting straight intercepts of a selected length (e.g. one kilometre) on maps of uniform scale (e.g. 1:250 000), or by using computers to integrate the grid squares within which coastline segments occur, taking each grid square as representing a specific coastline length.

Variations in geomorphology around the world's coastline were illustrated by Bird and Schwartz (1985).

coastal rock outcrops vary from tropical to arctic and from humid to arid environments. Climate also conditions coastal vegetation and fauna, which have produced features ranging from salt marshes and mangrove swamps to shelly beaches, coral reefs and stabilized dunes, and also the organisms that attack rock surfaces (the processes of bioerosion, p. 80). Coastal processes include the effects of rising and falling tides and associated tidal currents, and are influenced by oceanographic factors such as sea temperature and salinity, determined by climate and the patterns of ocean currents. The various processes are discussed in Chapter 2. Mention has been made of ancient coastlines, produced by past changes in the relative levels of land and sea, and these changes have continued to influence the evolution of existing coasts. Within historical times coastal evolution has also been modified by the effects of various human activities on the coast and in the hinterland.

Evolution of coastal landforms can be considered in terms of morphogenic (morphodynamic) systems, within which various factors influence the processes acting upon the coast (Cowell and Thom 1994). There is an input of energy (e.g. wind, tide, living organisms) and materials (e.g. water, rock, sediment) which interact to generate the coastal landforms, and there is feedback in the sense that the developing morphology modifies geomorphological processes, and thus becomes a factor influencing subsequent changes. These can be studied in terms of response to various coastal processes operating over specified periods: that is, as process–response systems. Attempts have been made to quantify the various inputs and to describe and analyse the interactions mathematically, but the ideal of a complete quantitative understanding of a coastal system is more easily advocated than achieved. It is realistic to formulate and attempt to solve specific problems, and establish empirical relationships between process and change that can be put to practical use in coastal management.

COASTAL EVOLUTION

The shaping of coastal landforms has been influenced by a range of morphogenic factors. These include geology, which determines the pattern of rock outcrops on the coast, on the sea floor and in the hinterland, and the movements of the earth's crust which result in uplift, tilting, folding, faulting and subsidence of coastal rock formations. Climatic factors have influenced the wind and wave regimes that shape coastal features, and the weathering processes that decompose and disintegrate

CHANGING COASTLINES

While some coastlines have changed little over the past 6000 years, most have advanced or retreated, and some have shown alternations of advance and retreat. A coastline advances where the deposition of sediment exceeds the rate of erosion, or where there is emergence due to uplift of the land or a fall in sea level; it retreats as the result of erosion exceeding deposition, or where

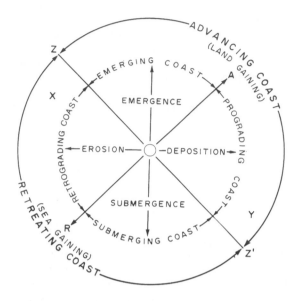

Figure 1.2 Analysis of coastline changes in terms of emergence and submergence, progradation and retrogradation, as proposed by Valentin (1952).

there is submergence due to land subsidence or a sea level rise (Figure 1.2).

Coastlines have changed at varying rates in response to coastal processes, with sudden changes during storms, earthquakes and volcanic eruptions and more gradual changes over quieter intervening periods. Coastline changes can be measured over various time scales, ranging from the past few thousand years down through recent centuries or decades to the annual or seasonal fluctuations and short-term changes related to the various tidal cycles or caused by particular weather events. Some changes are cyclic over varying periods; others continue as erosion or deposition proceeds. Temporal scales can be documented on both advancing and receding coastlines (Stapor et al. 1991, Cambers 1976).

Measurement of coastline changes can be made by comparing historical maps and charts, providing these were based on accurate surveys, with the configuration shown on modern maps, air photographs or satellite imagery (Carr 1962). Maps and charts of sufficient accuracy have been available for parts of western Europe and North America for the past two centuries, but for much of the world's coastline there is little information preceding the era of air photography in the past few decades. Information on global coastline changes over the past century was summarized by Bird (1985a).

On long-settled coasts changes have been determined from historical and archaeological evidence, as around the Mediterranean Sea, where it is locally possible to detect the advance or retreat of parts of the coastline over at least two thousand years (Kraft et al. 1988). Changes since present sea level was established (within the past 6000 years) may be determined from evidence of the preceding land surface intersecting the sea floor (p. 73) or from stratigraphical and sedimentological analyses of coastal depositional formations, using radiometric and other forms of dating as well as palaeontological and archaeological evidence (Carter and Woodroffe 1994).

Traditional methods of observing, mapping, and measuring changes on the coast and the processes that cause them have recently been supplemented by new techniques, including various electronic measuring instruments (Dugdale 1981) and the application of modelling (Fox 1985). Computers are used to process and extend field survey data, generating serial models of beach or coastal dune topography from which the pattern of gains and losses can be mapped and quantified (p. 131). Air photographs have been used for some time as an aid to the mapping and measurement of coastal changes, and colour photography has extended these studies to the nearshore sea floor. Satellite imagery has been used to trace coastline changes over the past three decades. Short-term changes, which range from a few minutes to a few hours (as on beaches or dunes during a storm), require monitoring by repeated field surveys, the use of micro-erosion meters, serial photo-recording or photogrammetry.

Some coastline changes have resulted from human activities, such as reclamation (also known as land claim), the making of new ground by enclosing or filling nearshore areas, which in places has advanced the coastline several kilometres (French 1997). The Netherlands have a long history of winning land from the sea by building dykes (sea walls) to enclose areas that were previously beneath the sea (at least at high tide) and draining these to form polder lands, thereby advancing the coastline seaward. New land has also been created on densely-populated coasts in southeast Asia, as in Tokyo Bay and Hong Kong, and Singapore has increased its land area by 10 per cent in recent decades by landfill.

Coastlines have also been modified by the introduction of structures such as groynes and breakwaters, intended to stabilize features that were changing in ways considered unacceptable, notably where erosion threatened seaside towns, ports, or other developed coastal areas. The dredging of harbour entrances and the dumping of material on the coast and offshore have also modified coastal topography. In consequence, many coastlines have become largely or entirely artificial, and the extent of these is increasing rapidly. Appropriate coastal management may succeed in maintaining or enhancing the coastal environment, but there have been mistakes that could have been avoided if those concerned had understood the principles of coastal geomorphology.

2

Coastal Processes

INTRODUCTION

Processes at work in coastal waters include winds, waves, tides and currents, which together provide the energy which shapes and modifies a coastline by eroding, transporting and depositing sediment. Although waves, tides and currents interact, one process augmenting or diminishing the effects of another, it is convenient to discuss them separately.

WAVES

Waves are undulations on a water surface produced by wind action. The turbulent flow of the wind blowing over water produces stress and pressure variations on the surface, initiating waves which grow as the result of the pressure contrast between their driven (upwind) and advancing (downwind) slopes. Waves consist of orbital movements of water that diminish rapidly from the surface downwards, until the motion is very slight where the water depth (d) equals half the wave length (L) (Figure 2.1). The depth at which waves become imperceptible is termed the wave base, and in theory erosion by waves could ultimately reduce the world's land areas to a planed-off surface at this level. Orbital motion in waves is not quite complete, so that water particles move forward as each wave passes, producing a slight drift of water in the direction of wave advance.

Wave height (H) is the vertical distance between successive crests and troughs, wave steepness the ratio between height and the length (H/L), and wave velocity (C) the rate of movement of a wave crest. Wave height is proportional to wind velocity, and wave period (T, the time interval between the passage of successive wave crests) to the square root of wind velocity. Wave dimensions are also determined partly by fetch

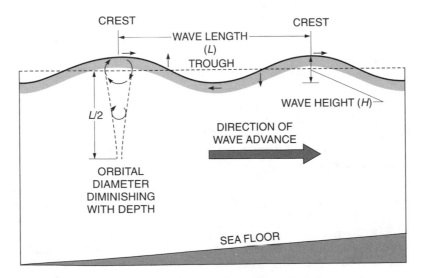

Figure 2.1 Wave terminology, and the pattern of currents as wave crests and troughs move shoreward.

(the extent of open water across which the wind is blowing) and by the duration and strength of the wind. Large waves are generated by severe storms, and in mid-ocean the largest storm waves, generated by prolonged strong winds over distances of at least 500 kilometres, can be more than 20 metres high, travelling at more than 80 kilometres per hour. Waves transmitted across the oceans from storm centres become long and regular, and are known as ocean swell. In coastal waters waves are diminished by friction with the shallowing sea floor, but locally generated storm waves can still be several metres high when they break on the shore. On the Atlantic coast of the United States, for example, occasional hurricanes generate waves up to five metres high when they break. Such storm waves can cause erosion or deposition well above the level of the highest tides.

Simple equations indicate the relationships between wave parameters. In deep water, wave velocity (C_0) is the ratio (L_0/T) of wave length (measured in metres) to wave period (measured in seconds). Wave length (L_0) in deep water (where $d > L/2$) can be used to calculate wave velocity (C_0) from the following formula, in which g is the gravitational acceleration (about 980.62 cm/sec^2 at latitude 45°):

$$C_0^2 = \frac{gL_0}{2\pi}$$

from which, since $L_0 = C_0 T$,

$$C_0 = \frac{gT}{2\pi} \quad \text{or } 1.56T \text{ in metres/second}$$

so that $L_0 = 1.56T^2$, providing a means of calculating wave length from measurements of wave period in deep water.

Measurements of nearshore waves can be made using a staff with graduated electric wires, a pressure transducer on the sea floor, or sonic devices mounted on a pier or platform. The problems of monitoring waves were discussed by Morang et al. (1997), and more detailed accounts of the nearshore wave field are given by Hardisty (1994) in relation to beaches and Sunamura (1992) on rocky shores.

Ocean swell

During storms strong winds generate irregular patterns of waves, varying in height, length and direction, which radiate from the generating area. The longest waves move most rapidly, and are most durable, so that as waves move across the ocean they become sorted into swell of more regular (gradually diminishing) height and (gradually increasing) length, which eventually arrives to break on a distant shore. There are major wave-generating storm regions in the Southern Ocean and in the northern parts of the Atlantic and Pacific Oceans.

Ocean swell consists typically of long, low waves with periods of 10 to 16 seconds. As they move towards the coast the wave crests gain in height and steepness, and as they enter shallow water they break to produce the surf observed on the shores of the Pacific, Atlantic and Indian Oceans. Storms in the Southern Ocean initiate the south-westerly swell that travels thousands of kilometres to arrive on the western and southern shores of Australia, New Zealand, the Americas and Africa. The long swell that breaks on the coast of California has travelled about 12 000 kilometres across the Pacific Ocean.

South-westerly swell generated by gales south of Africa is transmitted across the Indian Ocean to the southern coasts of India and Sri Lanka and the western coast of Thailand. It reaches the southern coasts of Sumatra and Java, and other Indonesian islands as far east as Timor. South-westerly swell originating south of Australia and New Zealand moves across the Pacific Ocean to coasts between Chile, California and Alaska, and the stormy waters south of South America produce south-westerly swell across the Atlantic Ocean to West Africa and western Europe (Portugal to the Hebrides). Occasionally a south-westerly ocean swell with wave periods up to 20 seconds arrives on the south coast of Britain, breaking heavily on the Loe Bar in Cornwall and Chesil Beach in Dorset, and this was probably generated in the vicinity of the Falkland Islands. Similar swell has been recorded on the Cornish coast about four days after hurricane disturbances off Florida.

As south-westerly swell moves across the oceans it fans out to produce a weaker southerly and south-easterly swell. Southerly swell occasionally reaches Iceland and the Aleutian Islands, and south-easterly swell arrives on the coasts of South America (Argentina to Recife in Brazil), south-east Africa and southern Arabia, south-east India, south-east Australia (eastern Tasmania north to Fraser Island) and the east coast of New Zealand. South-easterly swell is often augmented by the effects of south-east monsoon and trade winds in coastal waters.

Storms in northern latitudes generate similar ocean swell, especially in winter, when a north-westerly swell from the north Pacific arrives on shores between British Columbia, California and Central America. In the north Atlantic a north-westerly swell extends to the coasts of western Europe (Ireland to Portugal) and West Africa (Morocco to Senegal). It is frequently masked in high latitudes by locally-generated storm waves. The north-westerly swell is stronger in the northern winter, but fades in the summer months, whereas the south-westerly swell from the Southern Ocean is stronger in the northern summer (southern winter) and weaker in the northern winter. This leads to seasonal alternations, the winter north-westerly swell alternating with the summer south-westerly swell on the coasts of Portugal and California. These seasonal contrasts are well known in Half Moon Bay, California.

The north-westerly swell in the north Pacific diverges to form a weaker north-easterly (or northerly) swell on the north and east coasts of Japan, extending to China, Vietnam, the north and east coasts of the Philippines and northern New Guinea. In the north Atlantic there is similar modification of the north-westerly swell to produce a north-easterly swell from Cape Hatteras south to the eastern islands of the Caribbean and the north-east coast of Brazil. Again, the storm-generated waves radiating across these oceans may be augmented by waves produced by north-easterly winds by the time they reach these coasts.

Storm waves

Apart from ocean swell, nearshore wave regimes depend on climatic conditions in coastal waters. Storm waves, generated by strong wind action, arrive frequently on west-facing coasts in latitudes subject to frequent westerly gales, as on the Atlantic coasts of north-west Europe and the Pacific coasts of Canada and the north-west United States, and there are stormy coasts in Patagonia (southernmost Chile) and on the western seaboard of South Island, New Zealand. Although storms in the Southern Ocean generate the large waves that spread out across the Atlantic, Indian and Pacific Oceans, only small waves reach the coasts of Antarctica, even on parts that are ice free in summer.

Monsoon winds generate waves on the coasts of India and south-east Asia. In peninsular Malaysia the south-west monsoon (May to September) produces waves along the west coast, and the north-east (winter) monsoon generates strong wave action on the east coast, extending to south-east Thailand and the coast of Vietnam.

Large waves generated by occasional tropical cyclones (also known as hurricanes or typhoons) are accompanied by storm surges (p. 20) on the south-east coast of the United States, in the Caribbean, Madagascar and Mozambique, India and the Bay of Bengal, from Thailand to Vietnam, southern China and southern Japan, and in northern Australia. By contrast, the coastal waters of equatorial regions (such as north-eastern Brazil and Indonesia) are relatively calm, except where they receive ocean swell of distant origin, as in the Gulf of Guinea, southern Indonesia and the Pacific coast of Central America. Ocean swell dominates open coasts outside the stormy zones, although there may occasionally be strong locally generated wave action.

Nearshore waves

Ocean swell generally arrives as regular waves, breaking at intervals of 12 to 16 seconds on the shore. When ocean swell arrives from different sources there are variations in wave height as interacting sequences of waves break upon the shore. The idea that every seventh wave is larger is legendary, but occasional higher waves occur as the result of the merging of two or more sets of waves. Sometimes there is a steady increase in the height of successive waves to a combined phase maximum, followed by a diminution as the waves move out of phase. Known as surf beat, this interaction can produce maximum waves breaking at intervals of several minutes, accompanied by pulsations of current flow alongshore and onshore–offshore. Wave set-up is the raising of sea level close to the shore as the result of waves driving water in. It is roughly proportional to incident wave height, but also depends on shore gradient and beach texture, shingle absorbing more wave energy than sand.

Waves generated locally by winds (particularly onshore winds) blowing over coastal waters are typically shorter (wave period < 10 seconds) and less regular than ocean swell of distant derivation, and in stormy periods they are much steeper. They may be superimposed on ocean swell arriving in coastal waters, an onshore wind accentuating the swell and adding shorter waves to it, a cross wind producing shorter waves which move at an angle through the swell, and offshore winds flattening swell to produce relatively calm conditions in the nearshore zone.

Ocean swell is usually strong enough to produce waves that shape the coastline, and waves generated in coastal waters by onshore winds exceeding about 20 kilometres per hour (Beaufort Scale 4 and over) also break on the shore with sufficient energy to erode coastal rock formations and transport sediment along-shore (p. 23).

Geographical variation in nearshore wave conditions is related to fetch and coastal configuration. Coastlines protected by promontories, reefs or offshore islands receive swell (if at all) in a much modified and weakened form, so that locally wind-generated waves predominate. This is the case around landlocked seas, such as the Mediterranean and the Baltic, the Arabian Gulf and the Gulf of California, and in embayments with constricted entrances such as Port Phillip Bay in Australia. Around the British Isles wave regimes are determined largely by winds in coastal waters, ocean swell reaching the Atlantic coasts and penetrating the English Channel, the Irish Sea and the northern North Sea only to a limited extent.

On many coasts a particular wave direction is clearly dominant. The south coast of England, for example, is dominated by waves produced by the prevailing south-westerly winds, and much of the east coast of Australia has a prevalence of south-easterly wave action. Other coasts show greater variation: the north Norfolk coast in England has waves arriving from the north-west, north and north-east according to local wind conditions, which change with the passage of depressions and anticyclones. In some years the north-easterly waves are dominant, in others north-westerly waves prevail. The east coast of Port Phillip Bay, Australia, has seasonal variations in dominant waves, with westerly and south-westerly waves prevalent in summer and north-westerly waves commoner in the winter months. Correlation with meteorological patterns can be used to determine long-term dominant wave incidence (and resulting sediment movement) in such conditions.

Wave refraction

Ocean swell has parallel wave crests in deep water, but as the waves move into shallowing water they begin to be modified by the sea floor: the free orbital motion of water is impeded and the frictional effects of the sea floor retard the advancing waves. Sea floor topography thus influences the pattern of swell moving towards the coast, the angle between the swell and the submarine contours diminishing so that the wave crests become realigned until eventually they are parallel to the submarine contours.

This is known as wave refraction (Figure 2.2). Where the angle between the ocean swell and the submarine contours is initially large the adjustment is often incom-

plete by the time the waves arrive at the shore, so that they break at an angle (usually $< 10°$). Where the angular difference is initially small the waves are refracted in such a way that they anticipate, then fit the outline of a beach or of cliffs cut in soft formations, breaking synchronously along the coastline (Figure 2.2A). Sharp irregularities of the sea floor have stronger effects, a submerged bank retarding the waves while a submarine trough allows them to run on (Figure 2.2B). Islands or reefs awash at low tide interrupt the waves and produce converging patterns of wave refraction (Figure 2.2C), while waves that have passed through narrow straits or entrances are modified by diffraction, spreading out in the wider water beyond. Waves are also retarded as they impinge on headlands, but run on into the intervening bays. In broad embayments they develop gently curved patterns, the waves in the middle of the bay moving on through deeper water while towards the sides, in shallower water, they are held back. In narrow embayments the refracted wave crests are more sharply curved.

These curved patterns of wave crests in coastal waters can be observed from headlands or seen on air photographs (Figure 2.3). Given knowledge of the direction of approach of waves in deep water, their length or period, and the detailed configuration (bathymetry) of the sea floor from nautical charts, it is possible to construct diagrams confirming the patterns of refraction that develop as waves enter nearshore waters and approach the coast. Such diagrams are now generated with the aid of a computer program (Brampton 1977). Wave refraction diagrams can also be used to predict the direction of longshore currents produced when waves arrive at an angle to the shore.

The shaping of curved coastlines by refracted swell is obvious on ocean coasts, but wave patterns generated by local winds on landlocked seas, lakes or coastal lagoons can also produce curved coastlines. On the shores of the Mediterranean Sea, for example the curved outlines of Languedoc, the Venetian–Trieste coastline, and the Bay of Benghazi have been shaped by waves generated locally, and the same is true of the coasts of the Caspian Sea, the Sea of Azov, the Red Sea, the Arabian Gulf and much of the Gulf of Mexico. In Australia there are smoothly-curved shores in land-locked embayments such as Port Phillip Bay (Black and Rosenberg 1992), or coastal lagoons such as Lake Wellington (Gippsland Lakes), determined by the resultant directions of waves that come from various directions, and orthogonal to the maximum fetch. There may also be secondary smoothing by

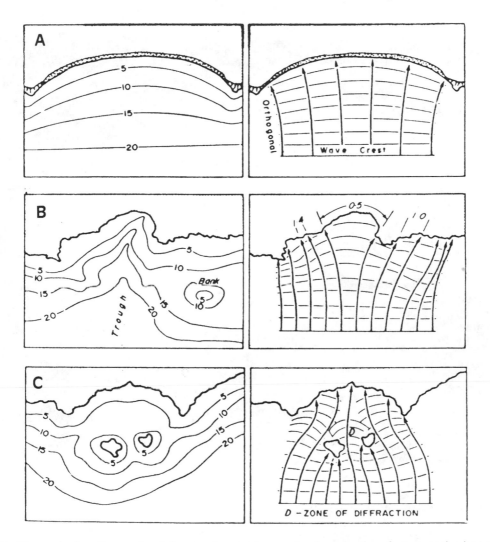

Figure 2.2   Wave refraction diagrams in relation to submarine topography. A – refraction of waves moving into a bay, B – refraction of waves over a trough (refraction coefficients indicated), C – refraction round and diffraction between offshore islands.

longshore currents generated by waves coming in on either side of the maximum fetch.

Wave energy

When orthogonal lines (rays) are drawn at equal intervals on wave refraction diagrams, at right angles to the alignment of waves in deep water, and projected shoreward they converge where they pass over a submerged bank or reef and diverge into where deeper water is traversed; in general, they converge towards headlands and diverge in embayments (Figure 2.2B). The spacing

of these orthogonals shows how wave energy is distributed when the waves reach the coastline, a shoreward convergence indicates a concentration of wave energy, whereas a divergence indicates weakening. This can be expressed as a refraction coefficient (R), equivalent to the square root of the ratio between the distance between neighbouring orthogonals in deep water (S_0) and their spacing on arrival at the coastline (S). Calculated from measurements on a wave refraction diagram, this coefficient is an indication of the relative wave energy arriving on each part of the coastline (p. 16). Such features as stream outlets or lagoon entrances are often found on coastal sectors (p. 236)

Figure 2.3   The curved outline of the beach in Seven Mile Bay, Tasmania, has been shaped by gently refracted ocean swell, the wave crests breaking to form surf which sweeps sand on to the shore. This is a swash-dominated beach.

of diminished wave energy (Bascom 1954), while erosion is intensified where convergent orthogonal lines indicate a focusing of wave energy. However, there are many exceptions to the generalization that erosion is concentrated on headlands and deposition in bays: on the Guyana coast, for example, there is accretion on convex parts of the coastline (Lakhan and Pepper 1997).

On coasts sheltered from strong wave action, particularly where the nearshore zone is broad and shallow, wave energy is attenuated and breaking waves diminished. Wave energy may be dissipated to such an extent that waves do not reach the shore. The grassy marsh coastline of western Florida receives very little wave action (except during hurricanes), and in the broad, shallow Gulf of Bo Hai, in northern China, there is usually almost no wave action along the low-lying shores of the Hwang-ho delta (Zenkovich 1967).

Breaking waves

As waves move into shallow water ($d < L/2$) they are modified in several ways. Their velocity C_s diminishes according to the formula:

$$C_s^2 = \frac{gL}{2\pi} \tan h \frac{2\pi d}{L}$$

The shallowing water also diminishes their length and period, and as they approach the shore their height increases, they steepen, the crests becoming narrower and sharper, and the intervening troughs wider and flatter. Orbital movements within each wave become increasingly elliptical, and shoreward velocity in the wave crest (used by surfboard riders) increases until it exceeds the wave velocity. When the orbital motion can no longer be completed, the oversteepened wave front collapses, forming a breaking wave (breaker). This sends forth a rush of water (swash) through the surf zone and up on to the shore, followed by a withdrawal, known as the backwash, by which water returns to the sea. These alternations can be documented photographically against a graduated scale, or measured with a dynamometer. In the nearshore zone the waves generate alternations of shoreward and seaward movement, the shoreward motion being generally stronger, except when breakers plunge and withdraw water and sediment seaward (p. 122).

The pattern of breaking waves varies with wave height and nearshore gradient, so that low waves break closer to the shore than high waves, and waves of a specific height break closer inshore where the gradient is steep than where it is gentle. There are complications where the nearshore topography is rugged, with reefs that disrupt wave action, or where bars have been formed by wave action in the nearshore zone (p. 199). Applications of this principle have been used to modify incident wave forms by re-shaping nearshore morphology (Bird 1996a). Underwater structures were built to diminish wave energy on an eroding coast near Niigata in Japan, and a 120 metre long boomerang-shaped artificial reef has been constructed off Cottesloe, Western Australia, with the aim of creating waves suitable for surfing.

The ways in which waves shape and modify the shore depend on their incidence and dimensions as they break, and the resulting patterns of water flow. Breaking waves are influenced by local winds (especially winds blowing offshore, which steepen wave fronts), by changes in nearshore water depth accompanying the rise and fall of tides or other short-term sea level changes, by currents (including those resulting from wave motion), by the gradient and topography of the sea floor, and by the general configuration of the coastline. Waves that arrive and break parallel to the coastline move water and sediment to and fro (onshore/offshore drift) while those that arrive and break at an angle to the coastline move water and sediment along the shore (longshore drift) (p. 116). Waves that break against hard rocky coasts or solid sea walls are reflected, and tend to move sediment seaward.

The energy of a breaking wave (E) depends largely on breaker height (H_b) and the density of the water (p), together with gravitational acceleration (g) according to the formula:

$$E = \frac{1}{8} pgH^2$$

However, there is considerable variation in the dimensions of waves reaching the shore, and the difficulty of estimating typical wave height can be overcome by using the concept of the significant wave ($H_{1/3}$), defined as the average height of the highest one-third of all waves observed over a period of 20 minutes, or the highest 33 in a train of 99 waves. The significant wave can be used to compare conditions in coastal waters at different times, or on different coastlines. On ocean coasts and those facing stormy seas significant wave height commonly exceeds three metres. Dolan and Davis (1992) devised a storm intensity scale (duration × maximum $H_{1/3}$) for the Atlantic coast of the United States with five categories up to major hurricanes.

A breaker coefficient (B) was defined by Galvin (1972) as

$$B = \frac{H}{GsT^2}$$

Where H is wave height at break point, g is gravitational acceleration and T is wave period. Waves breaking on beaches show a range of forms, within which Galvin recognized four types:

1. surging breakers, which are low and gentle waves until they sweep up a relatively steep beach,
2. plunging breakers, with fronts that curve over and crash (producing little swash but a strong backwash),
3. collapsing breakers that subside as they move towards the shore,
4. spilling breakers, which are short and high, and produce foaming surf as the swash runs up a beach of gentle gradient.

Although four types of breaking waves have been distinguished, the important point is whether they are constructive (washing sediment up on to the beach) or destructive (causing beach erosion), and this depends on whether the swash and backwash generated by breakers (whether parallel or at an angle to the shore) achieve net shoreward or seaward movement of beach material (p. 122). In general surging, spilling and collapsing breakers have strong swash (onshore flow) followed by a gentle backwash, producing net shoreward movement of sediment, whereas plunging breakers have a short swash and a relatively stronger backwash, so that they withdraw sediment from a beach. This distinction takes us back to the work of Johnson (1956), who suggested that waves which have steepness ($H_0 : L_0$) outside the breaker zone of less than 0.025 produce breakers which are constructive (moving sediment on to beaches), while those with higher ratios (especially > 0.25) become destructive (withdrawing sediment from beaches) (p. 122). However, wave profiles also vary in relation to the nearshore gradient (tan β), so that relatively gentle waves can be destructive where the nearshore slope is steep, and relatively steep waves may become constructive after traversing a broad nearshore slope of low gradient. Further discussion of interactions between wave characteristics and beach morphology is given in Chapter 5.

Water temperature variations have little effect on wave action in the tropics and temperate zones, but on cold Arctic and Antarctic coasts waves become ineffective when the nearshore water freezes. Nearshore ice fringes protect the coastline in winter, but can become erosive when waves disrupt them in the summer thaw. The length of the cold season determines the period of frozen nearshore water, and when the ice thaws and disintegrates waves drive it up on to a beach, shore platform or salt marsh. On the shores of the St Lawrence Gulf in Canada and around the Baltic Sea wave-piled ice fragments may accumulate on the shore as irregular aprons or ridges up to five metres high.

Rip currents

Backwash is the seaward flow of water that has been carried shoreward by breaking waves, but the augmentation (wave set-up) of nearshore sea level by shoreward movement of water, due to wave motion, must also return seaward, typically as localized rip currents (Figure 2.4). Within these, water flows back through the breaker line in sectors up to 30 metres wide, attaining velocities of up to 8 kilometres per hour before dispersing seaward. The shoreward movement of breaking waves and the seaward return currents form the main components of the nearshore water circulation.

Rip currents occur in distinct (though variable) patterns on many shores (particularly along beaches), related to variations in wave set-up. A light or moderate swell produces numerous small rip currents, and a heavy swell produces a few more widely spaced and concentrated rips, fed by stronger lateral currents in the surf zone. Rip currents may cut channels seaward through the nearshore zone (across any nearshore bars), and deposit fans of sediment at their seaward limits; in some places these channels have grown headward to form re-entrants in the beach. When waves arrive at an angle to the beach the rip currents head away diagonally through the surf instead of straight out

to sea, and cut oblique channels through the nearshore zone. Rising tides and onshore winds raise the water level along the shore, and thus intensify rip currents.

Wave currents

Waves that break parallel to the shore produce orthogonal swash and backwash on beaches, and mainly onshore–offshore movements of water and sediment, but there are usually lateral variations in breaker height, particularly where wave refraction has generated contrasts indicated by varying orthogonal spacing (p. 11). Any such lateral variations in breaker energy are resolved by divergent longshore current flow in the nearshore zone. As a rule the natural variability of wave dimensions results in energy changes that result in pulsations in longshore and onshore–offshore current flow.

It has been suggested that waves breaking parallel to the beach generate orthogonal edge waves, standing oscillations that develop at right angles to the coastline as the result of resonance between waves approaching the shore and waves reflected from it (Guza and Inman 1975, Hardisty 1994). Such edge waves are thought to produce longshore variations in breaker height as their crests augment incoming breakers. Augmented breakers form swash salients, sectors of higher water level from which there is longshore flow into intervening troughs, where outflowing rip currents develop. If edge waves are generated with the same (or doubled) period of incident waves the two motions could combine to produce a regularity in nearshore water circulation which may explain the rhythmic nature of beach cusps (p. 125) and certain sand bars (p. 200).

When waves arrive at an angle to the shore they deflect the nearshore water circulation, generating longshore currents. The effects of such wave-induced longshore currents are difficult to separate from the associated effects of oblique swash and orthogonal backwash produced when waves arrive at an angle to the shore and break on the beach, for both processes move sediment alongshore, the action of oblique wave swash causing beach drift (p. 116), while the wave-induced currents cause longshore drift in the nearshore zone. Whereas current velocities of at least 15 centimetres per second are needed to mobilize sand, agitation of sea floor sand by waves can lead to its entrainment by gentler currents.

Figure 2.4   Rip currents (arrowed) in the surf on Woolamai Beach, Phillip Island, Australia.

Dominant waves

The direction, height, and periodicity of waves approaching the shore have a strong influence on coastal outlines in plan and profile, particularly on cliffed coasts cut in soft formations or features formed by deposition. When wave conditions change there are corresponding adjustments in the morphology of depositional features such as beaches and spits. Wave conditions change frequently, and it is necessary to analyse several years records to establish characteristic annual and seasonal wave regimes. Direct observations of wave conditions are made from lightships and coastal stations, and these can be supplemented by analyses of meteorological data, when the direction and strength of winds are correlated with the pattern and dimensions of locally-generated waves, and by data on ocean swell regimes.

Dominant waves tend to determine such features as beach alignment and the net direction of longshore drift and sorting on beaches. Variations in coastal aspect can result in differing patterns of longshore drift in response to dominant waves.

High, moderate and low wave energy coasts

Coasts exposed to ocean swell and stormy seas (generally with deep water inshore) are known as high wave energy coasts, while those sheltered from strong wave action, bordering narrow straits, landlocked embayments, or island or reef-fringed seas with limited fetch, or where wave energy has been reduced by intense refraction, are termed low wave energy coasts. It is useful to recognize an intermediate category of moderate wave energy coasts.

High wave energy coasts can be defined as those with mean annual significant wave height ($H_{1/3}$, measured as the waves break) exceeding two metres, moderate wave energy coasts one to two metres, and low wave energy coasts less than a metre. High wave energy coasts typically have bold, rugged cliffs and long gently-curving beaches, as on the stormy Atlantic shores of Europe and the ocean coasts of southernmost Africa and southern and western Australia. Coastlines become more irregular on moderate wave energy coasts, while on low wave energy coasts, beaches are typically shorter, with such features as cusps, lobes and spits, as well as deltas and marshlands diversifying the shore, numerous shoals and bars, few cliffs (except on weak formations such as clay) and a generally intricate configuration. Examples are found in the Danish archipelago, around Puget Sound on the north-west coast of Washington state, and on the shores of many estuaries and lagoons. The coasts of northern Sumatra and western Malaysia, bordering the Strait of Melaka, are typical of tropical low wave energy coasts, with extensive mangrove fringes, coastal plains formed by confluent deltas, occasional beach-fringed segments and very few cliffs. Similar features are seen where near-shore and fringing reefs diminish incident wave energy, as on the north coast of Viti Levu, in Fiji. The Gulf coast of Florida, in the United States, has generally low wave energy, wave action being reduced by the broad, gently-shelving offshore profile, and shows an intricate coastline with only limited beaches interspersed with minor spits, deltas, and marshes and persistent offshore shoals. However, this generally low wave energy coast is occasionally subject to drastic modification by strong wave action during hurricanes, when beaches are over-washed, and the shore may develop new erosional features that persist during subsequent calmer conditions.

TIDES

Tides are movements of the oceans set up by the gravitational effects of the moon and the sun in relation to the earth. They are very long waves that travel across the oceans and are transmitted into bays, inlets, estuaries or lagoons around the world's coastline. Oceanic tides are indeed tidal waves, but this term has been widely misused as a synonym for tsunamis, which are large waves generated by tectonic events (p. 21).

The ebb and flow of tides produces regular changes in the level of the sea along the coast, and generates tidal currents. The lunar cycle produces semidiurnal tides (two high and two low tides in approximately 25 hours), well displayed around the Atlantic Ocean; the solar cycle produces diurnal tides (one high and one low tide every 24 hours), as registered on the Antarctic coast; and elsewhere the two are mixed, yielding unequal high and low tides (e.g. high high, low low, low high and high low), as around much of the Pacific and Indian Ocean coasts. Where the effects of lunar gravity are stronger, high and low tides occur about 50 minutes later each day.

Variations in tidal type are significant in terms of durations of low tide drying (less than 12 hours with semidiurnal tides, but more than 12 hours with diurnal and mixed tides) which affect shore weathering and intertidal ecology. Tidal currents are also generally stronger with semidiurnal tides because of the greater rapidity of tidal fluctuations, but in general the most important aspect of tides in geomorphology is their vertical range.

The rise and fall of the tide on a coast is measured by tide gauges, located chiefly at ports. The highest and lowest astronomical tides are those that occur at a particular point on the coast in calm weather over a period of at least a year. Tides recorded shortly after each new moon and full moon, when earth, sun, and moon are in alignment, and have combined gravitational effects, are relatively large, and are known as spring tides (the term is a little confusing, as these tides do not occur only during the spring season). The highest spring tides occur fortnightly, (actually at intervals of about 14.6 days). Maximum spring tide ranges occur about the equinoxes (late March and late September), when the sun is overhead at the equator. At half-moon (first and last quarter) the sun and moon are at right angles in relation to the earth, so that their gravitational effects are not combined; tide ranges recorded shortly after this are reduced, and are known as neap tides. The relationship between shore zones and tide ranges is shown in Figure 2.5.

There are also longer term variations of tide range in response to astronomical cycles of the relative positions of the sun, moon and earth. The moon, for example, returns to a similar position relative to the earth every 27.5 days, but its orbit is such that it returns to almost exactly the same position only once in 18.6 years. This long-term oscillation, based on the precession of the lunar orbit, is the nodal cycle (Pugh 1987a), which reached maxima in June 1950, February 1969 and October 1987: the next is due in May 2006. These maxima have been detected in sea level records in the Venice Lagoon (p. 42). Still longer astronomical tidal cycles have been identified, with maxima occurring in 1745 and 1922, and another expected in 2192, but these are of small dimensions (Cartwright 1974).

Tidal forces also cause minor fluctuations (up to 30 centimetres) in land levels as the gravitational pull of the sun and the moon are exerted on the earth's crust. Tidal oscillations of sea level recorded on the coast are thus resultants of the upward and downward movements of land and sea.

Tidal movements can also influence wave action. In general the transverse profile of the shore and near-shore zones is concave, and as the tide rises the water deepens, so that larger waves break upon the shore. Where tidal currents flow in one direction they can reduce the velocity and size of waves coming in from a contrary direction.

Tidal currents

Tidal currents, produced as tides rise and fall, alternate in direction in coastal waters, reversing as the tide ebbs; their effects may thus be temporary or cyclic, and can be measured using current meters (Morang et al. 1997) or traced with the aid of floating drogues or seabed drifters. In the open ocean tidal currents rarely exceed 3 kilometres per hour, but where the flow is channelled through gulfs, straits between islands, or entrances to estuaries and lagoons, tidal currents are strengthened, and may locally and briefly exceed 16 kilometres per hour. Modifications of tides transmitted into such basins are discussed in Chapter 9 (p. 240). Stronger tidal currents are generated by spring tides than neap tides because a larger volume of water is moved, and the strongest currents are generated where the tide range is large. Tidal currents attain 16 kilometres per hour in the Raz Blanchart, between Alderney and Cap de la Hague in north-western France, and are also strong through Apsley Strait, which separates Melville and Bathurst Islands off the north coast of

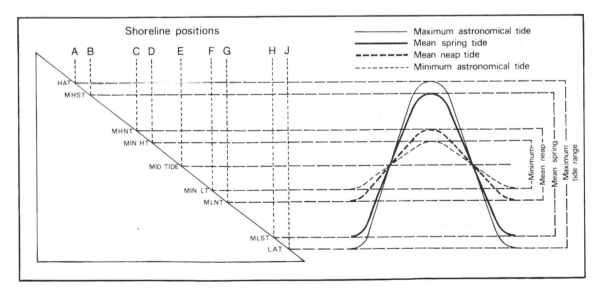

Figure 2.5 Tidal curves, tide ranges and shore zones. HAT – Highest Astronomical Tide, MHST – Mean High Spring Tide, MHNT – Mean High Neap Tide, MIN HT – Minimum High Tide, MIN LT – Minimum Low Tide, MLNT – Mean Low Neap Tide, MLST – Mean Low Spring Tide, LAT – Lowest Astronomical Tide. The maximum shore width (AJ) is at maximum and minimum astronomical tides, diminishing to mean spring tides (BH), mean neap tides (CG) and minimum tides (DF).

Australia. Tidal waterfalls form between some of the islands in the Buccaneer Archipelago, where tide ranges are very large (locally more than 10 metres) off the Kimberley coast in north-western Australia.

Tidal oscillations impinging on a coastline may set up longshore currents, as on the north Norfolk coast in England, where the longshore flow is westward during the two to three hours preceding high tide, then eastward for another two to three hours as the ebb sets in. Tidal currents have little direct effect on the morphology of coasts facing the open sea, but where they pass in and out of estuaries (p. 231) or through narrow straits, they can scour the bordering shores. At the entrance to Deception Pass, on Whidbey Island in Washington (north-west United States) strong tidal currents (up to 34 kilometres per hour) have truncated a spit. In general, tidal currents do not cause beach erosion or deposition, but they carry sediment along the coast in the nearshore zone, and this may eventually be delivered to beaches alongshore. Longshore currents generated by tides may either augment or reduce those produced by oblique waves, depending on the direction of approach, but the effects of tidal currents are otherwise generally subordinate to the effects of waves in the nearshore zone.

Macrotidal, mesotidal and microtidal coasts

Tide ranges can be defined as the difference between mean high and mean low spring tides where the tides are semidiurnal, or between mean higher high tide and mean lower low tide where the diurnal component introduces inequality in the levels of successive tides. Global data given in the Admiralty Tide Tables show that mean neap (or minimum) tides vary between 30 per cent and 80 per cent of mean spring (or maximum) tide values and that astronomical tide ranges are 20 per cent to 30 per cent higher than mean (or maximum) spring tide ranges. Coastwise comparisons can be conveniently made using mean spring (or maximum) tide range. Typically this shows more intricate variations along indented coastlines (as around the British Isles) than on open oceanic coasts.

Tide range in mid-ocean is small, of the order of half a metre, but is amplified where the tide is transmitted into shallow coastal waters, particularly in funnel-shaped gulfs and intricate embayments. Where the mean spring tide range is less than two metres, the environment can be termed microtidal, between two and four metres mesotidal, four to six metres macrotidal, and more than six metres megatidal. Large tide ranges are found on coasts where the tidal wave steepens across broad continental shelves and into gulfs and estuaries, especially where there is resonance between incoming and reflected tides, as in the Bay of Fundy, in eastern Canada. Microtidal ranges are typical of the more open coasts of the Atlantic, Pacific, Indian, and Southern Oceans and of landlocked seas such as the Baltic, Mediterranean, Black, Red and Caribbean. Because of local variations it is difficult to show global tide ranges on a textbook map, but the following paragraphs describe the broad pattern.

In Britain there is a microtidal sector in central southern England, between Portland Bill and Christchurch Bay, produced by a tidal circulation in the English Channel that yields large tides over on the north coast of France. Otherwise macrotidal conditions predominate in England and Wales, with very high (megatidal) ranges in the Bristol Channel (Avonmouth up to 14.4 metres), and from North Wales around the east coast of the Irish Sea (Liverpool 8.3 metres). The coasts of western and northern Scotland are generally mesotidal, increasing to macrotidal south from Montrose. The east coast of England is largely macrotidal, tide ranges increasing southward from Berwick upon Tweed (4.1 metres) to Boston (6.8 metres), diminishing in Norfolk and Suffolk to mesotidal (microtidal 1.9 metres between Caister and Lowestoft), then rising again to macrotidal southward from Clacton (4.7 metres).

The west coast of the Atlantic Ocean is generally microtidal, except for sectors in southern Argentina (Puerto Gallegos 10.4 metres), where there are strong currents in Magellan Strait, the Amazon delta, northern New England and the Bay of Fundy (Parrsborough 11.6 metres), the Gulf of St Lawrence (rising to 4.9 metres at Quebec), Leaf Basin, a ria opening into Ungava Bay (11.5 metres), and Frobisher Bay on Baffin Island to the north (9 metres). Maximum tide ranges in Minas Bay at the head of the Bay of Fundy attain more than 18 metres, and in Leaf Basin up to 20 metres. The coasts of Greenland are mesotidal, as are those of western and northern Iceland, but south-east Iceland is microtidal. Much of Norway is mesotidal but tide ranges diminish southward (Bergen 1.2 metres) and are negligible around the Baltic Sea. The coasts of Denmark and north-west Germany are generally microtidal, increasing to mesotidal in estuaries (Wilhelmshaven 4 metres) and along the Dutch coast (Westkapelle 4 metres), and continuing to increase through Belgium and northern France to the megatidal area of the Bay of Mont Saint Michel (Granville 11.5 metres). They then diminish round Brittany (Brest 6.1

metres) to mesotidal in the Bay of Biscay and Atlantic Iberia (Cadiz 2.7 metres) and then microtidal (Gibraltar 0.9 metres). The Mediterranean coasts are microtidal, tide ranges exceeding one metre only near Sfax (1.3 metres) in Tunisia. The coasts of West Africa are generally microtidal and mesotidal, the tide range diminishing south from Sierra Leone (2.6 metres) and remaining microtidal around southern Africa until it becomes mesotidal in Mozambique (Beira 5.6 metres) and the Strait of Madagascar. North from Kenya the Indian Ocean and Red Sea coasts are microtidal, as is the western side of the Arabian Gulf. Tide range increases to two metres at Kuwait and the east coast of the Arabian Gulf is mesotidal, as is the Pakistan coast (Karachi 2.3 metres). There is a small macrotidal sector in north-west India (Bhavnagar 8.8 metres), then a rapid diminution to microtidal conditions around India and Sri Lanka. Tide range increases in the Bay of Bengal, with macrotidal sectors on the Ganges delta and in Burma (Mergui 5.3 metres). The coasts of Thailand and peninsular Malaysia are generally microtidal, but increase locally to mesotidal and even macrotidal (Port Kelang 4.1 metres). Microtidal and mesotidal conditions predominate in Indonesia, Papua New Guinea, the Philippines, Japan and east Asia, with a macrotidal zone in the Yellow Sea (Inch'on, South Korea 8.2 metres) and another exceeding five metres in the northern part of the Sea of Okhotsk, where tides are augmented to 13 metres at the head of the Zaliv Shelikhova gulf. The Arctic coast of Russia is microtidal westward to the Barents Sea, where it becomes mesotidal (Murmansk 3.2 metres).

In the north Pacific Ocean tide ranges are small in the Bering Strait, but increase in the Gulf of Alaska where Anchorage (9.4 metres) and Port Essington (5.5 metres) are macrotidal. Mesotidal conditions predominate in British Columbia but become generally microtidal on the southern shore of the Strait of Juan de Fuca (Port Angeles 1.8 metres) and along the Pacific coasts of Washington, Oregon, California and Mexico. The Gulf of California has tide ranges of up to nine metres. They increase southward to macrotidal on the Panama and Colombian coasts (Balboa 5.0 metres), then diminish to mesotidal in Ecuador (Guayaquil 3.6 metres) and are microtidal through Peru and Chile to Cape Horn (Orange Bay 1.7 metres), except for local augmentation in the gulfs of southern Chile (Puerto Montt 5.8 metres).

On the Australian coast microtidal conditions prevail from Brisbane (1.8 metres) south to Cape Howe, then along the whole of the southern and western coast with the exception of certain embayments. Tide range increases into Spencer Gulf (Port Augusta, 2.0 metres) and Gulf St Vincent (Port Wakefield, 2.1 metres), but the narrow entrance to Port Phillip Bay impedes tidal invasion, so that mean spring tide range diminishes from Point Lonsdale (1.1 metres) at the entrance to about 0.6 metres within the bay. In north-western Australia the tide range increases behind a wide continental shelf to 5.8 metres at Port Hedland, 8.2 metres at Broome and 10.5 metres in Collier Bay on the Kimberley coast before diminishing to 5.5 metres at Darwin. The Gulf of Carpentaria is microtidal, but in north-east Queensland there is a mesotidal sector (Townsville 2.5 metres) increasing to macrotidal in the vicinity of Mackay (4.9 metres) then diminishing southward to Brisbane. New Zealand has generally microtidal coasts, increasing to mesotidal locally (Auckland 2.9 metres, Westport 3.0 metres), and the Pacific islands are microtidal. The coasts of Antarctica are also generally microtidal, with a few stations mesotidal (e.g. Breidbay 2.6 metres).

A large tide range implies a broad intertidal zone, more than 20 kilometres of sandflats and mudflats being exposed at low spring tides in the Bay of Mont Saint Michel, in France, and about 10 kilometres off Broome in north-west Australia. In Britain the intertidal zone is more than three kilometres wide in Morecambe Bay, in The Wash and on Maplin Sands on the Essex coast. Wave energy is expended in traversing such a broad shore zone, and waves which reach the coast at high tide have been much diminished by friction in the shallow water passage. Zones of weathering and ecological processes become wide and dispersed. Strong tidal currents shape elongated parallel banks and troughs on the sea floor, particularly where they flow through narrow seas, as in the Straits of Dover and Torres Strait, between Australia and Papua New Guinea (Off 1963). The strong currents through Deception Pass in Washington state have scoured the rocky floor, sweeping sediment out of each end of the narrow passage.

Movements of sea floor shoals close to the coast may modify wave patterns and result in coastline changes (p. 152). Where the tide range is small, as around the Mediterranean and Baltic Seas and along the south-west coast of Australia, wave action, intertidal weathering processes and biogenic activities are concentrated within a narrow vertical zone.

Changes in tide range may occur as the result of deepening or shallowing of the entrances to bays, inlets and estuaries. For example, the dredging of the approaches to Venice Lagoon in Italy has led to an increase in tide range within the lagoon.

Tidal bores

Where tides drive water into funnel-shaped gulfs or estuaries the narrowing and shallowing configurations cause the front of the tidal wave to steepen and form a narrow wave that moves rapidly upstream as a tidal bore. The velocity (c) is equivalent to $\sqrt{g(h + H)}$ where g is gravity, h river depth and H wave height. During spring tides in the Bristol Channel the rising tide becomes a tidal bore up to two metres high, moving up the Severn estuary at about 15 kilometres per hour. Similar bores develop in estuaries and inlets on other macrotidal coasts, as in Turnagain Inlet in Alaska and the Bay of Fundy in Canada. In the Amazon River the tidal bore (pororoca) moving upstream attains a height of up to six metres. The large tides on the north-west coast of Australia flow up several estuaries, but little is known of how far they run upstream and what dimensions they attain. The mascaret, a tidal bore that flowed upstream in the River Seine to Rouen, has been much reduced by dredging in the estuary downstream.

Tide-dominated coasts

The influence of tides on the coast depends largely on tide range, which determines the zone over which wave action can operate. On macrotidal and megatidal coasts the coastline is reached, only briefly and intermittently, by waves, wave action being withdrawn from the backshore for most of the tidal cycle. Brevity of wave attack impedes cliffing, although cliffs may be cut in soft rock or unconsolidated deposits (such as glacial drift or salt marshes), as on parts of the shore of the Bay of Fundy and in estuaries with a large tide range (p. 231). In such situations beaches are usually found at about high tide level, and the shaping of the beach profile, sorting of beach sediments and modification by longshore drift are all retarded by the vertical dispersal of wave energy as the tide rises and falls. Tidal currents play an important part in the shaping of sea floor topography exposed as the tide rises and falls: a morphology that can be termed tide-dominated.

The distinction between wave-dominated and tide-dominated morphology raises problems. Beaches on microtidal coasts are almost entirely wave-dominated features, but wave action is usually present in areas classified as tide-dominated (even though its effects are modified as tides rise and fall). Salt marshes, mangrove and seagrass banks and terraces are often considered tide-dominated, but waves certainly erode their margins and wash sediment into them as the tide rises.

Strong current action within salt marshes and mangroves is transverse to the shore and confined to the vicinity of tidal creeks (p. 209). Purely tide-dominated morphology may exist on the floors of megatidal gulfs, such as the Bay of Fundy, although even there wave action contributes to the shaping of shoal and ripple forms as the tide rises and falls. The problem of defining tide-dominated environments has complicated efforts to use sedimentary sequences as indications of past changes in tide range and tidal current velocities (Hinton 1998).

Extensive salt marshes have formed on low energy macrotidal and megatidal shores, as in Bridgwater Bay, in the Bristol Channel and on the shores of the Bay of Mont Saint Michel in north-western France. Wide mangrove swamps occur in similar situations in low latitudes, as on the shores of bays and estuaries in northern Australia where the tide range is large. Strong tidal currents that develop in inlets and constricted bay mouths can scour bordering beaches and interrupt longshore drift of beach sediments, sometimes curtailing spit growth, as on Sandy Point in Western Port Bay, Australia (p. 130).

On microtidal coasts, by contrast, wave energy is concentrated within a narrow vertical zone, facilitating cliff-base erosion and impeding the spread of salt marshes and mangrove swamps. Beach morphology, sediment transport and nearshore sea floor topography all become wave-dominated in these situations.

A change in sea level can modify tides. If sea level rises it is likely that there will be a slight increase in tide ranges around the world's coastline as the oceans deepen, the rise that actually occurs being modified as tidal amplitudes adapt to the changing coastal and nearshore configuration. If sea level falls tide ranges may diminish, except where the new coastal outlines result in increased local amplification.

STORM SURGES

Storm surges occur when strong onshore winds build up coastal water to an exceptionally high level for a few hours or days, and are most pronounced when they coincide with high spring tides. Strong onshore winds also generate large waves accompanying the raised sea level, overwashing beaches, flooding low-lying coastal areas and causing extensive changes in a short period. Beach erosion is usually severe during a storm surge, and if the coast consists of soft formations it may be cut back substantially to a new morphology. One of the best documented storm surges occurred in the North

Figure 2.6   A storm surge during a tropical cyclone produced large waves which dislodged and piled up beach rock on the shore near Port Hedland, Australia.

Sea on 31 January–1 February 1953, when a northerly gale raised the sea level up to three metres along bordering coastlines, flooding extensive areas in eastern England and low-lying parts of the Dutch and German coasts. A similar surge accompanied the passage of a deep depression across the North Sea in 1978 (Steers et al. 1979), and there have been many such episodes in this region in recent centuries, when coasts have been eroded and briefly submerged: the limits reached by storm surges are commemorated on marker posts or on the walls of buildings, as at the Harbour Inn, near Southwold in Suffolk.

Hurricanes raise the sea level temporarily along the Gulf and Atlantic coasts of the United States by as much as six metres, causing extensive erosion and damage to structures, and typhoons in the China Sea have similar impacts on the coasts of Japan and the Philippines. Major storm surges occur from time to time at the head of the Bay of Bengal, when large areas of the Sundarbans are devastated by flooding

and wave overwash. In northern Australia tropical cyclones generate surges in a region where heavy rains accompany the summer monsoon, and cause widespread flooding in coastal districts. Erosion is severe during such episodes. Beaches are depleted, and such features as the thrown-up blocks of sandstone at Quobba in Western Australia and on Grand Cayman (Jones and Hunter 1992), the piled-up beach rock near Port Hedland (Figure 2.6), or overwashed fans of beach sediment may persist for years or decades after such catastrophic events.

OTHER SEA DISTURBANCES

Apart from storm surges, exceptional disturbances of sea level occur during and after earthquakes, landslides or volcanic eruptions in and around the oceans. These produce tsunamis, very large waves which may attain heights of more than 30 metres by the time they reach the coast. They are most common in the Pacific Ocean, which is bordered by zones of crustal instability, and they are responsible for occasional catastrophic flooding and beach erosion on Pacific coasts. Charles Darwin observed a tsunami, in the form of three large waves caused by an earthquake in Chile in 1835. In 1946 a tsunami was initiated by an earthquake off the Aleutian Islands, and waves travelling southward caused extensive devastation in the Hawaiian Islands, 3700 kilometres away, 4.5 hours later. Large waves were recorded at many places around the Pacific Ocean, and at Scotch Cap, Alaska, the tsunami destroyed a lighthouse and radio mast on a cliff 30 metres above sea level. An earthquake off the coast of Chile in 1960 also produced giant waves: a tsunami 11 metres high devastated Hilo in Hawaii and a few hours later waves up to 40 metres high broke on the east coasts of Hokkaido and Honshu in Japan.

In 1883 the explosive eruption of Krakatau, a volcanic island in Sunda Strait, Indonesia, generated a tsunami up to 30 metres high on the nearby coasts of Java and Sumatra, sweeping away beaches, hurling coral boulders and blocks up on to fringing reefs and shore platforms and destroying the lighthouse at Anyer. In 1998 an earthquake 30 kilometres off the north coast of New Guinea produced a 10 metre tsunami that overwashed the sandy barrier at Sissano, west of Wewak, forming a coastal lagoon.

The effects of tsunamis on the coast, notably around the Pacific basin, may be visible for many decades after the event. In addition to changes effected on the coastline, very large waves may re-shape the nearshore profile and thereby change the pattern and dimensions of waves approaching the coast, resulting in subsequent erosion or accretion that would not otherwise have occurred. Like storm surges, they can throw large blocks and boulders up on to cliffs. Bryant et al. (1996) found such features on the south-eastern coast of Australia, and argued that occasional very large tsunami have strongly influenced coastal morphology there, even contributing to the shaping of coastal barriers. An exceptionally large tsunami may have occurred about 105 000 years ago in the south-west Pacific as the result of a massive submarine landslide in Hawaii, producing erosional and depositional features on bordering coasts. It has been suggested that features produced by this tsunami can be found on the south-eastern coast of Australia, but these have still to be assessed (Bryant et al. 1996).

Giant waves can be generated in restricted areas by landslides and rock falls. A massive rock fall into Lituya Bay, an Alaskan fiord, in 1960 swept a wave 15 metres high down the fiord at 160 kilometres per hour, washing over a spit at the entrance and dispersing beach sediments seaward. Disturbances of a similar kind have occurred in the vicinity of ice coasts as the result of iceberg calving in Greenland and around Antarctica.

These are unusual phenomena. On most coasts waves influence the shaping of beaches within a zone a few metres above and below present mean sea level, the extent of this zone depending largely on tide range. Rapid changes occur on beaches during storms and occasional storm surges, but there are also more gradual gains and losses, leading to the re-shaping of the beach in plan and profile, during long intervening periods of less boisterous weather. There is sometimes difficulty in deciding whether particular coastal features are the outcome of brief catastrophic events or the product of gradual evolution in response to everyday processes.

OCEAN CURRENTS

Ocean currents are gentle movements of water in response to prevailing wind patterns and density variations in the oceans, or resulting from differences in the salinity and temperature of the water. They have little effect on beaches or nearshore morphology, except where they bring in warmer or colder water, which modifies ecological conditions and thereby influences the distribution of such features as coral reefs or kelp beds, the presence of which can affect beach forms and shore dynamics. The Gulf Stream in the North Atlantic, for example, brings relatively warm water to the Atlantic coasts of Britain and Norway, which could otherwise be subject to shore ice processes in winter.

Wind-driven currents

Wind-driven currents are produced where winds move the surface water, building up sea level to leeward and lowering it to windward as wave action proceeds. Wind-generated currents are not as regular as the alternating tidal currents, but their effects are cumulative in the direction of the prevailing wind. Strong currents are produced when winds drive surface water into gulfs, through narrow straits, or in and out of estuary and lagoon entrances. These may strengthen or oppose the currents produced by tides in similar situations, and it may be difficult to separate the effects of the two.

River-mouth currents

Currents are also produced by discharge from river mouths. In tidal estuaries and at inlets to lagoons fed by rivers, fluvial discharge augments tidal ebb currents and diminishes tidal inflow (which may nevertheless be strong enough to move water in on a rising tide). River outflow may carry sediment into the sea, maintain or enlarge river outlets, and form a seaward jet, which refracts approaching waves and can act as a breakwater impeding or interrupting longshore currents and sediment flow. Fluvial discharge currents are strong off streams fed by melting ice and snow from coastal mountains, as in Norway and Alaska during the summer months.

Effects of currents

Any of these currents can move fine to medium sand (grain diameter 0.1 to 0.5 millimetres) when their velocity exceeds about 15 cm/sec, stronger currents being required to move coarser material. Currents generated by winds and tides can be strong enough to move sand or even gravel on the sea floor, either contributing to longshore drift, supplying material to a beach, or carrying it away offshore, but these effects are usually subordinate to the onshore–offshore and alongshore movements of beach sediment by wave action. Currents can prevent nearshore deposition, and erode channels or remove shoals, thereby deepening the nearshore water and permitting larger waves to break upon the shore. Sea floor shoals of sand or gravel may form adjacent to rocky headlands as the result of deposition by current flow, examples being the Skerries, off Start Point, and the Shambles, off Portland Bill, on the south coast of England. Material from such shoals may in due course be washed on to a beach, but the more immediate effect is to modify wave refraction patterns in nearshore waters and reduce wave energy reaching the nearby shore.

In general, tidal and other marine currents are more effective in shaping sea floor morphology than in developing beach configuration, and the early theory that long, gently-curving beaches on oceanic coasts were produced by marine currents sweeping along the shore has given place to the view that these outlines are determined by refracted wave patterns (p. 118). Nevertheless, changes in the topography of the sea floor, due to erosion by current scour or deposition from slackening currents, modify patterns of wave refraction and may thus indirectly affect beach outlines. Currents often play a part in removing material eroded

by waves from the coast, or in supplying the sediment that is subsequently built into beaches by wave action.

NEARSHORE WATER CIRCULATION

The combined effects of wind-generated waves, astronomically-generated tides, various forms of current flow and other disturbances of the sea produce a highly variable energy flux in nearshore waters. As has been noted, the several processes interact: a rising tide, for example, deepens nearshore water (the nearshore zone being generally concave upward in profile), thereby increasing the height and energy of waves that reach the shore, and a tidal current flowing in one direction can reduce the velocity and dimensions of waves moving in the opposite direction. Marine currents in the nearshore zone are the resultant of potential flows generated by winds, waves, tides and other forces, and there is much variation in current direction and velocity. In addition, wave variability results from the arrival of waves of differing height and length, generated from differing distances and directions, and there are often irregular wave patterns arriving in the nearshore zone. The outcome is a complex nearshore hydrodynamic system that moves sediment on the sea floor and onshore, offshore and alongshore, and influences the shaping of the coastline, including beaches, and the nearshore sea floor.

While it is possible to deduce the coastal features that would be shaped by a set of wind, wave and current processes continuing uniformly over a period of several hours or days, such an adjustment is rarely attained before there are changes in one or more of the driving components and the partially adjusted features begin to be modified. Stability is thus an ephemeral concept in terms of depositional coast morphology, where any attempt to define a stable coastline must be in terms of a stated time scale. On hard rock coasts there may be prolonged phases of stability when there is no response to nearshore processes, while on soft rock coasts, beaches, marshes and deltas there is almost continuous instability as nearshore processes frequently mobilize sediments.

WIND ACTION

In addition to the effects of waves and currents (which may be at least partly generated by wind action), coastal landforms may be shaped or modified by the wind. Strong winds deflate fine-grained sediment (sand, silt and clay) from beaches and tidal flats, lowering their surfaces, and causing the movement of rock particles onshore, alongshore or offshore. Sand blown from the beach or foreshore is transferred to the zone above high tide level, and deposited as dunes (Chapter 7), which may remain in position or be swept inland or along the coast by wind action. Wind-drifted silt and clay may be deposited down-wind from source areas such as intertidal mudflats that dry out sufficiently for the wind to mobilize this fine-grained sediment. Weathering of rocky shores (p. 76) may produce sediment fine enough to be carried away by the wind, which thus contributes to the lowering of shore rock surfaces.

Wind-blown rock particles may become airborne, or may bounce or roll as they travel down-wind: impacts with rock surfaces result in their rounding and attrition, while rock surfaces are scoured or smoothed by abrasion as the wind drives sand particles at and across them. Wind action also enhances evaporation, drying out wet rock outcrops on cliff faces and shore platforms, especially at low tide (p. 78).

OTHER PROCESSES

Other processes influencing coastal evolution include runoff after heavy rain or from the melting of snow or ice, which causes gravuring of cliff faces and gulleying of coastal slopes, forming downwashed fans, and the outwashing of sand from beaches. Weathering processes that have influenced the shaping of coastal landforms include physical weathering (by insolation, freeze-and-thaw or wetting and drying), chemical weathering (by solution, salt crystallization, mineral decomposition or base exchange) and biological weathering (by shore organisms, burrowing animals or root penetration).

Some shore processes are conditioned by the salinity of sea water. Salinity can be estimated from water density, but is usually measured either as total dissolved solids or instrumentally from the concentration of chloride ions (chlorinity), and is expressed in parts per thousand (ppt). In the oceans salinity averages 35 ppt salt concentration (mainly chlorides), but there are variations in salinity from almost fresh in parts of the Baltic Sea to hypersaline (> 35 ppt) in the Red Sea and Shark Bay in Western Australia. Sea salinity increases in warm and windy environments, especially along arid coasts where the salt concentration is augmented by high evaporation, and diminishes in cold areas, particularly near the mouths of rivers and along coasts with

melting glaciers and ice sheets. Salt water and sea spray have corrosive effects on shore rock outcrops and also produce distinctive habitats for marine and estuarine flora and fauna, which may influence weathering, erosion, transportation and deposition of sediment in the coastal environment. The effects of salt weathering are discussed in Chapter 4 (p. 78).

MODELLING COASTAL PROCESSES

Laboratory simulation of these various processes has been attempted using scale models such as water tanks in which waves, tides and currents can be generated and their combined effects assessed. The aim has been to test hypotheses concerning the ways in which these processes cause erosion, move sediment and promote deposition on the sea floor and along the coast. A wave tank was used by Bagnold (1940) to investigate beach processes. Such physical models have limitations because of the difficulty of scaling down materials and processes without modifying their physical properties (e.g. coherence, friability, expansion and contraction of sediments; viscosity and surface tension in water), but they have been useful in exploring potential responses to marine and nearshore processes (Silvester 1974).

Measurements of nearshore processes, using wave recorders, tide gauges and current meters, have been made at coastal laboratories, such as the Scripps Oceanographic Institution at La Jolla in California and the laboratory at Sochi on the Russian Black Sea coast, where piers are equipped for instrumental surveys and computerized recording of accompanying sediment movements and changes in coastal and nearshore morphology. Conclusions from such monitoring apply only to the nearby coastline, and there is a risk that insertion of numerous instrument-bearing structures may modify the natural processes, and cause changes that would not otherwise have occurred.

Mathematical modelling has been much used by engineers as a basis for computer simulations of coastal processes. It can be used to study the effects of integrated process systems (waves, tides and currents) on nearshore sediment flow, and the ways in which these processes and responses will be modified by the introduction of structures such as groynes or breakwaters. It is important to be sure that the information used is accurate and comprehensive, and to test predictions against what actually happens, in order to refine the model and improve subsequent forecasts.

CONCLUSION

These various processes cause the movement of sediment on the sea floor, and particularly in the nearshore zone and along the coast. They have contributed to the shaping of the existing sea floor and coastal landforms, and can be correlated with changes taking place with the sea at its present level.

Many coasts show features that developed when the sea stood at different levels in the past, or when it was rising or falling. The operation of waves, tides, currents, and the other processes that have been discussed has sometimes been at a higher level than it is now, and sometimes lower. There have been phases of still-stand when the relative levels of land and sea remained constant at particular altitudes, and phases when the sea was rising or falling relative to the coastal land. It is now necessary to examine the history of changing levels of land and sea.

3

Land and Sea Level Changes

INTRODUCTION

Around the world's coastline there are features that formed when the sea stood higher or lower relative to the land, especially during Quaternary times. Beach deposits or marine shell beds stranded above present high tide level are indications of an emerged coastline on which the sea was formerly at a higher level, while drowned valley mouths indicate a submerged coastline where the valleys were excavated by their rivers when the sea stood at a lower level. (The terms emerged and submerged are used in coastal geomorphology in preference to emergent and submergent.) It is sometimes difficult to determine whether a change in relative sea level (RSL) has been the result of upward or downward movements of the land, or an actual rise or fall in the level of the sea, or some combination of the two. Changes resulting from uplift or subsidence of the land are known as tectonic movements, while those due to an actual rise or fall of sea level are eustatic movements, and relative sea level is the resultant of these land and sea level changes.

A marine transgression occurs when sea level rises to invade the land, and a marine regression when sea level falls, exposing the former sea floor as a land area. Relative sea level has risen and fallen frequently through geological time. Few coastal areas have been tectonically stable for long, and many show obvious indications of continuing instability, notably earthquakes and volcanic activity. On the other hand, there have been phases of stillstand, when the sea has remained at or close to its level relative to the land long enough for recognizable coastline features to have developed. Some of these are above present sea level (emerged), others below (submerged), and on much of the world's coastline the past 6000 years have been one of these phases of stillstand with the sea at its present relative level.

This sea level depends partly on the volume of water in the oceans, which is determined by the balance of evaporation and precipitation produced by the hydrological cycle, and partly on the size and shape of the crustal depressions that contain sea water. Mean sea level provides a datum from which upward or downward movements of relative sea level can be measured.

MEAN SEA LEVEL

A definition of mean sea level must exclude short-term variations such as those caused by tides and waves, described in the previous chapter. As has been noted (p. 18) tides have a vertical range varying from almost zero to about 20 metres around the world's coastline, and tidal cycles vary from about 12 hours to 18.6 years, with some even longer, though minor, astronomical cycles. Tidal oscillations are in theory symmetrical about a long-term mean sea level, conventionally defined as the arithmetic mean of the height of calm sea surface (i.e. excluding waves and oscillations related to winds and atmospheric pressure variations) measured at hourly intervals over at least 18.6 years. In practice mean sea level at points on the coast is related to a national datum such as American Sea Level Datum, or Ordnance Datum in the United Kingdom (Kidson 1986).

There are also short-term fluctuations of mean sea level (several days to weeks) in relation to weather conditions, notably the variations in atmospheric pressure that accompany the passage of depressions and anticyclones, together with the effects of associated winds. Strong wind action produces storm surges (p. 20) that drive water shoreward and raise sea level temporarily in the nearshore zone by up to several metres, as well as producing very high waves that break upon the coast. Persistent onshore winds can maintain high water levels, especially in gulfs and semi-enclosed bay areas.

Sea level also varies seasonally in relation to temperature, pressure, and wind regimes. It rises about a centimetre with each millibar fall in atmospheric pressure. Analyses of monthly or seasonal data show that maximum annual sea levels occur at different times around the world's coastline: in September in the eastern North Atlantic, for example, and in April along the eastern seaboard of Australia (Guilcher 1965). In the South China Sea mean sea level is about 40 centimetres higher during the north-east monsoon (November to March) than during the south-west monsoon.

There are sea level changes related to longer-term atmospheric pressure cycles such as the North Atlantic Oscillation, which is correlated with sunspot cycles, and the two to seven year El Niño Southern Oscillation, which produces high pressure and low sea level over the south-east Pacific Ocean and low pressure and high sea level in the Indian Ocean, then reverses these (Fairbridge and Krebs 1962). Komar (1986) showed that sea level rose temporarily about 20 to 30 centimetres along the Pacific coast of North America during the 1982–83 El Niño Southern Oscillation.

It is necessary to exclude these variations and so determine mean sea level before deciding the extent of sea level changes over particular periods.

CAUSES OF SEA LEVEL CHANGE

Apart from these short-term variations in sea level, upward or downward movements of sea level can result from several causes, as indicated in the following sections.

Eustatic movements of sea level

Sea level rises when the volume of water in the ocean basins increases, and falls when it is reduced. These changes are world-wide because the oceans are interconnected. They have been termed eustatic, a term which was introduced (Suess 1906) when it was assumed that such changes were equivalent throughout the world's oceans, but it is now realized that there have been regional discrepancies in the amount of sea level rise or fall on particular coastlines (p. 31).

Through geological time the volume of water in the oceans has been gradually augmented by the arrival of small quantities of juvenile water supplied from the earth's interior, primarily from volcanic eruptions, implying a long-term rise in sea level, but it is unlikely that these accessions have had much influence on sea level changes during the Quaternary.

Steric changes

An increase in atmospheric temperature results in warming and expansion of the oceans, and sea level rises, whereas if the oceans cool they contract, and sea level falls (Warrick et al. 1993). The volume of sea water also diminishes as salinity increases, and rises as it freshens. Ocean volumes also vary with the density of sea water, related to temperature, salinity and atmospheric pressure. These are known as steric changes.

It has been calculated that a rise of 1°C in the mean temperature of the oceans would increase their volume so that sea level would rise about two metres. Estimates of Pleistocene variations of mean ocean temperature (based on palaeo-temperature measurements on fossils from ocean floor deposits) are within 5°C of the present temperature, so that this could only account for sea level oscillations of up to 10 metres.

Sedimentation

Sea level can also rise because of a gradual reduction in the capacity of the ocean basins resulting from the accumulation of sediment carried from the land to the sea, whether by rivers, melting glaciers, slope run-off, landslides, wind action or coastal erosion. This is a very slow process, termed sedimento-eustatic. Transference of all the land above present sea level into the ocean basins would raise the level of the oceans by more than 250 metres, but present estimates of denudation rates account for a sea level rise of only about three millimetres per century.

Tectonic movements

Relative sea level can change because of tectonic movements, upward or downward, of the earth's crust, changing the shape of ocean basins and raising, lowering or deforming the continents. These may be epeirogenic, orogenic, or isostatic movements. Epeirogenic movements are broad, rather uniform tectonic uplifts or depressions which have tended to raise continents and depress the floors of ocean basins, with warping restricted to a marginal hinge-line. Some coasts lie close to this zone of marginal warping, with uplift on the continental side and depression on the oceanic side, as on much of the coast of southern Africa.

Changes in the ocean basins

Epeirogenic movements have changed the shape and capacity of the ocean basins. An increase in the capacity of an ocean basin (measured below sea level) causes a lowering of sea level, and a reduction causes sea level to rise. It has been suggested that the floors of the ocean basins, notably the Pacific Ocean basin, have been subsiding intermittently, particularly during Quaternary times, increasing their capacity and resulting in successively lower sea levels marked by emerged stairways on bordering coasts. Depression of ocean floors has been accompanied by lateral movements of the continental plates at rates of up to five centimetres per year, which have also modified the configuration of the ocean basins. As tectonic movements within the ocean basins raise or lower sea levels all over the world this kind of change is termed tectono-eustatic.

Epeirogenic land movements

Epeirogenic movements have raised each of the continents during and since Mesozoic times, but the uplift has been punctuated by phases of stability during which planation surfaces have been formed, with land areas reduced to base-level by long-term weathering and erosion. A succession of such planation surfaces, separated by bold scarps initiated during each episode of epeirogenic uplift, was recognized in Africa by King (1962), who suggested that they were present on each of the continents. Some of the planation surfaces may have been cut by marine processes during phases of higher relative sea level, intermittent epeirogenic uplift having produced hard stairways of marine terraces in which the oldest was the highest and farthest inland, separated by ancient cliffs cut when the sea stood at each level, and typically remain horizontal over long distances. Such stairways could also be produced by intermittent eustatic lowering of sea level, or some combination of eustatic and epeirogenic movements.

Orogenic movements

Orogenic movements are more complicated deformations of the earth's crust associated with mountain-building, notably at the convergence of tectonic plates. They are active around the margins of the Pacific Basin, notably on the west coast of the United States, in Japan, the Philippines, Indonesia, Papua New Guinea and New Zealand. These are termed neotectonic regions, characterized by frequent earthquakes and volcanic activity, and their coastlines show evidence of irregular displacement by uplift, lowering, tilting, folding and faulting. Sea level changes on tectonically-active coasts have been at least partly due to the rising or sinking of the land margin, either gradually or as the result of sudden earthquakes. Such displacements are much in evidence around the Mediterranean, where there has been recent tectonic uplift in the Oman region of Algeria, the southern Peloponnese and western Crete, and the Izmir region of southern Turkey, and where submergence of ports such as Carthage in Tunisia, built in the classical Greek and Roman era, about 2000 years ago, indicates coastal subsidence. On the southern coast of North Island, New Zealand, orogenic movements have led to a juxtaposition of uplifted and emerged coastlines with downwarped and submerged coastlines, parts of the coastline having been determined by associated faulting.

In the vicinity of Wellington, New Zealand, parts of the coastline were uplifted by up to a metre during the 1855 earthquake, forming emerged shore platforms, and an earthquake at Hawke Bay, New Zealand, in 1931 drained a lagoon near Napier by up to two metres to form an emerged coastal plain. The 1964 Alaskan earthquake raised intertidal shore platforms out of the sea, for example on Montague Island, where a sudden uplift of 11 metres led to an advance of the coastline of 400 metres while parts of Homer Spit, in Kachemak Bay, sank nearly two metres (Shepard and Wanless 1971). Successive earthquakes have raised coastlines around Tokyo Bay, producing a sequence of emerged Holocene terraces, with shore platforms raised out of the sea by the 1703 and 1923 movements. During the Colombian earthquake of 1979, the coast around Tumaco subsided up to 1.6 metres (Herd et al. 1981). In the Rann of Kutch, on the border between Pakistan and India, subsidence caused by the 1819 earthquake resulted in the sea rising to submerge an area of about 500 square kilometres.

Similar neotectonic features have been reported from sectors of the Antarctic coastline, and radiocarbon dating of sediments indicating former sea levels along the coast of Argentina has revealed alternating sectors of orogenic uplift and depression (Codignotto et al. 1992).

Isostatic movements

Isostatic movements are adjustments in the earth's crust resulting from loading or unloading of the surface. Areas heavily laden by the accumulation of lava or sedimentary deposits show crustal subsidence, as in the vicinity of large deltas, where the load consists of sedimentary deposits accumulating at the mouth of a river. Parts of a delta not maintained by active sedimentation become submerged as this subsidence proceeds (p. 252).

Isostatic movements have also occurred in regions that were glaciated during Pleistocene times, as in northern North America and northern Europe. As ice accumulated the earth's crust was loaded and depressed, and when it melted, the unloaded crust gradually rebounded. On the shores of the Gulf of Bothnia sea level is falling at rates of up to a centimetre per year because of land uplift due to the melting of former glaciers and ice sheets. Recovery of depressed crustal areas continues for some time after the ice has gone, until a physical equilibrium is restored.

Sea level changes have also resulted from another form of isostatic movement, the loading of continental shelves during a marine transgression. The weight of the deepening water depresses the continental shelf, and the adjacent coast is downwarped as a result, a response known as hydro-isostatic subsidence (Bloom 1979). Sea level rise is thus augmented, the extent of hydro-isostatic subsidence varying in relation to the width and slope of the continental shelf (which determines the volume and weight of water load gained during a sea level rise) as well as to the structure and strength of geological formations in the area affected. Sea level rise during marine transgressions may thus have been greater on steep coasts with deep water close inshore than on gently-shelving coasts. Sea level rise can be similarly augmented where soft sediments, such as peat, are being compressed beneath the sea floor by the gathering weight of water as the sea rises.

Volcanic movements

Volcanic activity has also resulted in a relative rise or fall of sea level. In the nineteenth century the geologist Charles Lyell observed that the limestone pillars in the Roman Market of Serapis, in the town of Pozzuoli near Naples in Italy, showed a zone about two metres above present sea level that had been drilled by molluscs when the sea rose to flood the structure (Figure 3.1). Sea

Figure 3.1   Dark zones formed on columns by marine organisms during a mediaeval phase of higher sea level in the Roman Temple of Serapis, at Pozzuoli, near Naples, Italy.

level evidently rose as the land subsided after the pillars were built in Roman times, and fell when the area was uplifted, an oscillation thought to be due to evacuation and re-filling of a subterranean lava chamber associated with the nearby volcano of Vesuvius.

Glacio-eustatic movements

Major oscillations of sea level accompanied the waxing and waning of the earth's ice cover in response to alternations of cold (glacial) and mild (interglacial) climate during the Quaternary Epoch. These sea level changes are termed glacio-eustatic. During glacial phases the earth's hydrological cycle was interrupted when the climate cooled sufficiently for precipitation to fall as snow, which accumulated as glacial ice and persistent snowfields in polar and mountain regions. Retention of large amounts of water frozen on land depleted the oceans, and there was a world-wide lowering of the sea level. During the milder interglacial phases the trend was reversed, water released from melting snow and ice flowing back into the ocean basins to produce a world-wide sea level rise (marine transgression), the sea at times extending above its present level.

During the Last Glacial phase, late in Pleistocene times, sea level fell about 140 metres, but about 18 000 years ago the polar ice sheets and mountain glaciers began to melt, and there was a world-wide sea level rise. This was called the Flandrian transgression when it was first recognized by Dubois (1924), working in Flanders and along the north and north-west coast of France, but it was subsequently found to be world-wide, and has become known as the Late Quaternary marine transgression. It continued into Holocene times, and came to an end about 6000 years ago when the sea attained approximately its present level.

The earth is still in an Ice Age, in contrast to the ice-free conditions that have prevailed through most of geological time, apart from earlier glaciations in the Pre-Cambrian and Permian. If the remaining land-borne ice sheets, glaciers and snowfields on the earth were to melt they would release sufficient water to produce a further world-wide marine transgression, raising the level of the oceans by at least 60 metres (Donn et al. 1962), and causing widespread coastal submergence and the loss of existing coastal lowlands (including most of the world's major cities). Melting of floating ice, as in the Arctic Ocean and the ice shelves of Antarctica, will not increase the volume of the oceans, and will have no effect on sea level.

Human impacts on sea level

Sea level has risen on coasts where the land has been subsiding because of human activities such as groundwater extraction, which depletes the aquifers under and around coastal urban and industrial centres. As underground water is withdrawn, the sediments of the aquifers are consolidated and compressed by the weight of overlying rock formations (and buildings, if any), and the loss in volume results in subsidence of the land surface. This has contributed to a relative sea level rise in the Venice region of Italy, and around Bangkok in Thailand (p. 42).

Relative sea level has risen with the pumping of oil or natural gas from underground strata, as in southern California and the Ravenna region in Italy (p. 42). Similar submergence has followed the loading of coastal land with artificial structures, and some port and land reclamation schemes, including the construction of artificial islands, have caused subsidence and changes in local tide regimes, raising relative sea level, especially in bays and estuaries (Bird and Koike 1986). A local rise in relative sea level may follow the deepening of nearshore areas as the result of dredging, with increased tidal penetration. Shore walls and tidal barrages built in the Thames and Medway estuaries have resulted in higher high tides and increased penetration upstream of waves driven by storm surges, effectively raising relative sea level.

Geoidal changes

Sea level changes have also been influenced by shifts in ocean surface topography that result from geoidal changes, as Mörner (1976) pointed out. Evidence from satellite sensing has shown that the oceans have an undulating surface configuration, with bulges and troughs rising and falling up to 90 metres above and below a smooth geoid because of gravitational, hydrological and meteorological forces. These bulges and troughs are related partly to gravity patterns and other geophysical phenomena, including tides and the earth's rotation, and partly to climatic patterns and ocean circulations. In the North Atlantic, tide gauge records show that, when the effects of coastal tectonic movements were excluded, mean sea level rose between 1920 and 1950, the rise being faster on the American coast than in Europe, probably because of variations in sea surface topography (Pirazzoli 1989).

Similar changes have occurred as the result of variations in the distribution of oceanic bulges and troughs, especially during the Quaternary. Shoreward movement of bulges and troughs results in a sea level rise where high areas move coastward and a sea level fall as low areas move in.

Changes related to ocean currents

Variations in sea level also result from changes in transverse gradients that develop where ocean currents are strong. Off the east coast of the United States and in the China Sea the ocean is up to two metres above its general level in areas adjacent to the poleward currents of the Gulf Stream and Kuroshio, and any variation in the velocity or extent of these currents will lead to changes of sea level on nearby coasts. Moreover, the ocean envelope is related to the axis and velocity of the earth's rotation, and any variation in these will cause differential changes of sea level around the world's coasts.

MEASURING CHANGES OF LEVEL

Having outlined the reasons why relative sea level has changed, the evidence of emerged and submerged coastlines can be considered. It is necessary to relate the levels of coastline features that now stand above or below present sea level to a specific datum. Some studies have reported the levels of emerged or submerged coastlines above or below mean sea level, the datum used on topographic maps in many countries, but others have used low water spring tide level, the datum used on nautical charts of coastal regions, and others high water spring tide level, which is more easily determined and more readily accessible for surveying work on the coast. Each of these has advantages for particular purposes, the first fitting in with surveyed contours and bench marks on topographic maps, the second enabling chart soundings and submarine contours to be used without modification, and the third being the most practical in field work, but the discrepancies between them increase with tide range, and it is necessary to adjust levels to the same datum before comparisons are made. The most convenient method is to convert all measurements of emerged and submerged coastline features to mean sea level datum, and state the local spring tide range above and below this level.

Emerged or submerged shore features usually consist of benches or terraces, backed by former cliffs, and sometimes bearing beach deposits or marine shell beds. Their levels can be measured at the base of the abandoned cliff, but this is frequently obscured by beach deposits, dunes or sediment that has been washed down from adjacent slopes, and it is rarely possible to determine it accurately. For this reason, the levels of emerged or submerged coastlines are usually given approximately, e.g. five to eight metres above or below mean sea level.

CORRELATION AND DATING OF FORMER COASTLINES

Segments of emerged or submerged coastlines are often found at intervals along a coast, with gaps produced by dissection or tectonic displacement. They cannot necessarily be correlated simply in terms of their altitude above present sea level, for where the land has been tilted laterally along the coast, fragments of the same coastline will be found at different levels, and fragments of different coastlines may be found at similar levels.

Where marine deposits are associated with emerged coastlines, correlation may be possible in terms of distinctive fossils. Specific assemblages of minerals may occur in old beach sands, but lateral variations in mineral composition of present-day beaches are such that correlation in terms of this evidence can only be tentative.

A more reliable means of correlation is based on radiocarbon dating of samples of wood, peat, shells, bone or coral obtained from deposits associated with former coastlines. Radiocarbon analysis permits estimates of the age of samples less than about 50 000 years old, and is a means of distinguishing Holocene deposits (less than 10 000 years old) from those formed during Pleistocene times. Dates obtained from radiocarbon measurements are stated in years BP (Before the Present), the present being defined as the year AD 1950. The margin of error is usually indicated as ± the standard deviation of the measurements made (e.g. 5580 ± 200 years BP).

Other geochronological methods include potassium–argon dating, notably of Quaternary volcanic rocks, and uranium–thorium dating which has been used to date corals, other marine carbonates and dune calcarenites. Uranium–thorium dating of emerged coral reefs has enabled the Last Interglacial high sea level to be dated at 125 000 years BP. Use has also been made of fission track dating of minerals (notably zircon) in volcanic ash deposits, and thermoluminescence (T/L), based on glow curves relating intensity of light emission to temperature, in estimating dates of sand grains in emerged coastal deposits (Bryant et al. 1996), and lead isotope (^{210}Pb) ratios have been used to date salt marsh deposits within the past century (French 1996). These and other methods of dating are being developed to determine the age of emerged and submerged coastal deposits and of associated landforms.

EMERGED COASTLINES

On many coasts there are stranded beach deposits, marine shell beds and platforms backed by steep cliff-like slopes, all marking former coastlines that now stand at a higher level than when they originally formed. Many are well above present high tide level. These are emerged coastlines, and they owe their present position either to uplift of the land, or a fall in sea level, or some combination of movements of land and sea that has left the coast higher, relative to sea level, than it was before. Most are of Quaternary age, but some of the older and higher coastline features may have originated in Tertiary times, along with shore deposits that mark the limits of Tertiary and earlier marine transgressions. As well as providing data for the geological sequence of land and sea level changes they demonstrate landforms and associated deposits developed by marine action in the past.

Sequences of terraces or emerged beaches measured on one sector of coast may not be recognizable on adjacent sectors because terrace preservation requires a particular relationship between rock resistance, the intensity and duration of marine erosion at the higher coastline phase and the degree of subsequent degradation and dissection. As a rule, permeable sandstones, limestones (including coral) and gravelly formations retain terraces better than impermeable rocks that are more resistant (restricted terrace development at the higher coastline phase) or less resistant (greater destruction of terrace by subsequent denudation). Coastal terraces are rarely well preserved on soft or rapidly-weathering rock formations.

Emerged coastlines have been reported at various levels around the world's coastline. Some are essentially platforms or benches, thought to have been formed by marine action, and backed by a rising slope interpreted as a former cliff that has been degraded by subaerial processes. There are good examples in western Scotland, notably in the Oban district, and on the coast at Palos Verdes, California, where the modern cliffs intersect terraces at various levels. Solution notches have been cut in cliffs on limestone coasts

above present sea level, notably on emerged coral islands in the Pacific Ocean (Pirazzoli 1986). In Phang-nga Bay, southern Thailand, emerged notches in a limestone cliff indicate a sudden uplift or sea level fall (Figure 3.2): a gradual change would have produced a taller single enclave as the basal ledge was cut down.

In southern Britain the best preserved emerged beach is between five and eight metres above mean sea level (typically one to three metres above high spring tide level). There are lateral variations in level which may indicate that several stages of higher sea level are represented, but the junction between modern shore platforms and the cliff base also shows lateral variations of several metres in relation to present sea level (p. 81) and the emerged beach may have shown a similar height range. In the early literature these were sometimes called raised beaches, when it was assumed they owed their higher elevation to the uplift of the coastal land, but with the realization that there have been substantial variations in sea level the term emerged beach is preferred.

The emerged beaches of southern Britain consist of sand, rounded pebbles and cobbles and occasional boulders, often with some shelly material, and sometimes cemented into sandstone or conglomerate by carbonates or ferruginous precipitates. There are good examples on the coasts of Cornwall, notably in Gerrans Bay and Falmouth Bay (Figure 3.3), and along the south coast at Portland Bill and Black Rock, near Brighton. They are overlain by Late Pleistocene periglacial deposits and so probably date from a Last Interglacial phase of higher sea level. On the limestone coasts of the Gower peninsula in South Wales similar emerged beaches are overlain by glacial drift deposits, and are therefore also of Pleistocene age.

On some coasts there are emerged features with associated deposits which have yielded a Holocene radiocarbon age. These may result from one or more episodes of higher Holocene sea level or they may be due to tectonic uplift within the past 6000 years. In places there are indications that the higher sea level was also warmer, indicated by the fossil oysters on a Holocene terrace 10 metres above sea level at Vestervøy in south-east Norway. There are also low-level emerged terraces of Holocene age on uplifted coral reefs, notably in the Pacific Ocean.

Beach ridges (p. 141) that show a seaward decline in the levels of ridge crests have been cited as evidence of an emerging coast, but as the dimensions of beach ridges are also related to the heights of the waves that built them a seaward decline of ridge crests could result

Figure 3.2   Notches on a limestone cliff in Phang-nga Bay, Thailand, the upper one formed during a higher relative sea level phase, the lower one at present mean sea level.

from diminishing wave heights. Nevertheless, there are emerged beach ridges in western Scotland that have certainly been raised by isostatic uplift. On the island of Islay in western Scotland beach ridges are found at various levels up to 37 metres above sea level, stranded on a wide Pleistocene emerged shore platform (McCann 1964).

Stairways of emerged features are found on several coasts. In North America, the Atlantic coastal plain bears depositional terraces which have been correlated with successive alternations of sea level since Miocene times, the sea attaining levels of 13.5, 6, 7.5, 4.5, and 0 metres respectively, with intervening low sea level phases (Oaks and Coch 1963). On the Pacific coast similar terraces have been dislocated by the San Andreas fault near San Francisco, and warped and uplifted terraces are found near Los Angeles and on Santa Catalina Island (Emery 1960). In Chile, coastal terraces have been displaced tectonically in a region subject to earthquakes, and as the Chilean coast runs parallel to the main axes of Andean uplift there is a possibility that terraces found at the same height

above sea level have been raised epeirogenically (Fuenzalida et al. 1965).

In the south-east of South Australia a series of stranded beach and dune ridges runs roughly parallel to the coast at successive levels up to more than 60 metres above present sea level. They were tilted transversely when the Mount Gambier region rose and the Murray-mouth area subsided during Quaternary times (Figure 3.4). The levels of the successive beach ridges result from land movements as well as sea level oscillations.

On the north-east coast of New Guinea (Huon Peninsula) a remarkable series of emerged Quaternary coralline terraces gains elevation and multiplies across an axis of uplift in the vicinity of the Tewai gorge, where the highest attains more than 600 metres above sea level. Radiocarbon dating of the younger terraces has been supplemented by uranium–thorium dating of the earlier and higher terraces to indicate that the sequence was initiated about 220 000 years ago, and that the terraces formed during successive stillstands, the last of which was about 6000 years ago. Similar

Figure 3.3   Emerged gravelly beach on the coast of Falmouth Bay in Cornwall.

results were obtained from studies of the emerged coralline terraces that form a stairway on the tilted coral cap of Barbados (Chappell 1983). On the north coast of New Guinea, near Aitape, localized uplift has raised a mangrove swamp deposit 50 metres above sea level.

Coasts bordering recently deglaciated areas have also been subject to continuing (isostatic) uplift of the land, and show sequences of differentially elevated coastlines, as on the southern shores of Hudson Bay in Canada, where Holocene beaches have been raised by up to 315 metres (Fairbridge 1983). On Skuleberget in north-east Sweden a hilltop beach on a coastline 8600 years old has been raised 286 metres above the present level of the Gulf of Bothnia by the Holocene isostatic rebound. Other Late Quaternary emerged coastlines have been traced around the Baltic Sea, the most notable being the Litorina Sea coastline, prominent as far west as Denmark, and possibly extending to Scotland and north-east England. Similar features occur around Puget Sound in the north-western United States.

Planation surfaces cut across tilted and folded geological formations have been found at various levels on each of the continents, some of them dating from Tertiary or Mesozoic times. Some are emerged marine terraces with associated beach deposits, formed during Pleistocene and earlier phases of higher sea level. In south-west England, for example, there is an extensive planation surface (known as the Trevena Platform) between 90 and 130 metres above sea level, rising to a break of slope which is thought to be a former coastline. St Agnes Beacon, in Cornwall, rises sharply above this platform, and has old beach and dune deposits at about 130 metres indicating that this hill was once an island in the sea (Bird 1998). In south-east England a planation surface rising to a possible former coastline about 210 metres above present sea level bears marine deposits that were assigned an early Pleistocene (Calabrian) age on palaeontological grounds, their contained fossil assemblage matching marine deposits of this age elsewhere. Similar platforms have been described at various levels (notably rising to breaks of slope about 130, 60, 30 and 15 metres above sea level) in

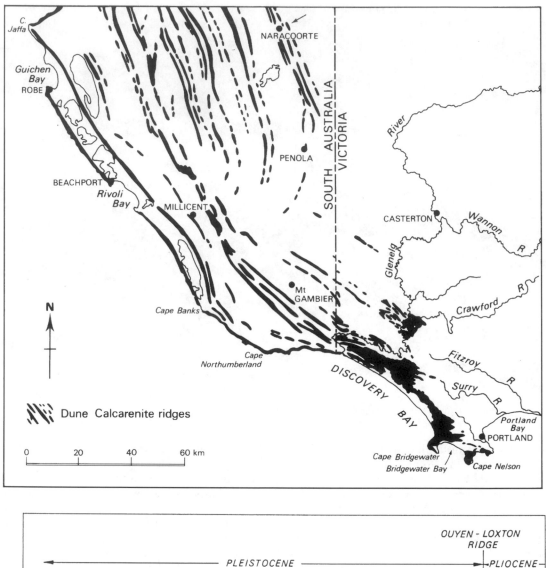

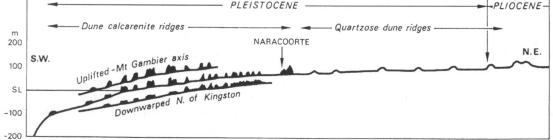

Figure 3.4   Emerged dune calcarenite ridges in the south-east of South Australia and western Victoria rise to an axis of uplift near Mount Gambier. A cross-section south-west to north-east through Naracoorte shows the ridges and intervening swales (former coastal lagoons) which occur at a higher level on the Mount Gambier axis and a lower level to the north, towards the Murray valley.

the British Isles, but confirmatory evidence of contemporary beach deposits is generally lacking.

While some planation surfaces are emerged sea floors, shaped by marine processes when the sea stood at higher levels, others may have been formed by subaerial processes, especially rivers, and some may be the outcome of periglacial or desert weathering and erosion. Whatever their origin, they have little influence on coastal geomorphology beyond determining the form and altitude of cliff crests. Thus in south-west England there are stretches of even-crested cliffs cut into platforms at various levels above the sea. Along the north coast of Cornwall cliff recession has cut into the 90–130 metre platform, whereas along the south coast the flat-topped cliffs of the Lizard Peninsula border a platform about 60 metres above sea level. Similar features are seen in south-west Pembrokeshire.

SUBMERGED COASTLINES

Submerged coastlines formed when the sea stood at various lower levels, during stillstands that punctuated falling sea levels and ensuing marine transgressions. Sea floor morphology has been charted using depth soundings and radar, and evidence of submerged coastlines has been found in the form of sharp breaks of slope, thought to be submerged cliffs and shore platforms, at various levels offshore. Stairways of submerged terraces have been mapped off various coasts, one of the clearest being the paired sequence of seven terraces at successive levels down to 110 metres below present sea level, bordering Tsugaru Strait, between Hokkaido and Honshu, Japan (Emery 1961). A submerged coastline at a depth of about 18 metres has been reported off the coasts of the United States (Shepard 1973).

Some submerged coastlines maintain consistent levels, and could have been produced when sea level was lowered by eustatic movements, or as the result of tectonic subsidence. Stearns (1974) described a sequence of submerged coastlines around the Hawaiian Islands down to a depth of 1100 metres, regarding those down to 140 metres as having formed during the Late Quaternary glacio-eustatic oscillation of sea level, while those at greater depths were the outcome of earlier tectonic subsidence. Submerged beaches have been found at various levels around the Australian coastline. In Bass Strait a well-defined break of slope indicates a submerged coastline at a depth of 60 metres

below present sea level, possibly formed during the Last Glacial low sea level phase in Pleistocene times (Jennings 1959).

Other submerged coastlines have been tilted, folded or faulted by tectonic movements that have disrupted the sea floor, as off southern California (Emery 1960). Local tectonic subsidence has led to partial submergence of prehistoric structures on the shores of the Golfe du Morbihan in Brittany.

Beach deposits have been found along some submerged coastlines, as on the floor of the Gulf of Mexico (Curray 1960). In Australia evidence of a submerged coastline came from sediment with nearshore fossils, dredged from a depth of 130 metres off the New South Wales coast (Smith and Iredale 1924). Dune calcarenite ridges that formed behind Pleistocene coastlines now submerged can be traced off the coasts of Western Australia and in Encounter Bay, South Australia, while sea floor contours near Flinders Island show the outlines of submerged parabolic dune topography.

Solution notches have been found at various levels below present sea level off limestone coasts on Crete and elsewhere in the Mediterranean, formed in low stillstands of sea level followed by sudden subsidence or a rapid sea level rise (Pirazzoli 1986). Coastal submergence in these areas is also indicated by the drowning of port structures built about 2000 years ago.

Coastal submergence is also indicated in the southeast North Sea where field patterns seen on the sea floor near Pellworm off Schleswig-Holstein, Germany are considered to be part of a landscape inundated by a storm surge in 1362. The notion that a storm surge can result in permanent submergence of a coastal landscape needs refinement, for storm surges cause only temporary inundation. Permanent submergence is more likely to be due to land subsidence here.

Much evidence of submerged coastlines has doubtless been lost because smoothing of the sea floor by wave action during successive episodes of marine transgression has obliterated or concealed them. Submerged beaches may be hidden beneath a veneer of finer deepwater sediments. Moreover, the emerged sea floors during low sea level phases became landscapes shaped by terrestrial processes (rain and rivers, frost and ice, and wind action), and coastline features that had formed during the preceding sea level fall may have been removed or buried by subaerial erosion and deposition. Because of these effects, remnants of the older submerged coastlines are unlikely to persist on the present sea floor. Those that have been found probably date from the Late Quaternary.

SEA LEVEL CURVES

When the levels of emerged and submerged coastlines are graphed against their geological age, relative sea level curves are obtained, showing the fluctuations of sea level relative to the land over a specified period. Attempts to draw curves showing absolute sea level movements (i.e. the pattern of sea level rise and fall that would be registered on a tectonically stable coastline) on a global scale have been abandoned because this is now seen as a theoretical abstraction. Where coastal submergence or emergence has taken place it is difficult, if not impossible, to separate land movements from a rise or fall in sea level, but it is useful to produce curves showing the relative levels of land and sea over time on particular sectors of the coast (Bloom 1965).

Some sea level curves show the elevation of former coastlines against their age, determined by radiocarbon or other methods of dating. Others have been drawn on the assumption that the large scale oscillations of sea level and climate during Quaternary times were related to variations in solar radiation. Estimates of these variations, based on astronomical theory, are projected back through time to show a sequence of warmer and colder periods during the Quaternary, which are correlated with interglacial and glacial phases to give a time scale for sea level fluctuations. An alternative approach used oxygen isotope ratios measured from analyses of fossil foraminifera in sedimentary cores obtained from the floors of the oceans, where sedimentation has been very slow, and deposits marking the whole of the Quaternary era are only a few metres thick. The ratio of the oxygen isotopes O^{16} and O^{18} in fossil foraminifera is an indication of the temperature of the environment at the time they formed and thus, compared with present ocean temperatures, an indication of the scale of warmer or colder climates in the past. Foraminifera obtained from successive levels in the stratified sequence of ocean floor sediments yield evidence of the sequence of past changes of climate, with warmer and colder phases that can be correlated with interglacial and glacial phases in the Pleistocene period. It is assumed that each climatic oscillation was matched by a glacio-eustatic oscillation of sea level, a warmer ocean indicating a higher sea level, and a cooler ocean a lower sea level, and this can be adapted to a graph of changing sea levels calibrated with vertical dimensions (Fairbridge 1961).

Shackleton and Opdyke (1973) analysed oxygen isotopes and palaeo-magnetism in ocean-floor cores from the Equatorial Pacific, representing a sedimenta-tion sequence over the past 870 000 years. Palaeo temperatures within the upper 2.2 metres (deposited in the past 128 000 years) were used to deduce a sea level curve which showed maxima correlative with those derived from studies of terrace sequences in Barbados and New Guinea. This indicates a high interglacial sea level (the Eemian or Ipswichian stage) about 80–120 000 years ago, a Last Glacial minimum about 20–25 000 years ago, and the ensuing Late Quaternary marine transgression.

LATE QUATERNARY SEA LEVEL CHANGES

Sea level curves indicate that about 80 000 years ago sea level around much of the world's coastline was slightly higher than it is now, but it then began to fall, and remained low during the Last Glacial phase, between 80 000 and 6000 years ago (known as the Devensian in Britain, the Würm in Europe and the Wisconsinan in North America). The extent of lowering of sea level during this Last Glacial phase has been estimated from the volume of water abstracted to form the late Pleistocene glaciers and ice sheets, which indicates that the sea fell more than 100 metres below its present level. An estimate based on extrapolation of the pre-Holocene floor of the Mississippi valley (which can be traced beneath a thick alluvial fill) out to the edge of the continental shelf, suggested a lowering of 140 metres (Fisk and McFarlan 1955).

As a result of sea level lowering during the Last Glacial phase the world's continental shelves emerged as wide coastal plains, and coastlines advanced towards their outer edges. The British Isles became a peninsula of Western Europe, there were wide coastal plains off what is now the Gulf and Atlantic coasts of the United States, and Australia was enlarged and linked to Tasmania and New Guinea (Figure 3.5). Rivers extended their courses to the lowered coastlines, incising valleys across the continental shelves and dissecting earlier terraces.

Reference has been made to the fact that about 18 000 years ago the earth's climate was still very cold, glaciers and ice sheets were close to their maximum extent, with the sea about 140 metres below its present level. Then the earth's climate became warmer, and the ice cover started to melt. As water returned to the oceans there was a world-wide rise of sea level, known as the Late Quaternary marine transgression, between 18 000 and 6000 years ago, when the rising sea submerged the Last Glacial landscapes of

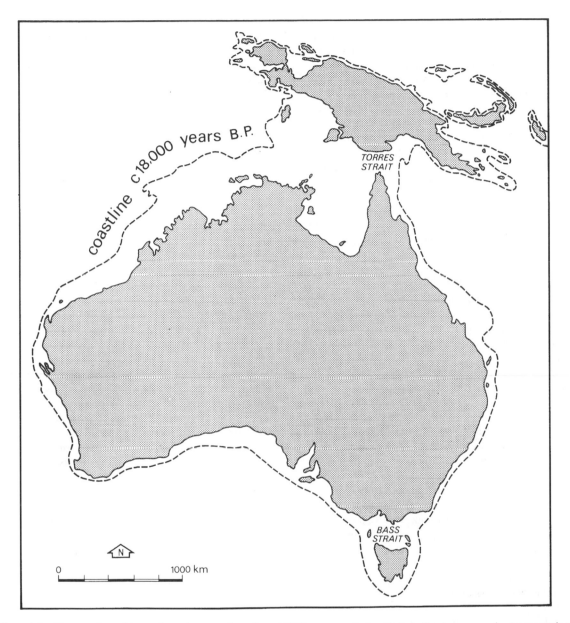

Figure 3.5   The coastline of Australia and New Guinea about 18 000 years ago, before the Late Quaternary marine transgression rose to submerge the continental shelf and separate Australia from Tasmania by the formation of Bass Strait (9000 years ago), and from New Guinea by the formation of Torres Strait (8000 years ago).

the continental shelves and brought the sea to roughly its present level.

Evidence of stages in Late Quaternary marine submergence has been obtained from borings in coastal plains and deltas (notably on the Gulf Coast of the United States and in the Rhine delta), where fresh water peat and relics of land vegetation have been encountered well below the present sea level, overlain by marine deposits formed as the sea rose. These stages have been determined by dating (mainly by radio-carbon ratios) such materials as shells, wood or peat associated with these deposits found at specific depths below present sea level. In attempting to relate dated materials to contemporary sea levels it is necessary to take account of the extent to which stratigraphic horizons have been lowered, relative to present sea level, by subsequent compaction of underlying sediments (especially peat and clay) and by the crustal subsidence (an isostatic response to sedimentary loading) which has taken place in many deltaic regions.

When the dates are plotted against former sea levels a graph is produced which traces the sequence of sea level changes relative to the land for particular coastal regions. The resulting curves show that the sea level rise was rapid, averaging just over a metre per century between 18 000 and about 6000 years ago. Most studies have deduced an oscillating rise, with pauses and occasional slight regressions, particularly where there are Holocene stratigraphic sequences of alternating peats and marine sediments. A spasmodic sea level rise, with minor advances and retreats, would have aided the shoreward sweeping of sea floor sediment described on p. 104. However, some studies have indicated a smooth and steady increase accompanying the warming of global climate, and have judged the supposed oscillations as statistical aberrations. Further research is needed to decide which is the correct view.

There are certainly discrepancies between Holocene sea level curves from different parts of the world's coastline (Figure 3.6), some of which result from the complicating effects of land uplift or depression in coastal regions, and some from regional variations in the scale and sequence of sea level changes. The contrast between Late Quaternary sea level history on the Atlantic coast of the United States, where the marine transgression is still proceeding slowly, and southeastern Australia where the marine transgression slackened or came to a halt about 6000 years ago, is well known, as is the fact that post-glacial isostatic recovery has produced a falling sea level in Scandinavia and northern Canada.

Submergence of Late Pleistocene landscapes as this marine transgression proceeded may be the basis for the story of Atlantis, a land area lost beneath the sea. Archaeologists have been inclined to the idea that the submergence of Atlantis, mentioned by Plato as a lost land off the west coast of Europe but now thought to be near Crete in the Mediterranean, was due to the tsunami generated by the explosive eruption of the Thera volcano (Santorini) about 3500 years ago, but although this would account for widespread damage it would not explain the permanent submergence of a former land area (James 1995).

Similar legends recur in the folklore of coastal people. In south-west England the lost land of Lyonesse is thought to have lain in the area of the Isles of Scilly, while on the Welsh coast stories of drowned cities and palaces probably originate from imaginative interpretations of the sarns, bouldery ridges of Pleistocene glacial drift that extend out from the coast and across the sea floor, submerged by the Late Quaternary marine transgression (Steers 1964). Coastal tribes of Australian aborigines tell stories suggesting that their ancestors retreated from lands now submerged on the sea floor.

Processes of marine erosion and deposition began to shape existing coastlines as the sea approached its present level. Areas that had been uplands on the emerged coastal plain during the Last Glacial low sea level phase became islands or reefs offshore, their outlines modified by wave erosion and deposition. Valleys that had been incised across the sea floor were submerged and largely filled with sediment. It is sometimes possible to trace their offshore alignments in existing sea floor morphology, and they can be located by seismic surveys, which detect the sediment fill. Along the present coastline the landforms described in later chapters, that took shape as low-lying areas, were submerged to form embayments and inlets, valley mouths became estuaries, and higher ground persisted as steep coasts and promontories. Cliffs were cut as the rising sea encountered rock outcrops of varying resistance; sand and gravel were deposited to form beaches, spits and barriers; wind-blown sand accumulated as coastal dunes; river sediments began to fill estuaries, and in some places to build protruding deltas; and corals and other marine organisms built reef structures.

Much remains to be done to elucidate the sequence and effects of Late Quaternary sea level changes on coastal landforms. An example of a problem that remains to be solved is the presence of glacial erratics on the shores (and in some low-level emerged beaches) of south and south-west England and Brittany. One of these erratics, on the shore platform near Porthleven in Cornwall, is Giant's Rock, a large boulder of microcline gneiss of a kind unknown in Britain (Figure 3.7). It has been generally assumed that erratics of this kind have been deposited from icebergs that melted when they became stranded on the shore, but they must have been emplaced when the sea stood at or a little above its present level. The shores of the English Channel are well beyond present-day iceberg limits, and if the climate was colder sea level should have been somewhat lower than it is now. One possibility is that icebergs were at that stage much larger, and floated further south, during stages when the Scandinavian and North Atlantic ice sheets were

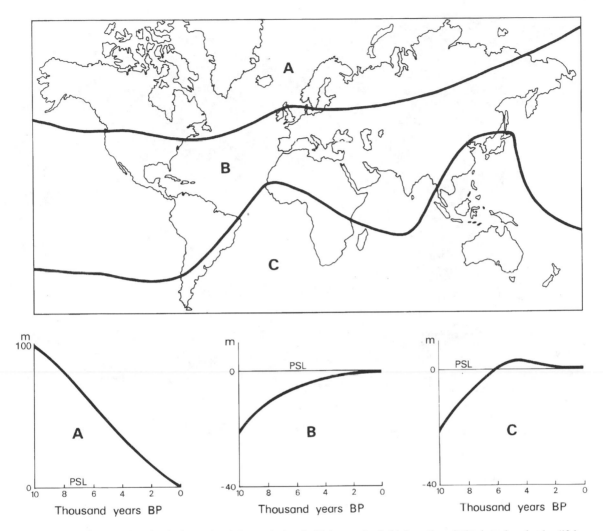

Figure 3.6 Major variations in relative sea level change during the Holocene. A – in high northern latitudes, where land uplift has resulted in a falling sea level, B – in middle latitudes, where the sea has risen at a slackening rate to attain (but not exceed) its present level, C – in much of the southern hemisphere and south-east Asia, where the sea rose above its present level between 3000 and 6000 years ago and has since fallen back. There are many other local variations related to local land uplift or subsidence, and to geodetic changes in ocean level.

melting rapidly and disintegrating, possibly before the Gulf Stream became established to bring relatively warm water to the English Channel. Alternatives are that the Antarctic ice sheet was much reduced, raising ocean levels while northern hemisphere ice persisted, or that there have been sea level oscillations in the English Channel independently of the global glacio-eustatic response to the retreat of ice sheets (Bird 1998).

Another problem is posed by accordant valley mouths (i.e. where streams descend to present sea level) which may result from uplift of coastal land off-setting the Late Quaternary sea level rise that would otherwise have submerged the valley mouth. Examples of this occur on the coasts of southern Norway and eastern Scotland, where land uplift has occurred as a result of isostatic recovery following glaciation. Alternatively, as in the chines of the Isle of Wight, stream downcutting may have been matched by cliff recession with the sea at its present level.

Figure 3.7 Giant's Rock, near Porthleven, Cornwall, a large erratic boulder (1.8 metres high, 2 metres wide and 3.6 metres long) thought to have been deposited on the coast from a melting iceberg in Late Pleistocene times. For discussion of relative sea level implications see p. 40.

MODERN SEA LEVEL CHANGES

Around much of the world's coastline the sea level has been relatively stable during the past 6000 years, apart from minor oscillations, of the order of a metre. This has been a period of Holocene stillstand, when the relationship between land and sea level has been more or less stable. There are, nevertheless, sectors where the coastal land has continued to rise or fall within this period. Emerging coasts (as around the Gulf of Bothnia) border shallowing seas, and often show progradation as the result of sea floor sediments being carried shoreward by wave action, whereas submerging coasts border deepening seas and often show increasing erosion as waves grow larger and sediment is lost to the sea floor.

Continuing submergence is obvious in the Venice region of Italy, where sea level has risen by up to 30 centimetres since 1890, largely because of land subsidence following groundwater extraction. The frequency of city flooding by high tides (aqua alta) has increased from four times a year in the 1900s to 100 times in 1997 (Pirazzoli 1983). Bangkok is one of several other coastal cities where sea level has risen because of land subsidence correlated with groundwater extraction, and near Port Adelaide an annual relative sea level rise of up to a centimetre has been attributed partly to extraction of groundwater and partly to wetland drainage (Belperio 1993). Coastal subsidence south of Los Angeles, California, has resulted from the pumping of oil from underground strata, and there have been similar effects near Galveston in Texas and on the Bolivar coast in Venezuela. In the Ravenna region of Italy a sea level rise of up to 1.3 metres occurred between 1950 and 1986, due to subsidence resulting from subsurface compaction following the extraction of natural gas as well as groundwater. This drawdown has been most severe over the Ravenna Terra gas field north-east of the city, and there has been increased sea flooding along the Adriatic coastline where it intersects the subsidence bowl.

Evidence of modern sea level changes has been obtained from analyses of tide gauge records, from geodetic surveys and from certain biological indicators.

Tide gauge records

Some measurements of changes in sea level have been based on data from tide gauges, using calculations of annual and decadal means and running means to determine changes in mean sea level during the period for which the records are available. Most tide gauges are instruments installed for the purposes of navigation and harbour operations rather than to provide scientific information. They register actual fluctuations of the sea surface, including the effects of wind stress, waves, and changes in atmospheric pressure, and also register local effects, such as current swirl. Many are located at ports, on structures where there may be local sea level anomalies resulting from wave reflection and ponding, and where there is a risk of damage and disturbance in the course of ship movements and port activities. Their datum levels are sometimes poorly maintained, and if a disturbance or accident results in the subsidence of a harbour structure the tide gauge records this spuriously as a rise in mean sea level.

Some caution is therefore necessary in measuring sea level changes from available tide gauge records. More scientific instrumentation is being developed around the world's coastline, and this will in due course provide improved monitoring of sea level changes in the future.

Geodetic surveys

An alternative source of information on sea level changes in recent decades has been provided by repeated geodetic surveys of land areas, based on precise levelling. In Sweden and Finland, for example, successive geodetic surveys have confirmed the fall in sea level indicated by tide gauges along the shores of the Bothnian Gulf, which results from continuing isostatic uplift of Scandinavia following deglaciation (Figure 3.8). In the United Kingdom there is evidence from tide gauges that mean sea level has risen faster in the south-east and south-west (Sheerness and Newlyn) than in the north (North Shields and Aberdeen) (Woodworth 1987), but geodetic confirmation of transverse tilting is still awaited.

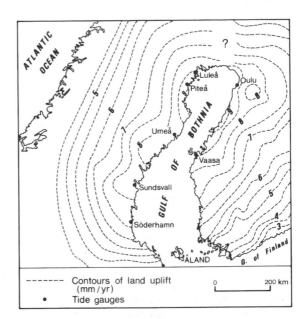

Figure 3.8 Rates of land uplift around the Gulf of Bothnia, based on Swedish and Finnish geodetic surveys.

Geodetic surveys are elaborate and expensive, and it will be some time before they can contribute much to knowledge of global sea level changes. Monitoring from satellites will eventually provide an integrated global tidal facility, with more accurate information on the rise and fall of land and sea levels around the world, using measurements based on the earth's centre as a datum. It will then be possible to distinguish the effects of land uplift or subsidence and sea surface rise or fall on vertical changes in sea level relative to the land registered in tide gauge records (Goldsmith and Hieber 1991).

Biological indicators

Evidence of sea level changes may also be obtained from repeated surveys of the levels of marine organisms such as oysters, barnacles, mussels, algae and kelp, where these are found at specific tidal levels in vertical zones encrusting cliffs, rocky shores, sea walls and pier supports. Biological zones are correlated with the depth and duration of marine submergence with the sea at its present level (Lewis 1964), and can therefore be expected to move up or down in relation to a rise or fall of sea level. A sea level rise should be indicated by upward movement of such zones on cliff faces, stacks and rocky protrusions, as well as on artificial structures, accompanied by landward movements across shore platforms and intertidal outcrops.

Measurements of the upper and lower levels of oysters and barnacles on concrete piling at Miami Beach, on the subsiding coast of south-east Florida, showed that these horizons moved 15 centimetres upward between 1949 and 1981, consistent with the relative rise of mean sea level registered on nearby tide gauges (Wanless 1982). It may be possible to detect such migrations with reference to historical photographs where mean sea levels can be determined in relation to fixed features such as steps or decking, and it would be useful to document existing levels of zoned organisms as a basis for future measurements. The most suitable organisms are readily identifiable plants or animals that occupy particular parts of the intertidal zone with consistent, well-defined upper or lower boundaries that can be correlated with particular stages of the tide, preferably at or close to mid-tide level. Indicator organisms should be able to migrate upwards as sea level rises by rapidly colonizing higher levels that are either untenanted, or occupied by other plants or animals that can be displaced or overrun.

An example from south-eastern Australia is the calcareous tubeworm *Galeolaria caespitosa*, which occupies a well-defined intertidal zone and is best developed on sites sheltered from strong wave action and abrasive sand movements. It forms either a thin layer, or cauliflower-like encrustations similar to the trottoirs, accretionary ledges of coralline algae that protrude from some Mediterranean rocky shores (p. 91). The vertical range of *Galeolaria* is typically 20 to 40 centimetres, with an upper limit close to mid-tide level, irregular on sites exposed to strong waves and variable swash and spray, but horizontal on sheltered sites such as the inner sides of harbour walls (Bird 1988). Monitoring of the upper limit of *Galeolaria* in Port Phillip Bay, initiated in 1988, indicated a rise of a few millimetres over the following decade, possibly in response to a sea level rise.

Sea level changes may also be indicated by studies of changes in zoned patterns of vegetation in salt marshes and mangrove swamps, but it is difficult to obtain precise measurements of sea level changes from such evidence.

RECENT CHANGES OF LAND AND SEA LEVEL

Evidence of recent changes of land and sea level from tide gauge records was discussed by Pirazzoli (1986),

who analysed tide gauge records from 229 coastal stations that had been maintained for at least 30 years, and found that 63 (28.5 per cent) showed a mean sea level rise exceeding two millimetres per year, 52 (22.5 per cent) between one and two millimetres per year, and 47 (20.6 per cent) less than one millimetre per year, the remaining 65 (28.5 per cent) having shown a mean sea level fall. As over 70 per cent of the records showed a positive trend a global sea level rise seemed likely, but the geographical distribution of the 229 stations was uneven, with strong northern hemisphere mid-latitude clustering, and only six in the southern hemisphere. Reference has been made that most tide gauges are located at ports, and may not be reliable indicators of mean sea level changes.

Emery and Aubrey (1991) extended this review, using 664 tide gauge stations, 65 per cent of which had kept records for at least 30 years. Analysis of 98 key stations showed that sea level had fallen only on sectors where the coast had been rising tectonically or isostatically. The Global Sea Level Observing System (GLOSS) is developing a much more representative network of 287 tidal stations around the world's coastline, and will provide more accurate information in the next few decades (Pugh 1987b, Woodworth 1991), both globally and regionally (Pugh 1991).

Sea level changes over the past few decades should be considered against the background of factors listed previously, which are known to have influenced Holocene sea level trends. Climatic fluctuations are thought to be responsible for the major oscillations of water level that have taken place in the Great Lakes in North America, during the past century, and to have contributed (along with dam construction in tributary rivers) to the changes in level recorded on tide gauges around the Caspian Sea which fell by 2.67 metres between 1930 and 1975, and has since been rising (Figure 3.9), with the onset of more humid conditions in the surrounding area (Kaplin and Selivanov 1995).

Tectonic movements have certainly influenced changes of land and sea level in recent decades. Sea level has continued to fall on coasts where isostatic land uplift has followed Late Quaternary deglaciation, as in parts of Scandinavia, Northern Canada, and Alaska, and where the land is rising tectonically, as in northern New Guinea, some Indonesian and Philippine islands, and parts of the Japanese coast (Pirazzoli 1991). Features characteristic of such emerging coasts include prograding beaches and marshlands and rivermouth rapids, as in the Fotlandsvatnet in south-west Finland (Ristaniemi et al. 1997) and the Selfloss River

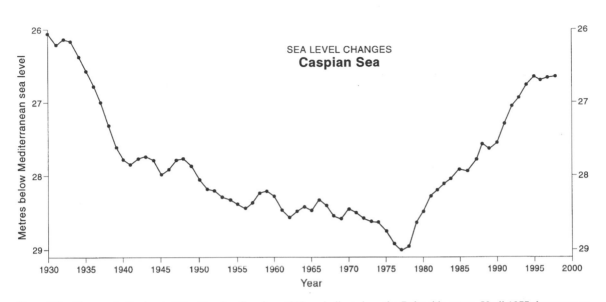

Figure 3.9 Changes in the level of the Caspian Sea since 1930, as indicated on the Baku tide gauge. Until 1977 there was an intermittent lowering of sea level, but subsequently there has been a rise of about two metres, probably because of a trend towards a more humid climate in the region. Information supplied by Dr S. Lukyanova of Moscow State University.

in Iceland. Emergence may result in the incision of streams in deltas and coastal plains, the drainage and dissection of salt marshes and mangrove swamps and their seaward progradation, the abandonment and degradation of cliffs, and shoreward drift of sea floor sediment to widening beaches, as on the shores of the Gulf of Bothnia.

Sea level has been rising on coasts where the land margin is subsiding, as on the Gulf and Atlantic seaboards of the United States, the south and south-east coasts of Britain, the Netherlands and north Germany, north-eastern Italy, and several other areas (Milliman and Haq 1996). The floor levels of Roman London are now about four metres below high-tide level in the Thames estuary, indicating an average rate of submergence of about two millimetres per year in this area over the past two millennia. At Newlyn in Cornwall, mean tide level in the 1990s was about 20 centimetres higher than that recorded for 1915–21, when the mean sea level at this site was calculated to establish a datum for British topographical surveys: a rise of just over two millimetres per year. It is possible that the records from this tide gauge station have been affected by hydro-isostatic subsidence (p. 30) of the adjacent Land's End peninsula. Continuing subsidence in southern and eastern England may be responsible for the persistence of broad estuaries in Suffolk and Essex, the wide mouth of the Thames estuary, and embayments such as Chichester Harbour and Southampton

Water, in each case submergence having exceeded the rate of sedimentary infilling. By contrast, mean tide level at Dundee in Scotland fell by just over two millimetres per year in this period, any sea level rise being more than offset by the effects of the ongoing isostatic uplift following Late Quaternary deglaciation (p. 30). If a sea level rise is taking place around Britain, it has probably been in progress since the Little Ice Age phase of colder climate recorded in the mid-eighteenth century. The possibility that tectonic deformation has continued in Britain since Late Tertiary times (the past five million years) cannot be ruled out.

Figure 3.10 shows the location of sectors of the world's coastline that have been subsiding in recent decades, as indicated by evidence of tectonic movements, increasing marine flooding, geodetic surveys, and tide gauges recording a rise of mean sea level greater than two millimetres per year over the past 30 years. Some are in areas of isostatic downwarping around major deltas; others are at least partly the outcome of human activities, notably groundwater extraction, as in the Venice and Bangkok regions, and oil extraction, as in Southern California.

The balance of the evidence suggests that a global sea level rise is in progress (Pirazzoli 1996), but more extensive monitoring is required to confirm its dimensions. It is possible that the so-called contemporary world-wide marine transgression of between 1 and 2 mm/year has been over-estimated, or that more com-

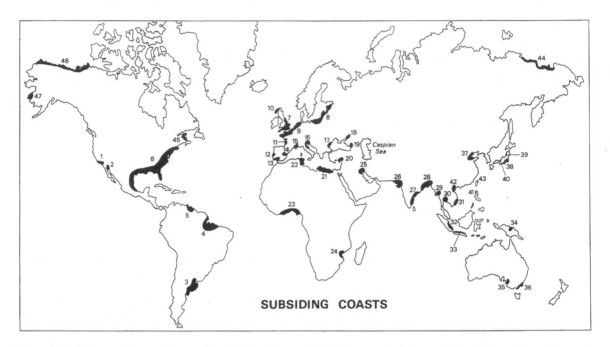

Figure 3.10 Sectors of the world's coastline that have been subsiding in recent decades, as indicated by evidence of tectonic movements, increasing marine flooding, geomorphological and ecological indications, geodetic surveys, and groups of tide gauges recording a rise of mean sea level greater than two millimetres per year over the past three decades.

KEY TO MAP: 1, Long Beach area, southern California; 2, Colorado River delta, head of Gulf of California; 3, Gulf of La Plata, Argentina; 4, Amazon delta; 5, Orinoco delta; 6, Gulf and Atlantic coast, Mexico and United States; 7, Southern and eastern England; 8, The southern Baltic from Estonia to Poland; 9, North Germany, the Netherlands, Belgium and northern France; 10, Hebrides, Scotland; 11, Loire estuary and the Vendée, western France; 12, Lisbon region, Portugal; 13, Guadalquavir delta, Spain; 14, Ebro delta, Spain; 15, Rhône delta, France; 16, Northern Adriatic from Rimini to Venice and Grado; 17, Danube delta, Romania; 18, Eastern Sea of Azov; 19, Poti Swamp, Soviet Black Sea coast; 20, South-east Turkey; 21, Nile delta to Libya; 22, North-east Tunisia; 23, Nigerian coast, especially the Niger delta; 24, Zambezi delta; 25, Tigris–Euphrates delta; 26, Rann of Kutch; 27, South-eastern India; 28, Ganges–Brahmaputra delta; 29, Irrawaddy delta; 30, Bangkok coastal region; 31, Mekong delta; 32, Eastern Sumatra; 33, Northern Java deltaic coast; 34, Sepik delta; 35, Port Adelaide region; 36, Corner Inlet region; 37, Hwang-ho delta; 38, Head of Tokyo Bay; 39, Niigata, Japan; 40, Maizuru, Japan; 41, Manila, Philippines; 42, Red River delta, North Vietnam; 43, Northern Taiwan; 44, East Siberian coastal lowlands; 45, Maritime Provinces, Canada; 46, Mackenzie delta and northern Alaska; 47, Yukon delta, Alaska.

The coast of the Caspian Sea is included, as sea level has risen by more than two metres since 1977 (Figure 3.9).

plete global studies will show that there have been geographical variations in the nature and scale of sea level change during the past century similar to those indicated by Holocene sea level curve discrepancies.

FUTURE SEA LEVEL CHANGES

During the next few decades monitoring of sea level changes, using the Global Sea Level Observing System (GLOSS), will clarify the pattern of upward and downward movements of the sea relative to the land, and is likely to demonstrate a world-wide sea level rise resulting from predicted global warming by the enhanced greenhouse effect. The nature and effects of a world-wide sea level rise will be discussed in Chapter 12.

4

Cliffs and Rocky Shores

INTRODUCTION

Three-quarters of the world's coastline is cliffed and rocky. The geomorphology of cliffs and rocky shores has been discussed in textbooks by Trenhaile (1987) and Sunamura (1992), and summarized briefly by Griggs and Trenhaile (1994). Cliffs are strongly influenced by the geology of coastal regions, particularly the structure and lithology of rock formations that outcrop on the coast and their response to weathering and erosion processes. Rock formations of varying age, from Pre-Cambrian (more than 560 million years old) to Holocene, outcrop on the world's coastlines, but most cliffs have been shaped during Pleistocene and Holocene times, mainly the past 6000 years, when the sea has stood at or close to its present level.

Climate has been an important influence on the weathering of coastal rock outcrops, which results from physical, chemical, and biological processes, related partly to subaerial conditions and partly to the presence or proximity of the sea. Rocks are decomposed or disintegrated by such processes as repeated wetting and drying, solution by rain water, thermal expansion and contraction, freeze-thaw alternations and shore ice effects, all related to temperature, precipitation, and evaporation regimes in the coastal environment. Rock debris falls to the cliff base as talus, which must be consumed or removed by wave action if cliff recession is to continue; if it persists as a protective beach, and marine erosion at the cliff base is halted, the cliff becomes degraded by subaerial processes.

Where the rocks are very resistant (e.g. massive granites) steep and high coastlines have changed little, if at all, as the result of marine erosion over the past 6000 years, but less resistant formations have been cut back as cliffs, some bordered by irregular rocky shores, others with smoother shore platforms at least partly exposed at low tide. Elsewhere there are cliffs or coastal slopes that plunge into deep water. A distinction is sometimes made

between hard rock cliffs and soft rock cliffs, also known as earth cliffs (May 1972). There has been little cliffing on emerging coasts, such as those of the Gulf of Bothnia in the Baltic Sea, where wave erosion has been withdrawing seaward across an emerging sea floor.

The simplest cliffs are found where marine erosion has attacked the margins of a stable land mass of coherent rocks, removing a wedge of material to form a steep (usually >30°, and often vertical) slope fronted by a shore platform. On many cliffed coasts there are complications introduced by the lithology and structure of outcropping rock formations, the degree of exposure to wave attack, the effects of subaerial weathering (physical, chemical and biological) on the coast, and the history of changing land and sea levels (Emery and Kuhn 1982). Even on relatively weak rock formations, the cutting of cliffs and shore platforms takes time, and implies that the sea has remained at or close to the same level in relation to the land for a prolonged period. However the small cliffs known as clifflets or microcliffs, cut into salt marshes and mangrove swamps (Chapter 8), can form quickly.

A distinction may be made between cliffs, steep to vertical coastal slopes that expose hard or soft rock outcrops and bluffs (which may formerly have been cliffs), more rounded and subdued in profile, with rock outcrops concealed by a weathered mantle, soils and vegetation. Cliffs are often (but not always) actively receding, their gradient depending on relative rates of cliff crest and cliff base recession. Bluffs are usually more stable (although some are retreating by way of intermittent and local slumping, and could be termed wasting bluffs). There is the difficulty that some cliffs, especially in North America, have been named bluffs: Scarborough Bluffs on the north coast of Lake Ontario, Canada, are receding cliffs. Coastal slopes which are partly vegetated and partly rocky may be termed cliffy bluffs.

Cliffs become bluffs as the result of subaerial degradation, notably where there has been coastal emergence, the accumulation of a protective beach, or the building of a protective artificial structure. Savigear (1952) traced stages in subaerial degradation (including periglaciation) of former cliffs cut in Old Red Sandstone to rounded slopes on the south Wales coast eastward from Pendine. Bluffs become cliffs when marine erosion is intensified by coastal submergence or the loss of a protective beach.

Steep coastal slopes that are not actively receding, but descend steeply to the shore, are seen on the north Devon coast bordering Exmoor (Figure 4.1), and the coast of the French and Italian Rivieras. Much of the New Zealand coast is similarly bold (Cotton 1974). The building of roads along steep coasts often results in instability, both during their construction when the slope is excavated and debris spills down, perhaps to the shore, and when the road is disrupted by slumping or rock falls.

In addition to cliffs and bluffs there are other steep coasts consisting of slopes that have been shaped primarily by subaerial processes, and are dissected by short, steep stream valleys. Some descend to basal cliffs and perhaps shore platforms, as on the steep coast of the Otway Ranges in south-eastern Australia, others to rocky or boulder-strewn shores, as on the steep coast of Macalister Range in north-eastern Australia. Both are relatively straight steep slopes, possibly upfaulted along the coastline. Off glaciated coasts the intertidal zone is often bouldery, with coarse residual material where moraines or boulder clay have been washed over by the sea, and waves have swept away the finer sediment.

CLIFF MORPHOLOGY

Vertical cliffs are best developed on homogeneous or well-stratified rock formations, notably sandstones and limestones (including emerged coral reefs). Coherent silty sediments such as brickearth (loess) stand in vertical cliffs on the shores of Pegwell Bay in Kent, and along the south-eastern shore of the Sea of Galilee, and there are vertical cliffs cut in glacial drift deposits along the Holderness coast in north-east England. The simple association of vertical cliffs

Figure 4.1   The steep coast of North Devon near Lynmouth, where vegetated slopes descend to a rocky and bouldery shore. The shore widens off the mouth of the River Lyn across a gravelly intertidal delta, which was enlarged during a severe flood in 1952.

fronted by wide seaward-sloping shore platforms is well illustrated on the chalk coasts bordering the English Channel (Figure 4.2). The chalk, an Upper Cretaceous formation, is a stratified limestone up to 500 metres thick, with some minor lithological variations, including nodular layers of hard siliceous flint which weather out to form shingle beaches (p. 95).

Cliff morphology is often related to variations in lithology and structure, picked out by marine erosion as the cliff base is cut back. The more resistant parts of coastal rock formations protrude as headlands, or persist as rocky stacks and islands offshore, whereas the weaker elements are cut back as coves and embayments. Resistance in this context means the strength and hardness of rocks attacked by the physical forces of marine erosion, or their durability in the face of physical, chemical and biological weathering processes. There are many examples around the world's coastline: the Pembrokeshire coast between St David's Head and Strumble Head has a succession of bays and coves cut out in relatively weak Ordovician sedimentary outcrops and intervening headlands on igneous intrusions (Steers 1953). Old Castle Head in south Pembrokeshire protrudes because hard rocks outcrop at the cliff base. A distinction may be made between discordant cliffs, cut across the strike of several geological formations, as on the Dorset coast, and concordant cliffs, which run parallel to the strike, along a single formation, such as the rugged cliffs on Whin Sill dolerite on the Northumbrian coast.

Rock resistance depends on several factors. Massive rocks are generally more resistant to erosion than rock formations divided by many joints, bedding planes, cleavage planes and fractured zones, which facilitate cliff dissection. It is possible to recognize categories of rock resistance ranging from very resistant quartzites and sandstones, massive granite and indurated metamorphic rocks down through moderately resistant slates, shales, grits and basalts to weak limestones (including chalk) and sandstones to very weak mudrocks (defined as rocks containing at least 90 per cent silt and clay, and formerly known as siltstones, shales and clays) and unconsolidated sands (Clayton and Shamoon 1998), but these categories have not yet been correlated with coastline configuration.

Figure 4.2 Chalk cliffs and an intertidal shore platform at Seven Sisters on the Sussex coast. The chalk here dips gently seaward.

Solid and massive formations are generally eroded more slowly than formations that disintegrate readily, such as friable sandstones, rocks with closely-spaced joints and bedding-planes or rock formations shattered by faulting. Most rock formations have planes of division that are weakened by weathering processes and penetrated by marine erosion, influencing the outline in plan of a cliffed coast. These include bedding planes, cleavage planes, joints and faults, overthrusts, and zones of soft or shattered rock. They are excavated by weathering processes and wave scour to form crevices, clefts, inlets and caves along the coast. Shore rocks are attacked by quarrying, the hydraulic pressure of breaking waves forcing air and water into fissures, which are gradually enlarged to form clefts and gullies. Intricate dissection by wave action along planes of weakness (including joints and faults) has produced the irregular and crenulate coastal outlines of the Tintagel area in north Cornwall, with many steep-sided coves, inlets, caves and gullies, all closely related to the intricacies of local geology (Wilson 1952). Many Cornish headlands are on strongly metamorphosed Devonian greenstones, between bays cut back into less resistant slates and shales.

On the coasts of Kent and Sussex vertical cliffs have been cut in chalk where the strata dip gently seaward (Figure 4.2), but in the Isle of Wight and on the Dorset coast, where the chalk outcrop dips steeply and is often strongly folded, there are more irregular features related to minor variations in lithology, such as ledges and reefs on the more resistant layers (e.g. the Melbourn Rock) and gentler slopes on softer marly horizons within the chalk. Where the dip is landward, as at Ballard Down in Dorset (Figure 4.3), and at the eastern (Culver Cliff) and western (Tennyson Down) ends of the Isle of Wight the cliffs undercut the chalk escarpment, and may be described as escarpment cliffs (Bird 1995, 1997a).

Contrasts related to lithology and structure are also well displayed on cliffs cut in Jurassic formations along the Dorset coast, where high, bold promontories occur on the limestones and bays have been excavated in outcrops of clay, shale or soft sandstone. On the north coast of Devon the cliffs near Hartland Point are vertical, cut by stormy seas across Carboniferous sandstones and shales compressed into tight zigzag folds along vertical axes running at right angles to the coastline. In detail the harder sandstones protrude and the weaker shales have been cut out, while on a larger scale there are minor headlands on anticlinal zones that have proved slightly more resistant, and coves and inlets along the transverse synclines.

More resistant formations, such as the granites of the Land's End peninsula in Cornwall, have been dissected by wave action along horizontal, inclined and vertical joint planes to form headlands and clefts, which in places have a castellated appearance related to the cuboid jointing. Distinctive cliffs have developed on columnar basalts, as in the Giant's Causeway in Northern Ireland, and there are vertical cliffs and stepped rocky shores on columnar dolerites at Pillar Point in south-eastern Tasmania. Old Red Sandstone

Figure 4.3   Escarpment cliff in landward-dipping chalk at Ballard Down in Dorset.

is another resistant formation which forms high cliffs on the north coast of Scotland between Duncansby Head and Skirza Head and on the west coast of the Orkney Islands, where cliffs up to 60 metres high are retreating by way of frequent rock falls and removal of the fallen debris by wave action. Also in Scotland the rugged cliffs of St Abb's Head show an impressive array of crags, clefts, caves, gullies, reefs, stacks and skerries, formed by dissection along numerous planes of weakness in Silurian slates and Devonian volcanic rocks, while rapid variations in cliff and shore morphology are related to intricate outcrops of sandstone, limestone and volcanic rock on the nearby coast at Dunbar.

Very high cliffs (rising more than 300 metres from the sea) are known as megacliffs. They are found on the Pacific coasts of Peru and Chile where coastal land has been uplifted tectonically, and on volcanic islands in the Atlantic Ocean, notably the Canary Islands, Madeira and Tenerife. High vertical cliffs border lava flows in Skye, eroded by the sea along joint planes, but their morphology is similar to (and passes laterally into) that of lava cliffs produced by glacial and periglacial processes inland, and fronted by grassy colluvial aprons.

Profiles of cliffs and bluffs

Cliff profiles are related to the structure and resistance of outcropping rock formations, the effects of subaerial weathering and denudation, and undercutting by marine erosion. The presence of harder or softer elements in the cliff face can influence the profile, as on Golden Cap in Dorset (Figure 4.4), where harder outcrops form ledges and the softer sands and clays gentler (sometimes vegetated) slopes and areas of subsidence (Bird 1995). Figure 4.5 shows structural ledges as steps on horizontal sandstones separated by weak clays. On the south-west coast of the Isle of Wight sandstone horizons in the clay-dominated Wealden Beds produce relatively steep cliffs where they outcrop at the cliff crest, as at Barnes High, and cliff-face ledges where they outcrop in the cliff profile, as at Sudmoor Point. In general, bolder cliffs occur on coasts exposed to strong wave action from the open sea, and more

Figure 4.4 Slope-over-wall profile at Golden Cap, Dorset. There is a capping of Upper Greensand, the slope is cut in Middle Lias sands and clays and the basal cliff in Lower Lias shales and marls.

Figure 4.5   Structural ledges on the cliff at Womboyn, New South Wales, Australia, corresponding with the upper surfaces of flat-lying sandstones separated by layers of soft clay.

subdued slopes on relatively sheltered sectors. On the western and south-eastern coasts of Australia cliffs have been cut into Pleistocene dune calcarenite, a calcareous sandstone formed by the lithification of coastal dune deposits (p. 191). These cliffs vary in profile with the inclination of dune bedding, with bold headlands where the dip is steep and seaward and gentler rugged slopes where it is landward.

On the high wave energy coast near Port Campbell in Australia vertical cliffs have been cut by marine erosion in horizontal stratified Miocene sedimentary rocks, and the huge waves that break against these during storms have cut out ledges along the bedding-planes at various levels up to 60 metres above high tide (Baker 1958). These structural ledges are rarely more than five metres wide, and are the product of present-day storm wave erosion; they are not emerged shore platforms. The power of storm waves is illustrated where large boulders have been thrown up on to the cliffs, as at Quobba in Western Australia and on Grand Cayman (Jones and Hunter 1992).

Cliffs on more sheltered sections of the coast, where strong wave action is intercepted by headlands, islands, or offshore reefs, or attenuated by a gentle offshore slope, may show profiles partly formed by subaerial weathering and erosion as well as those shaped by marine attack. Cliffs in these situations may develop slope-over-wall profiles, consisting of a coastal slope formed by rainwash and soil creep, the lower part of which is kept steep and fresh by wave attack during occasional storms. Cliffs cut in soft Tertiary sandstone on the coast of Port Phillip Bay in Australia show alternations in profile, being degraded by subaerial weathering and runoff during occasional downpours and cut back and steepened after dispersal of basal fans by wave action when storms accompany high tides. The removal of basal sediment produces instability which is transmitted up the cliff face until the cliff crest recedes (p. 67).

The profiles of steep coasts on similar rock formations show intricate variations related to exposure to wave attack. The bold profiles of cliffs of massive sandstone facing the ocean in the Sydney district contrast with the gentler, often vegetated, slopes on the same geological formations on the sheltered shores of Sydney Harbour and Broken Bay. The degree of cliffing developed in these situations is closely related to the local fetch, which limits the strength of attack by local wind-generated waves. On the west coast of Kintyre in Scotland cliffs and bluffs become bolder where they face higher wave energy arriving through the strait between Islay and the Antrim coast in Northern Ireland.

If wave action ceases at the base of a cliff, either because of a sudden fall in relative sea level, of the kind that has occurred on coastal sectors uplifted by earthquakes, as in the vicinity of Wellington and Napier on the North Island of New Zealand, the coastal profile begins to change. When marine erosion has ceased, accumulation and persistence of basal talus or the accretion of a wide beach causes cliffs that had been steep and receding to decline (degrade) to more gentle slopes, which in humid regions soon acquire a soil and vegetation cover. This is well illustrated on the North Island of New Zealand at Matata, west of Opotiki, where a widened beach fronts degraded bluffs. At Newhaven in Sussex the cliffs became protected when the beach prograded after a harbour breakwater completed in 1891 began to intercept shingle drifting alongshore. The cliffs are capped by Tertiary sands and clays, show an upper vegetated convex bluff, a middle vertical chalk cliff, and a basal vegetated concave talus slope. These mark stages in the evolution of a convex-over-concave vegetated bluff similar to that seen on valley-side slopes inland. The coast at Flamborough Head in Yorkshire shows the development of an upper convex slope and a basal concave slope (Figure 4.6), and similar changes have taken place on cliffs of Pliocene mudrock at Byobugaura on the east coast of Japan, following the building of sea walls along the shore 15 metres in front of the cliffs in 1967 (Sunamura 1992).

When marine erosion of a cliffed coast is halted and subaerial processes become dominant the sea cliff is degraded to a coastal bluff with slopes determined by the geotechnical properties of the rock outcrop. Such slopes are usually between 8° and 10° on soft clays and steeper on more resistant formations. On the north Norfolk coast a line of bluffs, formerly Pleistocene cliffs, were cut off by the development of spits, barrier islands, and marshlands formed during and since the Holocene marine transgression, when waves reworked glacial drift deposits left behind on what is now the floor of the North Sea. In Australia, enclosure of

Figure 4.6 Composite cliff profile at Flamborough Head, north-eastern England. The upper convex slope is on a capping of glacial drift, the cliff is cut in chalk, and there is a grassy basal talus slope behind the shingle beach.

former embayments by the growth of coastal barriers has been followed by the degradation of the former sea cliffs on the enclosed coast, as behind the Gippsland Lakes region in Victoria (Bird 1978a).

Cliffed coasts undoubtedly existed before the Last Glacial phase of low sea level, but during that phase, in the absence of marine attack, they were degraded by subaerial denudation. As a rule, the bluffs thus formed were undercut, as the sea attained its present level, and were rejuvenated as receding cliffs, but it is possible to find sectors where the Pleistocene bluffs have been preserved, usually behind areas of Holocene deposition. Thus the sandstone cliffs east of Hastings in Sussex pass into grassy bluffs at Cliff End, and these can be followed along the northern side of Romney Marsh (where they mark the cliffed coastline that existed prior to the formation of the Dungeness foreland), until they pass into the Lower Greensand cliffs of the Folkestone coast. In Australia the Port Campbell coast consists of vertical cliffs, except at Two Mile Bay, where these pass into Pleistocene bluffs to the rear of an emerged shore platform that has not yet been removed by the sea (Bird 1993a).

Slope-over-wall profiles

Some steep coasts consist of slopes which descend to basal cliffs. Such slope-over-wall profiles may develop in various ways. On soft formations a coastal slope formed by subaerial processes (runoff, slumping) may descend to a basal cliff cut back by wave action. Alternatively, the upper part of a cliff may be a slope cut in weathered rock while the lower part is a vertical cliff in more coherent unweathered rock or a seaward-dipping bedding plane, joint or fault plane may form a structural slope on the inclined surface of a relatively resistant formation which has been undercut and cliffed at the base by wave action, as in Cornwall on Porth Island near Newquay and Glebe Cliff south of Tintagel.

On the limestone cliffs of south Pembrokeshire thinly-bedded strata have weathered into a convex upper slope over a vertical wall of more massive lime-stone, as at Arnold's Slade and Mount Sion. On Flamborough Head in Yorkshire a grassy upper slope (bevel) cut in soft glacial drift deposits declines to a vertical cliff cut in firmer chalk, with a basal talus apron (Figure 4.6). On the Dorset coast between St Alban's Head and Durlston Head a slope in the softer Purbeck Beds descends to a cliff cut in Portland Limestone.

Where coastal bluffs have formed as the result of subaerial weathering and erosion on former cliffs pro-tected by a wide beach, a coastal barrier or an artificial structure such as a harbour breakwater, the removal of the protective feature will result in such bluffs being undercut by cliffs, forming a slope-over-wall profile until cliff recession consumes the whole of the formerly degraded slope.

Slope-over-wall profiles have also formed on coasts in high latitudes where cliffs cut in relatively resistant rock were degraded by periglacial freeze-and-thaw and solifluction (the movement of frost-shattered rubble down slopes) during cold phases of the Pleistocene when sea level was lower. The Pleistocene cliffs then became bluffs mantled by an earthy solifluction deposit with angular gravel (known as Head), which extended out on to what is now the sea floor in a broad, diminishing apron (Figure 4.7). During and since the

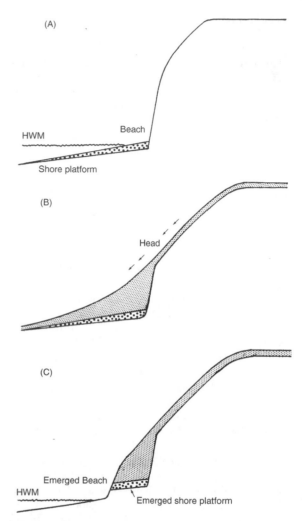

Figure 4.7 The evolution of a slope-over-wall coast profile where a Pleistocene cliff (A) became degraded by periglacial solifluction to form a Head-mantled slope (B), the base of which has been undercut (C). The Pleistocene beach is now seen on an emerged shore platform at the base of the modern cliff.

Late Quaternary marine transgression these peri-glaciated coastal slopes became vegetated, and were undercut by marine erosion, forming a steep or vertical basal cliff, exposing solid rock formations and produ-cing a slope-over-wall profile. The vegetated coastal slope, mantled by frost-shattered debris, is thus a legacy of past periglacial conditions, for it cannot be explained in terms of processes now at work.

Coastal landforms of this type are found on the western and south-western coasts of Britain, where

Figure 4.8   Slope-over-wall cliff coast profile at Dodman Point, Cornwall, with a steep slope formed by Pleistocene periglacial solifluction undercut by Holocene marine erosion, as shown in Figure 4.7.

vegetated slopes (typically 20°–30° but locally up to 45°) descend to steeper, rugged, rocky cliffs (Figure 4.8). The coastal slope may be a bevel of almost uniform gradient (especially where it follows seaward-dipping bedding, cleavage, joint or fault planes), but more often it is convex in form, like a hog's back, and occasionally it is concave, where the lower slopes of the Head deposit are preserved. Hog's back coasts are well developed on the Exmoor coast in north Devon, where the geological strike is parallel to the coast, forming coastal escarpments where the dip is landward (Figure 4.3).

The proportion of relict periglaciated slope to actively-receding cliff depends on the degree of exposure of the coast to wave attack. On sheltered sectors of the south coast of Cornwall (as on the eastern side of the Lizard peninsula near Coverack) the relict slope is well preserved down almost to high tide level. On the more exposed north coast of Cornwall, open to Atlantic storm waves, the relict slope has been largely, and in places completely destroyed by Holocene marine erosion. Little if any of the coastal slope remains on the coast of Watergate Bay in north Cornwall, which has high, almost vertical receding cliffs. The periglaciated coastal slope is better preserved on the sheltered eastern sides of major headlands, as on Dodman Point (where the slope of solifluction deposits becomes concave behind Vault Beach), than on their more exposed western shores.

There were probably several alternations of peri-glaciation and marine cliffing on these coasts as sea level oscillated during Pleistocene times. In places the coastal slope may include several facets, the gradient of which diminishes upward between successive breaks of slope. Savigear (1962) suggested that each facet could have been initiated as a vertical marine cliff which became degraded to a coastal slope, the upper facets being older and therefore more degraded than those at lower level, the latest being the active (Holocene) marine cliff undercutting the composite slope. Multi-faceted cliff profiles on the coast of Cornwall (as on Beeny Cliff) could be interpreted as the outcome of such a sequence, but generally the evidence for several stages of slope evolution has been destroyed by the intense Late Pleistocene periglaciation which shaped the coastal slopes that have been subsequently undercut by the sea.

Periglacial disintegration of rock outcrops and the formation of Head deposits by solifluction on coastal slopes left some residual massive rock formations (mainly quartzite), which survived frost-shattering, protruding from the debris-mantled coastal slope as tors or buttresses. These are seen as tors near the top of coastal slopes and buttresses where they jut out from the slope at lower levels. In Cornwall there are good examples in the Dodman district (Figure 4.9). There has also been slumping of the periglacial drift mantle on coastal slopes, forming hummocky vegetated topography, as at Dizzard Point on the north coast of Cornwall (Bird 1998).

On the north coast of Cornwall where islets and stacks such as those at Bedruthan Steps and Tobban Horse near Porthtowan retain segments of the Head-mantled slope it is possible to estimate the extent of cliff-base recession during the last 6000 years, since the sea rose to its present level: typically at least 100, and locally as much as 200 metres.

Periglaciated slope-over-wall profiles are found between 60°N and 39°N on the coasts of Scotland, western and southern Ireland, Wales, Brittany, western France and northern Spain (as at Cabo Finisterra) and south to Cap Mondego in Portugal. They are not well developed in equivalent latitudes of the Pacific coast of North America, partly because of the rarity of suitably resistant coastal rock formations, but more because the periglacial zone narrows and fades out south of the limit of Pleistocene glaciation on the west coast of North America. The contrast in coastal landforms is thus the outcome of differing Late Quaternary environmental histories (Williams et al. 1991). Slope-over-wall profiles are found on the coasts of Korea and eastern Russia, and on the shores of islands in the Southern

Figure 4.9 A buttress of resistant quartzite protruding from the Head-mantled slope on the coast at Boswinger, near Dodman Point in Cornwall.

Ocean, as on Macquarie Island (55°S) and the Auckland Islands (51°S) (Fleming 1965).

On rocky sectors of the Antarctic coast and on parts of the Canadian and Siberian Arctic coast, degradation of coastal slopes by periglacial activity still continues. Head deposits are being formed where rocky outcrops are shattered by recurrent freezing and thawing, producing solifluction deposits that move downslope, accumulating as a basal (often concave) slope apron that may be dispersed by wave action in the sea. This has led to rapid slope recession in Sakhalin on the Russian Pacific coast (Zenkovich 1967). On the shores of the St Lawrence estuary, sectors of rocky cliff and shore platform developed alongside narrow straits where the fetch is limited, are the outcome of frost shattering and shore ice plucking, the waves washing away the products of cold-climate rock disintegration. Similar ice-plucked features have been described from the rocky coast of Disko, in Greenland (Nielsen, 1979). Rocky cliffs on Antarctic coasts, and in Greenland, show scoured features resulting from etching by strong winds, especially where they mobilize sandy material.

Humid tropical bluffs

Coastal slopes with a vegetation cover are extensive in humid tropical regions, where they may show little if any basal cliffing. These profiles result from the intensive subaerial weathering and denudation characteristic of the humid tropics, so that many coastal rock formations have been weakened as the result of rapid and deep decomposition by chemical weathering, and do not form steep cliffs. Instead there are forested bluffs, steep coastal slopes with a soil and vegetation mantle as in north Queensland, where the Great Barrier Reef prevents strong wave action reaching the mainland coast, and within the Indonesian archipelago, where weak waves across limited fetches achieve little more than the removal of material that moves down the slope and would otherwise accumulate on the shore (Bird and Ongkosongo 1980). Vegetation extends down to a trim line marking the upper limit of wave swash (Figure 4.10).

Cliffed coasts are thus comparatively rare on humid tropical coasts in south-east Asia and India, on the east

Figure 4.10   Tropical bluffs on the north-east coast of Australia showing rain forest descending almost to the high tide line.

and west coasts of tropical Africa, in Brazil and on the Pacific coast of Central America. Cliffing occurs where wave action is relatively strong, on exposed promontories, and is more extensive on coasts receiving ocean swell, especially where the coastal rock formations are weak, as in Paraiba, north-eastern Brazil, where 21 per cent of the coastline is retreating cliffs cut in sandy clays (Guilcher 1985a). Yampi Sound in northern Australia is bordered by low crumbling cliffs of deeply-weathered metamorphic rock, from which protrude bolder promontories of quartzite, a type of resistant rock that has been little modified by chemical weathering. Marine erosion therefore works upon coastal rock formations, the resistance of which is related to prior weathering under humid tropical conditions rather than their original lithology. Tricart (1962) described the persistence of a dolerite headland at Mamba Point in Liberia, where adjacent outcrops of thoroughly weathered granite and gneiss formed vegetated bluffs. Coastlines with humid tropical bluffs are generally fairly stable, but may be receding where there are recurrent landslides, which become rapidly revegetated.

Ice cliffs

A special kind of cliff is found where glaciers or ice sheets have advanced beyond the edge of the land and end in a wall of ice descending to, and below, the sea. These ice cliffs are well developed in Antarctica, where the Ross Ice Shelf terminates in cliffs up to 35 metres high. In summer icebergs calve from these cliffs and in winter the adjacent sea freezes, a fringe of sea ice persisting until winds and the spring thaw break it into floes. Melting in late spring is hastened by the growth of diatoms which darken the ice surface and increase its solar heat absorption. Protruding snouts of glaciers which reach the sea form floating ice tongues, somewhat richer in morainic debris than ice shelves. This debris is released and deposited on the adjacent sea floor, and some of it may subsequently be washed up on beaches. Wave action is weakened by the predominance of winds blowing offshore from the ice cap and impeded when sea ice forms.

Similar features are seen on ice cliffs in Alaska, and where glaciers still occupy the heads of some fiords, notably in Greenland, northern Canada and southern Chile. A concealed rocky coastline is protected from wave attack as long as the ice fringe persists. The effectiveness of marine processes depends on the seasonal duration of sea ice, which exists for 100–150 days per year in the St Lawrence estuary, and for longer and shorter periods in lower and higher latitudes.

Permanently frozen ground (permafrost) influences coastal evolution in northern Alaska and the Arctic coasts of Canada and Russia, where the summer thawing of interstitial ice contributes to rapid degeneration of tundra cliffs, consisting of peat and morainic sediments, which slump into the sea and are swept away by wave action. Tundra cliffs can retreat tens of metres in a few weeks (Zenkovich 1967).

Cliff dissection

As cliffs recede weathering and erosion penetrate zones of weakness such as faults, joints or outcrops of less resistant rock, cutting clefts and crevices which may develop into caves and blowholes or deep, narrow inlets, with stacks as the outcome of irregular cliff retreat. Cliffed coasts showing an array of such features are found on carboniferous limestone in south Pembrokeshire and on Old Red Sandstone on the west coast of the island of Jura in Scotland. Caves, archways and stacks have been formed by the dissection of chalk cliffs in Freshwater Bay, in the south-west of the Isle of Wight, and the Port Campbell limestone coast in Australia has spectacular examples of these dissection features.

Caves

Caves form where marine erosion has penetrated zones of weakness, and are common in rock formations that have numerous joints or faults, or segments of weaker rock. A famous example is Fingal's Cave on the Scottish island of Staffa, a cave 20 metres high and 70 metres long cut out along planes between columns of Tertiary basalt. On limestone coasts the sea may penetrate and widen fissures that originated as subterranean solution caverns prior to the Late Quaternary marine transgression. The caves of Bonifacio in southern Corsica, are believed to have formed in this way, and on the Port Campbell coast in Victoria subterranean caves excavated along vertical joints in Miocene limestone have been etched out by marine erosion to form deep inlets in the cliffed coast, with large caves at their heads, as at Lochard Gorge (Bird 1993a).

Blowholes

Where the hydraulic action of incoming waves and the compression of trapped air within a cave puncture the roof, a blowhole develops, and water and spray may be driven up through it as fountains of spray. Blowholes formed in this way are seen on Porth Island, near Newquay in Cornwall and on the Old Red Sandstone coast near Arbroath in eastern Scotland. An Australian example occurs at Kiama in New South Wales. On the coast of Cornwall some blowholes have grown into large circular cavities linked to the sea through archways, as at the Devil's Frying Pan near Cadgwith and the Lion's Den on the Lizard peninsula (Figure 4.11) (Bird 1998).

Figure 4.11 The Lions Den, a round hole formed by the collapse of a cave roof on the coast near Lizard Point, Cornwall.

Gorges

Gorges are spectacular steep-sided clefts or inlets formed where the roofs of caves have collapsed or where rock has been excavated by wave action along lines of weakness at an angle to the shore in coastal rock formations. Examples can be found on many cliffed coasts. They are known as geos or yawns on the coast of Scotland and zawns on the granitic Land's End peninsula in Cornwall, where some have been at least partly shaped by tin mining. Elongated inlets have been excavated between joint planes on the rocky coast of Le Puits d'Enfer in western France. Deep gorges of this kind are incised into the bold sandstone cliffs of the Jervis Bay district in New South Wales, and the limestones of Port Campbell in Victoria. On the south Pembrokeshire coast flat-topped cliffs cut into carboniferous limestone have been dissected along joints and fault planes to form narrow, steep-sided rocky inlets such as Huntsman's Leap (Figure 4.12). A distinction is made between gorges cut out entirely by marine erosion and stream-cut valleys, which can also form steep-sided inlets where their mouths have been submerged by the sea.

Natural arches

Where powerful wave action has excavated caves along joints and bedding-planes on a cliffed coast, some may extend through headlands to form a natural arch (Figure 4.13). An example is the Green Bridge of Wales on the Pembrokeshire coast, an arch of massive carboniferous limestone beneath which thinly-bedded strata have been cut out along joint planes (Figure 4.14). The Porte d'Aval is a slender natural arch on the cliffs of hard chalk at Etretat on the Normandy coast of France. Several natural arches have been called London Bridge, notably at Torquay in Devon and Portsea on the Victorian coast in Australia. On the Port Campbell cliffs in Australia a natural arch was lost when a storm removed the trunk from Elephant Rock, and another when the inner arch of London Bridge collapsed in 1990 (Figure 4.15).

Tunnels are elongated natural arches. On the north coast of Cornwall, Merlin's Cave is a tunnel 100 metres long excavated by marine erosion along a thrust plane between slates and volcanic rocks on Tintagel Island.

Stacks

Dissection of promontories on a cliffed coast may also isolate stacks (residual islets standing offshore),

Figure 4.12 Huntsmans Leap, a cleft cut out along joint planes in carboniferous limestone on the cliffed coast of Pembrokeshire in south-west Wales.

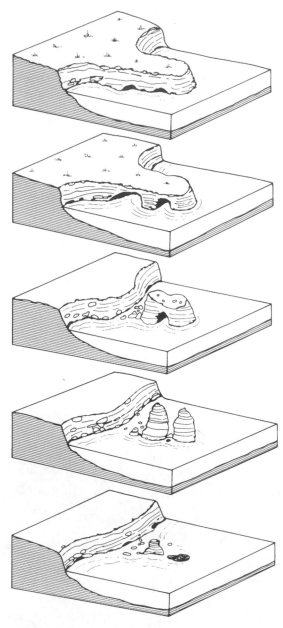

Figure 4.13 The dissection of a rocky promontory by caves to form natural arches which collapse leaving stacks that are gradually reduced by erosion.

formed either where a natural arch has collapsed, or where a transverse inlet has been cut along a zone of weakness through the headland. There are stacks on the receding cliffs cut in Permian sandstone on the Devon coast, notably at Ladram Bay, and in northernmost Scotland, where huge cliffs of Old Red Sandstone face powerful wave action from the north Atlantic. The Old Man of Hoy is a spectacular 140 metre stack in front of towering cliffs of Old Red Sandstone on the west coast of the Orkney Islands, and Elegug Rocks are high stacks off the south Pembrokeshire limestone coast (Figure 4.16). On a smaller scale, Old Harry Rocks are a group of stacks that have become separated from The Foreland, on the chalk promontory of Ballard Down in Dorset, as the result of excavation along joint planes. There is historical evidence of stages in the dissection of this headland along vertical joints to form stacks, which have been reduced by erosion: Old Harry lost his wife in a storm in 1896, and new stacks were formed by breaching of a long, narrow headland in 1920–21 (May and Heeps 1985). In general, the shaping of stacks is strongly influenced by weathering and erosion along steep or vertical joint planes, which thus dominate the bordering cliffs, but in many cases the seaward sides of a stack are steeper than the landward sides because they are exposed to stronger wave

attack. Such features are seen in the Twelve Apostles, a group of stacks cut in Miocene limestone on the Port Campbell coast in Victoria, Australia, and on the Kamchatka coast, where rapid cliff recession has left stacks up to a kilometre offshore (Zenkovich 1967). Flat-topped stacks occur where they have been cut

Figure 4.14   The Green Bridge of Wales, a natural arch cut in carboniferous limestone on the Pembrokeshire coast in south-west Wales.

into a coastal plateau, usually a Pleistocene terrace formed by marine erosion at a higher relative sea level, as on many of the rocky stacks off the Oregon coast.

Some stacks are residuals of more resistant rock left standing as marine erosion proceeds. The Needles are residual ridges of hard chalk, compressed and indurated in a zone of intense monoclinal folding at the western end of the Isle of Wight. Similar tall stacks of harder silicified chalk stand in front of the cliffs of Etretat in Normandy, one of which, the 25 metre high Grande Demoiselle de Fontenailles, collapsed during a storm in 1902 (Paskoff 1994). Off the Oregon coast many of the stacks are of hard lava in volcanic necks, the surrounding less resistant volcanic ash and sedimentary deposits having been removed by marine erosion. An example is the 72 metre high Haystack Rock, off Cannon Beach.

Most stacks have formed in Holocene times, being residuals left by cliff recession since the sea rose to its present level. This is certainly the case where the coast had been subject to Pleistocene glacial or periglacial

processes, and stacks retain deposits or facets inherited from cold climate weathering (p. 56).

Hanging valleys and coastal waterfalls

Some cliffed coasts show hanging stream-cut valleys truncated by cliff recession, their floors well above present sea level, with rivers pouring out as waterfalls cascading down to the shore, as at Ecclesbourne Glen, on the Sussex coast. On the north-west Devon coast there are opportunities to study the relationships between cliff recession and valley incision, which has produced a series of coastal waterfalls, as at Speke's Mill Mouth. At Tintagel in Cornwall Trevena Brook pours out over the cliff from a valley truncated by cliff recession, and on the Dorset coast there are several waterfalls, including one at St Gabriel's Mouth near Golden Cap.

Deep and narrow coastal ravines known as chines on the south-west coast of the Isle of Wight and in Bournemouth Bay have been cut by small streams in

Figure 4.15   The inner natural arch at London Bridge, near Port Campbell in south-eastern Australia (top), a month before it collapsed on 15 January 1990, leaving an outer arch (bottom).

Figure 4.16   Elegug Rocks, stacks cut in carboniferous limestone on the Pembrokeshire coast in south-west Wales.

valleys truncated by rapid cliff recession (Figure 4.17). There are waterfalls along the chines and down the cliff face. Numerous waterfalls pour down the steep slopes of the tectonically uplifted coast south of Jackson Bay in Fiordland, in the South Island of New Zealand.

Accordant valley mouths, with unaggraded stream channels opening to the shore, are rare, occurring only where cliff recession has matched the rate of stream incision or where coastal uplift has prevented valley-mouth submergence during the Late Quaternary marine transgression. Gullies cut into recently deposited volcanic ash and agglomerate on the shores of Anak Krakatau, an Indonesian volcano (Figure 4.18), are also accordant, downcutting having proceeded at a similar rate to cliff recession in these unconsolidated sediments (Bird and Rosengren 1984).

Dry valleys truncated by the recession of chalk cliffs on the Sussex coast, have produced the undulating crest of the Seven Sisters along an almost straight coastline (Figure 4.2). The undulating cliff crest crosses successive valleys lined with deposits of frost-shattered chalk (Coombe Rock) produced by Pleistocene periglacial action. These dry valleys, truncated by cliff recession,

decline seaward, and in places intersect the shore platform, producing either a re-entrant or a sector where the chalk passes beneath a sand and gravel fill, as at Birling Gap, near Beachy Head (Figure 4.19).

Steep-sided incised valleys cut by glacial meltwater, notably water overflowing from ice-dammed lakes, have been identified on the coast of Cardigan Bay in Wales, where the Irish Sea ice sheet abutted the coast in late Pleistocene times and enclosed lakes in the valleys of the Teifi and Nevern Rivers. These lakes rose in level and overflowed, cutting valleys across coastal promontories, as on Cemmaes Head and Dinas Head.

Hinterland topography

Cliffs are also influenced by the geomorphology of the immediate hinterland, in particular the topography intersected as the cliff recedes. Flat-topped cliffs have been cut into a coastal plateau, as on the Atlantic coast of Cornwall, particularly south of Portreath, where the even-crested cliffs of Reskajeage Downs rise about 100 metres above sea level. Cliffs cut into

Figure 4.17   Whale Chine, on the south coast of the Isle of Wight, formed where a valley cut into soft Cretaceous sands and clays has been truncated by cliff recession.

Figure 4.18   Gullied slopes of volcanic ash below the rim (R) of the crater of Anak Krakatau, Indonesia have been truncated by cliffs (C) cut by marine erosion. A more resistant lava flow (L) protrudes into the sea on the left. Reproduced by permission of Neville Rosengren.

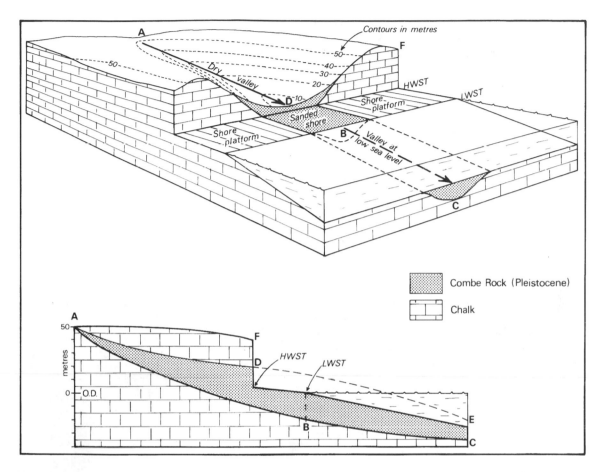

Figure 4.19 Evolution of a sanded zone in the shore platform cut in chalk at Birling Gap, Sussex, as the result of valley incision during a Pleistocene low sea level phase, infilling with Coombe Rock (a periglacial drift deposit) and truncation by cliff recession.

high ground recede more slowly than cliffs eroding into low-lying topography on similar rock formations, so that interfluves tend to become promontories between valley-mouth embayments. This is illustrated on the Yorkshire coast north of Flamborough Head, where cliffs have been cut into a landscape of plateaux and incised valleys on Jurassic formations, forming headlands and bays. Similar features are seen on the northern Oregon coast, and on the Australian coast near San Remo in Victoria. A crenulated outline of ridge headlands and valley bays has developed on the margins of dissected hilly country on Cretaceous sandstones. Where the hinterland slopes away from the coast, cliffs diminish in altitude as they recede, as on Beachy Head in Sussex and in the vicinity of Childers Cove near Warrnambool in Australia.

OUTLINES IN PLAN OF CLIFFED COASTLINES

Where a relatively resistant coastal rock formation is backed by parallel weaker outcrops penetration of the outer wall by marine erosion is followed by the excavation of coves and embayments in the softer rock. In such situations there has often been prior incision of valleys parallel to the coastline, cut in the softer outcrops behind a coastal ridge on harder rock. A well-known example of this is seen on the Dorset coast east of Weymouth, where several stages of dissection can be seen. Stair Hole, near Lulworth, is an early stage, a narrow breach in the outer wall of steeply-dipping Jurassic limestone. Close by, Lulworth Cove has a wider entrance through the limestone wall, and an almost circular bay carved out between converging valleys cut in soft Cretaceous sands and clays, backed by a high ridge of chalk. Farther east a much broader embayment has developed along the clay lowland corridor in front of the chalk ridge at Worbarrow Bay, while to the west the Jurassic and Wealden rocks have been removed almost completely, leaving only the small residual peninsula at Durdle Door and some outlying reefs such as Mupe Rocks on the harder Jurassic limestone. The coastline is the outcome of partial Late Quaternary marine submergence of limestone ridges and clay vales produced by subaerial denudation of the Jurassic formations, followed by marine cliffing (Bird 1995).

The outline in plan of a cliffed coast tends to become simplified and smoothed with the passage of time, except where there are marked contrasts in the structure and lithology of coastal rock formations, which may perpetuate irregular outlines as the cliff recedes. Where the recession of cliffs cut in a relatively weak formation such as glacial drift or dune calcarenite uncovers outcrops of harder basement rock at or near sea level, the latter emerge as headlands while the softer rocks are cut back as embayments. Such coasts, termed contraposed coasts, are exemplified by the west coast of Eyre Peninsula in South Australia, where receding cliffs cut in Pleistocene dune calcarenite extend behind and between exhumed promontories of harder Pre-Cambrian rock. In north-west Brittany granite has been exposed to form headlands as the Pleistocene periglacial deposits (known as Head deposits) that formerly concealed them were cut back by the sea, leaving intervening embayments backed by cliffs and bluffs cut into Head deposits.

Where the coastal outcrops are comparatively uniform in relatively soft rock, a receding cliffed coast develops an outline in plan related to the prevailing wave patterns. Where the waves are refracted by offshore topography or adjacent headlands cliffed coasts develop gently-curved outlines in plan, much like those on depositional coasts; indeed a curved cliffed coast may pass smoothly into a curved depositional coast, as in the asymmetrical embayment at Waratah Bay on the south-eastern coast of Australia (p. 119). Similar curved outlines have developed on the Tertiary cliffs of Bournemouth Bay in southern England in relation to refracted waves approaching from the south-west.

COASTAL LANDSLIDES

Mass movements have occurred on weak and weathered rock outcrops and in unconsolidated sediment on steep and cliffed coasts. Instability develops with an increase in shear stress (e.g. by greater loading) or a decrease in shear strength (i.e. weakening of the rock formation) reaching a threshold level and is relieved by movements down-slope. Such down-slope movements can take place in various ways: falls, slides and flows (Brunsden and Prior 1984). There are rock falls and toppling, where masses of rock collapse from the cliff face (Figure 4.20), translational slides where rock masses slip down a seaward dipping plane, rotational slides which collapse seaward down a curved plane to form back-tilted rock masses, mudslides where masses of coherent silty or sandy clay move irregularly down across a sheared surface and mudflows where highly lubricated fine-grained sediment moves downslope as

Figure 4.20   Subsiding and toppling of the cliff face cut in soft Miocene limestone near Port Campbell in south-eastern Australia.

a slurry. Mudslides and mudflows form lobate tongues that spread out across the shore and into the sea. They occur especially in the winter or the rainy season, when they may be accompanied by surface runoff, including streams that cut gullies and deposit basal sediment fans. Groundwater seepage from permeable formations has also contributed to instability, coastal landslides being most active during wet periods when the rock formations become saturated (shear stress increased by loading), or when frozen rock thaws and becomes incoherent after a cold spell (shear strength diminished).

Saturation loading of weak or porous coastal formations with groundwater has led to cliff collapse in the Head deposits on the Cornish coast at Downderry and Gunwalloe. By contrast, desiccation in dry summers leads to flaking and falling of clay particles from cliffs cut in Jurassic clays on the Dorset coast. Slumping may also occur as a consequence of changes in volume as ground temperatures rise and fall diurnally (cliff falls often occur at night), and can be initiated by the widening of crevices in the cliff face as the result of penetration by the roots of cliff-top vegetation and the pressure exerted as such roots

grow larger. Burrowing animals, notably rabbits, can contribute to slumping and erosion of cliffs and bluffs cut in soft rocks.

Near Newport, Oregon, the loading of a cliff by buildings has been followed by landslides which have damaged or destroyed many houses and led to the demolition of a condominium (Viles and Spencer 1995). The weight of cliff-top buildings also contributed to retreat on the Val de Lobo cliffs in southern Portugal (Alveirinho Dias and Neal 1992), and was partly responsible for the collapse of Holbeck Hall Hotel on a steep coastal slope at Scarborough in Yorkshire in 1993.

Coastal landslides are common where geological formations dip seaward. On the coast of Lyme Bay in Dorset several large coastal landslides, backed by receding cliffs, have occurred where masses of gravel-capped upper greensand, dipping seaward, have subsided over slumping Jurassic clays at Black Ven and Cains Folly (Figure 4.21). The town of Lyme Regis is threatened by the recession of upper cliffs behind enlarging coastal landslides. Measurements on a mudslide in Worbarrow Bay, Dorset, using erosion pins and electronic recording showed various types of

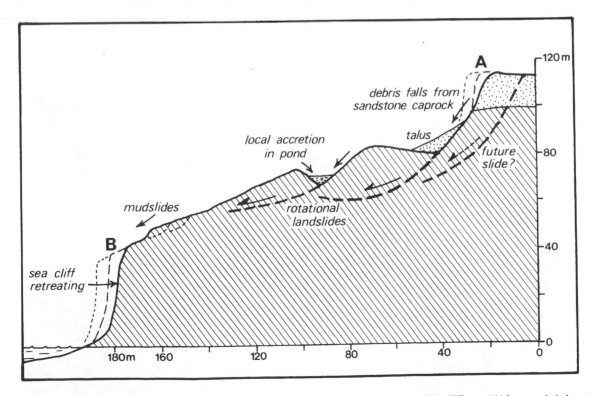

Figure 4.21   Recurrent landslides and basal cliffing on the coast near Lyme Regis in Dorset. The cliff crest (A) has receded about the same distance as the basal cliff (B) during the past century.

movement, some in frequent brief (slip-stop) episodes, some gradual, some in sudden surges (e.g. three metres in 20 minutes) and some apparently random (Allison and Brunsden 1990). This is one of several places on the south coast of England where pebbles from capping gravel deposits move gradually down through the landslide and are eventually deposited on the beach. Further east the profiles of Gad Cliff, Hounstout Cliff and St Alban's Head show massive cappings of Portland limestone which have disintegrated, toppled and subsided over Kimmeridge clay to form hummocky topography, undercut by basal cliffs.

A seaward dip in the coastal rock formations facilitates landslides and rock falls, the undercut rock sliding down bedding-planes into the sea, leaving the exhumed bedding-planes as a coastal slope. On the south coast of England coastal landslides are common where the permeable chalk and upper greensand formations dip seaward, resting on impermeable gault. Water seeping through the permeable rocks moves down the inclined clay surface, and if marine erosion has exposed the junction at or above sea level, lubrication of the interface leads to slipping of the overlying rocks. A spectacular landslide occurred in this situation on the Devon coast near Axmouth in 1839, and similar landslides have occurred on the south coast of the Isle of Wight, and between Folkestone and Dover in Kent, where a railway built along the undercliff is damaged from time to time by falling rock. The physical effects of wetting and lubrication of the clay surface are accompanied by chemical processes, for when the seeping water, rich in dissolved calcium carbonate, reaches the glauconitic gault, base exchange occurs, calcium ions displacing potassium ions so that alkalinity increases and the clay is deflocculated. The upper layers of the clay become a soft wet slurry that flows out at the base of the cliff and hastens the undermining of the chalk and upper greensand (Hutchinson et al. 1980).

On the south coast of the Isle of Wight masses of chalk and upper greensand have broken away from an upper escarpment and subsided to form the Ventnor Undercliff between St Catherine's Point and Luccombe. To the west successive landslides have led to rapid recession of a cliff cut in lower greensand, truncating and largely consuming a valley at Blackgang Chine, behind a slumping undercliff terrace on soft clays and sands (Figure 4.22), while on the

Figure 4.22 Undercliff terrace formed by landslides at Blackgang, Isle of Wight.

south-west coast of the Isle of Wight there are a series of broad amphitheatres cut in the Wealden Beds, backed by cliffs that overlook hummocky subsided terraces (Hutchinson et al. 1991).

On the Oregon coast there have been major landslides at Humbug Mountain, forming lobate slumped promontories that temporarily prograded the coastline, before being consumed by marine erosion. There was massive slumping on the high cliffs on the coast of Hawke Bay north of Napier in New Zealand caused by earthquakes in 1931 and 1990, and on the bluffs at Anchorage, Alaska, as a result of the 1964 earthquake. Cones of slumped talus that extended up to 600 metres into the sea have been cliffed and cut back subsequently by wave action.

West of Moonlight Head in Victoria, Australia a series of amphitheatres backed by cliffs in Tertiary sandstone are fronted by tumbled rocks where landslides have occurred on the underlying Palaeocene clay (Bird 1993a). Similar lobes at Fairlight in Sussex form natural breakwaters that intercept eastward drifting shingle, and this has also occurred on the shores of Lyme Bay, where shingle has accreted updrift

of landslide lobes instead of moving eastward to Chesil Beach (p. 104). Below the Lyme Bay cliffs are several arcuate festoons of boulders that commemorate the seaward limits of former lobes from which finer sediment has been dispersed by wave action (Figure 4.23).

Attempts to stabilize landslide-prone cliffs include artificial slope grading, the insertion of drainage systems and the planting of vegetation, the most satisfactory results being where a combination of these procedures is used. The use of explosives to blast cliff faces, notably where it is deemed necessary to remove dangerous overhanging rock often increases long-term instability by shattering and weakening the rock formation. Limestone cliffs near Llantwit Major, in south Wales, remained unstable after they were blasted in 1969 (Williams and Davies 1980).

CLIFF RECESSION

Cliffs are cut back mainly during storms, largely by wave action. The cliff base is undercut by the hydraulic

Figure 4.23 A loop of boulders below the cliff at Golden Cap, Dorset, marks the limits of a former landslide lobe that has been cut back by marine erosion.

pressure of wave impact and the abrasive action of water laden with rock fragments (sand and gravel) which during storms are hurled repeatedly at the cliff base. A basal abrasion notch is thus formed, and as it grows the cliff face becomes unstable, and an over-hanging rock mass eventually collapses (a rock fall). After a storm the backshore is littered with debris that has fallen from the cliff. Agitated by wave action, this becomes broken and worn down (a process known as attrition), and is either retained as a beach (which protects the cliff base from further abrasion), or carried away along the shore or out to sea by the action of waves and currents.

The vertical cliffs of chalk in south-east England and northern France are maintained by marine erosion in this way. May and Heeps (1985) noted that marine erosion cut basal notches and marine processes removed basal talus, but found that cliff recession was largely due to rock falls caused by saturation after heavy rain, and particularly to freeze-and-thaw action in winter, especially when mild weather follows a cold winter, as in 1940, 1947, 1963 and 1979. A massive rock fall on Beachy Head in Sussex early in 1999 was attributed to the expansion of saturated chalk during a spell of very cold weather. Freezing and thawing can cause toppling and slumping, produc-ing half-conical basal talus fans of broken rock, and leaving a white scar on the cliff face (which is otherwise grey, as the result of weathering and colonization by algae). The fallen rock is gradually consumed by weath-ering and corrosion (the chalk being dissolved by rain-water and aerated spray and surf) and removed or dispersed by wave action (off chalk coasts the English Channel is typically grey-green or opaline with chalk in suspension as well as solution). The vertical cliff profile thus restored is then further undermined by basal wave erosion until it collapses again. On a smaller scale, fans of frost-shattered rock form each winter at the base of chalk cliffs on the island of Rügen, on the German Baltic coast, and are consumed by the sea in the summer months.

On soft formations, such as clays, unconsolidated sands, deeply weathered rocks or unconsolidated glacial or periglacial drift deposits, cliffs and steep coastal slopes recede by recurrent slumping, particu-larly after wet weather or the thawing of a snow cover. Intermittent recession of this kind is seen on the soft Tertiary sands and clays of the Bournemouth coast and the northern shores of the Isle of Wight, on Jurassic clays and shales on the Dorset and Yorkshire coasts, on coal measures clays and sands in Pembrokeshire and on glacial drift deposits, as in eastern England, the Danish archipelago, New England or the islands of Puget Sound. Cliff recession is dramatized where buildings are undermined and lost, and little is left of coastal churches at Dunwich in East Anglia (Parker 1978) and Trzesacz in Poland, both damaged by recession of cliffs cut in glacial drift.

Recession of the cliff crest often takes the form of irregular breakaways, when lumps of rock become dislodged and fall away from the cliff. On the Dorset coast, slumping has occurred where cliff-top paths have become breakaways (Figure 4.24). Cracks develop behind, and parallel to, receding cliffs as a prelude to calving or slumping, and deformation and fracturing parallel to the coastline is seen on many cliff-top roads and damaged buildings close to cliff edges, as on the Holderness coast. Slumping coastal slopes are irregular, and basal debris fans are undercut by the waves, forming a slope-over-wall profile. The vertical cliff grows in height as it is cut back, but there is soon

Figure 4.24   Breakaway on the cliff crest at Black Ven, near Lyme Regis in Dorset (cf. A in Fig. 4.21).

further slumping. The cliffs thus recede as the result of alternating marine and subaerial erosion. Figure 4.21 shows such a sequence on the coast of Lyme Bay.

Coastal slopes and cliff faces in soft formations are dissected by gullies cut by runoff after heavy rain or melting snow, as on the Wealden clays and sands on the south-west coast of the Isle of Wight. A cliff face of soft Tertiary sandstone at Black Rock Point, on the shores of Port Phillip Bay, Australia, has been dissected and cut back by gullying, accompanied by the exudation of fine sediment by groundwater seepage. It became steeper and smoother in profile after cliff-top stabilization diverted runoff during heavy rains (Figure 4.25) (Bird and Rosengren 1987).

Weathering processes are particularly active on limestone coasts. On cliffs cut into Pleistocene dune calcarenites, seeping groundwater dissolves carbonates and washes out fine particles, forming cracks and crevices in the cliff face, and precipitation of carbonates from seeping groundwater hardens the surface, which is often darkened by algal colonization. Alveolar weathering produces cavities (tafoni) and honeycombing, and cliff faces may also be scoured by wind action, producing caves in the sandstone. At lower levels the cliff face becomes intricately pitted as the result of muricate weathering, which includes the effects of recurrent wetting and drying, salt crystallization, corrosion by rainfall and sea spray, and bioerosion by marine organisms (p. 80). Shore vegetation and fauna often include species, the growth and metabolism of which lead to decomposition or dissolution of coastal rock outcrops, especially limestones (p. 78). Bioerosion can also be achieved by terrestrial organisms, notably penetration of and widening of joints and fissures by the roots of plants growing on the cliff face. Nesting holes excavated by birds, particularly sand martins, contribute to the erosion of soft sandstone cliffs, while burrowing bees have dissected the soft sandy cliff near Redend Point in Dorset.

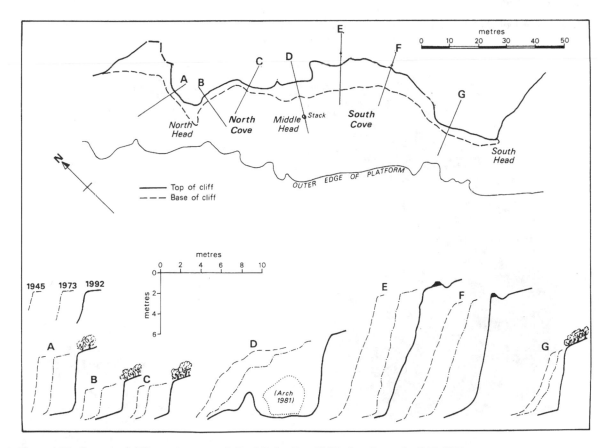

Figure 4.25 Surveys of cliff recession at Black Rock Point, Port Phillip Bay, Australia 1945–1992.

The effects of some organisms may be protective or even constructional. A dense growth of kelp or a firm encrustation of algae may protect a rocky shore from abrasion).

Superficial induration of the cliff face by the precipitation of cementing materials washed down the cliff or brought to the surface by seepage can slow erosion, at least until the hardened crust is breached and dissected. The vertical cliff face at Demons Bluff on the coast of Victoria, Australia, has been hardened in this way, and shows slumping in sectors where the indurated crust has been breached (Figure 4.26).

On the Pacific coasts of Oregon and Washington there are steep bluffs carrying scrub and forest which appear fairly stable, but are in fact receding as the result of intermittent and localized slumping, the slump scars being quickly revegetated in this cool and moist environment as the debris fans are consumed by the sea (Byrne 1964).

Rates of cliff recession

Measurements of cliff recession are most readily made at the cliff crest, but it is also possible to measure basal cliff retreat or the recession of the whole profile, the changing gradient depending on the relative rates of retreat of the cliff base and at the cliff crest. The rate at which a cliff retreats depends on the resistance and structure of the outcropping formations, the energy of incident waves, tide range, the presence of a rocky shore or shore platform which affords some protection to the cliff base, and the frequency and consistency of basal wave attack. As has been noted (p. 18) wave energy is concentrated in a narrow vertical zone on microtidal coasts and dispersed where the tide range increases. On the megatidal shores of the Bay of Fundy in eastern Canada, there are sloping cliffs in soft glacial drift, cut during the occasional brief episodes when high spring tides bring wave action to the cliff base.

Cliff recession is generally episodic, during phases of strong wave attack, and varies along the coastline with the resistance of coastal rock formations, aspect in relation to wave regimes and the nearshore profile and the incidence of rock falls and landslides, when the crest of the cliff may retreat, but the cliff base may advance, at least temporarily, by the accumulation of a lobe of slumped material. Some changes are instantaneous; others take place over time scales ranging up to

Figure 4.26   Cracks in the mudrock cliff at Demons Bluff, Victoria, Australia, which recedes as masses of rock fall from the cliff face.

centuries. Most existing cliffs have been cut back within the past 6000 years, when the sea has stood close to its present level, and on some coasts it is possible to measure the extent of this recession with reference to remains of the preceding subaerially weathered land surface intersecting the outer edge of the shore platform. Gill (1973) used this method to show that the cliffs on the Otway coast in south-eastern Australia had receded 105 metres on mudstones and 53 metres on sandstones over this 6000 year period (Bird 1993a).

Rates of cliff recession (and of changes on shore platforms) can be monitored by repeated profile surveys (Figure 4.25), by photogrammetric analysis (e.g. Jones et al. 1993), or by measurements on inserted pegs or micro-erosion meters which can be linked to computers (Stephenson 1997). Micro-erosion meters have been used to measure changes on cliffs and shore platforms in Yorkshire (Robinson 1976), at Kaikoura in New Zealand (Kirk 1977) and on the Otway coast in Australia (Gill and Lang 1983), but are difficult to apply where there is lowering by detachment of blocks of stratified rock, as on many chalk and sandstone coasts.

Cliff recession is usually expressed as averages in metres per year, but the actual retreat of the cliff crest is episodic and localized as each rock fall occurs. Measurements of cliff recession may be linear, areal, or volumetric, and are usually expressed as annual averages. Sunamura (1992) listed worldwide linear cliff recession rates (his Appendix 2) and found average rates of recession of:

- 1 mm/yr on cliffs cut in granite
- 1 mm–1 cm/yr on limestone
- 1 cm/yr on shale
- 10 cm–1 m/yr on chalk and Tertiary sedimentary rocks
- 1–10 m/yr on glacial drift deposits
- at least 10 m/yr on volcanic ash

Cliff-crest recession in glacial drift on sectors of the Polish coast has averaged a metre a year, with up to five metres in a stormy year (Zenkovich 1967), and recession of up to 100 metres per year has been measured on Wasque Point, a stormy headland on the south coast of Martha's Vineyard Island in Massachusetts (Kaye, 1973). May and Heeps (1985) found rates of cliff-top recession ranging up to more than one metre per year at various sites on chalk coasts, and an areal loss of 3264 square metres over 12 years on chalk cliffs in Sussex, 97 per cent of which was removed

in winter periods, and 42 per cent in two particularly cold, wet and stormy winters. On the Yorkshire coast Robinson (1977) observed quarrying by waves at the base of Liassic shale cliffs unprotected by a beach, and abrasion by waves armed with rock particles formed a basal notch by recession at up to 4.7 centimetres per year. Cliff erosion rates were 15 to 18.5 times higher where there was backshore beach material, moving to and fro on a cliff-base abrasion ramp sloping at $> 2.5°$.

It is necessary to take account of variations in cliff height and calculate the volume of rock removed as a cliff is cut back. During the 1953 North Sea storm surge Williams (1956) recorded 12 metres of recession on cliffs 12 metres high and 27 metres of recession on cliffs three metres high on a mile-long sector of the Suffolk coast at Covehithe (Figure 4.27). Because of the intermittency of such events, and lateral variations in their effects, measurements of average annual rates of cliff retreat can be misleading. Along the Port Campbell cliffs in Australia the mean rate of cliff recession is only a few centimetres per year, achieved by a series of occasional localized rock falls, when segments up to 200 metres long and 12 metres wide have suddenly collapsed into the sea (Baker 1943). One such collapse in 1939 near Sentinel Rock left a freshly exposed scar that is still discernible as a less weathered cliff sector 60 years later, its base aproned by fallen debris being slowly consumed by marine erosion. Evidence of such slumping is present on about five per cent of the length of the Port Campbell cliffs, suggesting a recurrence interval of many decades for such major events (Bird 1993a).

The cliff crest and cliff base at Scarborough Bluffs on the north shore of Lake Ontario have retreated at an average rate of up to 0.5 metres per year, mainly during episodes of storm wave attack in high lake level phases and as the result of recurrent slumping (Carter and Guy 1988). Slope-base recession on Calvert Cliffs, Maryland, by wave undercutting is more than one metre per year, and by freeze-thaw alternations about 0.5 metres per year (Wilcock et al. 1998). Cliff recession slackens and cliff profiles may become degraded (i.e. slope angle diminished) as shore platforms widen and become protective. This can be seen on the coast at Kimmeridge in Dorset, where cliffs are less steep behind wider sectors of structural shore platform on horizontal limestone (Figure 4.28).

Sea walls have been built to halt cliff recession, particularly at seaside resorts, some of which now stand forward from the coastline because of continuing cliff recession on the adjacent unprotected coastline. Often

Figure 4.27   Slumping cliffs at Covehithe in Suffolk.

the cliff beyond the end of a sea wall is cut back more rapidly as deepening nearshore water allows larger waves to break along its base. Examples of this are seen on glacial drift coasts at Withernsea and Mundesley in eastern England and Ustronie Morskie on the Polish coast.

Accumulation of large quantities of fallen rock or deposited beach material on the backshore serves to protect the base of a cliff from wave attack because storm wave energy is expended on the beach. Smaller quantities of beach material that can be mobilized and hurled on the cliff base during storms accentuate abrasion by waves. Herein lies the risk of removing beach material from the shore. Dredging of shingle from the nearshore area at Hallsands in south Devon for construction work at Plymouth dockyard during the 1890s led to accelerated beach and cliff erosion and storm wave destruction of the fishing village of Hallsands, which used to stand on a coastal terrace near the base of the cliffs (Robinson 1961). The cliffs

at Anzio in Italy bear remnants of the walls of Nero's Palace, which was built in front of them nearly 2000 years ago, when their recession had halted because of the formation of a wide protective beach, correlated with a lowering of sea level. In recent decades sea level has been rising and this beach has been removed by erosion, so that little now remains of the palace (Bird and Fabbri 1987). On the Holderness coast in eastern England Pringle (1981) correlated phases of accelerated cliff erosion and steepening with the passage of low sectors of beach (ords) in the southward-drifting beach system.

It has been suggested that some retreating cliffs maintain their general form (with cyclic variations) over time scales of about a century, showing a dynamic equilibrium between the landforms and the processes causing their recession (Cambers 1976, Brunsden and Jones 1980). Recession of vertical cliffs by intermittent rock falls and basal rejuvenation provides a similar sequence.

Figure 4.28   A structural shore platform on a horizontal layer of Jurassic limestone at Broad Ledge, near Kimmeridge in Dorset, the outer edge undergoing dissection by waves.

SHORE PLATFORMS

Many cliffs are bordered by shore platforms that extend across the intertidal zone. They can be formed in various ways, and the term shore platform is preferred to wave-cut platform because it describes the feature without implying its mode of origin. Some are seaward-sloping shore platforms (Figure 4.29A), planation surfaces that extend from the high tide line at the base of the cliff to a level below and beyond the low tide line. It has been suggested that a shore platform of this kind may be bordered by a wave-built terrace constructed by deposition at its outer margin, but although this can be found on some lake shores such a nearshore terrace is rarely formed off sea coasts (Dietz 1963).

Others are structural shore platforms (Figure 4.28) that coincide with the upper surfaces of resistant rock formations which may dip seaward, landward, or alongshore. There are also more or less horizontal shore platforms, standing close to the high tide level

(submerged only briefly at high tides), typically with a sharp drop or low-tide cliff (often >3 metres) at the seaward edge. These are termed high tide shore platforms (Figure 4.29B). Similar subhorizontal shore platforms are found close to low tide level (submerged for most of the tidal cycle), particularly on limestones, and generally these also end seaward in a sharp drop: they may be termed low tide shore platforms (Figure 4.29C). Shore platforms that are narrow (e.g. up to about 10 metres wide) are called benches and may be structural ledges coinciding with the outcrop of a resistant rock formation. On some coasts the shore topography includes sloping ramps and flat segments, as well as steps or micro-cliffs, uprising ridges and stacks, and dissecting features such as grooves and clefts. In high latitudes, particularly in Scandinavia, there are strandflats (p. 91), almost flat shore platforms that grade landward into gently shelving coastal plains. Some shore platforms are much dissected, and as dissection increases they grade into irregular rocky or boulder-strewn shore topography. Finally, there are

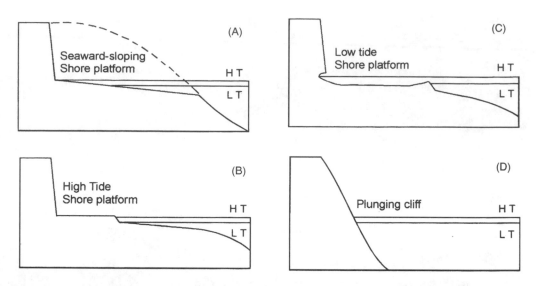

Figure 4.29   Shore platform types. A – seaward-sloping shore platforms, B – subhorizontal high tide shore platforms, C – subhorizontal low tide shore platforms, D – plunging cliffs with no platform on the shore.

coasts where cliffs descend into deep water close inshore (plunging cliffs, Figure 4.29D), without any platforms or intertidal rocky areas.

There is the problem of how smooth and continuous a planed rock surface should be for the term shore platform to be applied. Many shore platforms are dissected by minor grooves and channels or diversified by small rising outcrops, but as these increase in dimensions it may be more appropriate to use the terms incipient shore platform (with higher residual ridges and stacks) or dissected shore platform (with wide transverse gullies and furrows parallel to the coastline). It is difficult to define these quantitatively, but one suggestion is that the term shore platform should be used only where the platform surface is sufficiently well formed, intact and undissected for a four-wheel drive vehicle to be driven across it. Where less than 10 per cent of the shore is a smooth platform and the rest irregular the term rocky shore could be used.

The various kinds of shore platform and rocky shore may vary along the coast in relation to geology, aspect and morphogenic factors, such as tide range and wave regime. Contrasts in shore profile related to geology are evident where there are varied geological outcrops along a coastline and other factors are relatively uniform, and contrasts related to morphogenic factors may occur on a particular rock outcrop, especially fringing headlands and islands. Such contrasts have been analysed along the southern New South Wales

coast (Bird and Dent 1966) and along the Victoria coast (Hills 1971) in south-eastern Australia.

Before discussing the various kinds of shore topography it is necessary to examine the processes at work on rocky shores.

PROCESSES ON ROCKY SHORES

Shore platforms have formed on coasts where cliffs have receded in response to processes already described. As the cliffs recede, various erosion and weathering processes have shaped and modified shore topography.

Abrasion

Wave abrasion occurs when waves move rock fragments (sand, shingle, cobbles) to and fro across the shore. It results in the cutting of a smooth (or slightly grooved) seaward-sloping ramp or shore platform, often truncating the geological formations that outcrop on the shore (Figure 4.30). Waves armed with rock fragments are powerful agents of abrasion, for without such fragments they achieve abrasion only on soft formations, such as clay and shale. Rock fragments may be of local derivation, eroded from the cliff or quarried from the shore platform, or they may have been

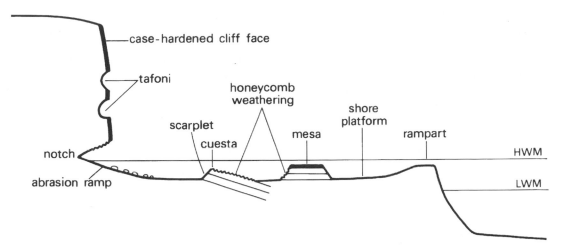

Figure 4.30 Cliff and shore platform features.

brought in by longshore drift from adjacent parts of the coast, or shoreward drift from the sea floor.

The effects of wave abrasion can be seen on shore platforms where the more resistant elements of rock persist as reefs and stacks above a smoothly abraded rock surface. Abrasion ramps slope gently upward to the base of a receding cliff, where there is an abrasion notch. The lowering of a ramp or platform by abrasion can be measured by repeated profile surveys, or with the aid of micro-erosion meters (Stephenson 1997). The potency of wave abrasion is seen where rock fragments that have become trapped in a crevice on a shore platform are repeatedly moved by wave action in such a way as to excavate smoothly-worn circular potholes, the rock fragments becoming well-rounded in the process (Figure 4.31). Pebbles rolled to and fro as the tide rises and falls also cut grooves across a shore platform at right angles to the coastline or along the outcrops of bedding planes, joints, faults or weaker formations, particularly where these run at an angle to the coastline. Wave abrasion is much reduced where marine vegetation has colonized the shore, protecting the rock surface and attenuating wave action.

Wave quarrying is a form of abrasion which occurs when loosened rock fragments are excavated from a rock surface by wave action. It can produce a smooth slope where it lays bare a bedding-plane or other flat or gently sloping surface on relatively resistant rock (a structural shore platform), but more often it produces irregularities, with pits, pools, clefts and cavities left where rock material has been removed. When wrack and kelp are torn off by storm waves their holdfasts may retain rock fragments, which are cast up on the shore.

Figure 4.31 Potholes on a dune calcarenite shore near Rye in south-eastern Australia, scoured by boulders circulated by wave action at high tide.

Wave abrasion is effective below low tide level, but diminishes rapidly offshore as the water deepens. The depth to which abrasion is effective is an important factor in the evolution of sea floor topography. Sand grains in sea floor sediments off the California coast (which has high wave energy) have been shown to become well worn and rounded in depths of up to 10 metres below low tide level (Shepard 1973). In deeper water they show less evidence of wear, and often remain angular. The wearing of these particles implies abrasion of the rock surface over which they are moved to and fro by wave action, so that sea floor abrasion is effective to a depth of 10 metres. Abrasion produces broad gently sloping platforms that rise gradually towards the coastlines. On a tideless coast such an abrasion platform could develop shoreward from a depth of 10 metres to the cliff base, and with an average inclination of 1° would be about 500 metres wide. Their possible width increases with tide range, and would be about 800 metres with an average slope of 1° where the tide range is five metres. Platforms up to several kilometres wide at various levels, that occur above and below present sea level, are often considered to be the product of marine planation (p. 37), but can only be explained as abrasion platforms if they developed during a slow marine transgression, the land margin being planed off by waves as the sea rose. There are many gently-sloping unconformities in the geological record, indicating that such abrasion has proceeded in the course of past marine transgressions.

Weathering

Weathering processes are at work on coastal rock outcrops which are not protected by a soil or vegetation cover in the intertidal zone, and at the cliff base up to above mean high tide level. Intricate pitting of rock surfaces (muricate weathering) results from repeated wetting and drying, which can lead to physical disintegration (decrepitation), releasing rock fragments which are swept away by wave sluicing. A distinction is made between wave abrasion of rock outcrops and wave sluicing of unconsolidated weathered rock material. The wetting and drying process is most effective in semi-arid zones where frequent desiccation in dry weather alternates with wetting by occasional rain as well as the highest tides, swash and sea spray. Rock formations disintegrate and are sluiced away by wave action down to the level of permanent saturation in coastal rock formations, which is generally just below high-tide level.

The process of wetting and drying is often accompanied by salt crystallization in the zone splashed by sea spray. Salt is deposited on rock surfaces and in cavities, and as it crystallizes, absorbing water, its volume increases, creating pressure that widens fissures and detaches rock fragments. Salt spray weathering can be observed on tropical shores where abrasion by waves is generally weak (because of low wave energy and the rarity of gravelly material) and the strong sun rapidly dries off rock surfaces wetted by saline spray. Tricart (1959) considered that salt spray weathering was a dominant process shaping shore features on the tropical coast of Brazil. It is difficult to separate the physical effects of wetting and drying from the physico-chemical effects of salt crystallization from drying spray on sea coasts, where wetting and drying is accompanied by salt corrosion.

Sandstone outcrops wetted by sea spray and rainfall, and then dried out, become pitted and honeycombed as sand grains are loosened by the decomposition of the cementing material that formerly bound them. Other fine-grained rock formations, such as siltstones, mudstones, shales, schists, phyllites and basalts, show similar superficial decomposition and disintegration. Wetting and drying and salt crystallization are not effective at lower levels, where the rock formations are permanently saturated by sea water or ground water. The effects of the two weathering processes are seen on basalt boulders on an emerged shore terrace near Lara on the west coast of Port Phillip Bay, Australia, which have been undercut marginally and planed off locally at the level of groundwater saturation. On the south Devon coast, Mottershead (1998) measured changes on a schist outcrop where weathering had lowered a surface that had been covered by an oil spill and left small pinnacles capped by the oil layer. Using relics of the oil layer as a reference level, he deduced that the surface was lowered at the rate of 0.625 millimetres per year over a seven year period (Mottershead 1989).

Solution

Rock outcrops on cliffs and shore platforms can be dissolved by rain water, sea spray and sea water. Solution occurs particularly on limestones, including emerged coral and algal reefs and dune calcarenites (Figure 4.32). Rain water absorbs carbon dioxide from the atmosphere (carbonic acid) and can dissolve limestone (mainly calcium carbonate) much more effectively than pure water, but the widely-reported notion

Figure 4.32 Karstic weathering has formed lapies on a limestone coast near Sorrento in south-eastern Australia. The shore platform ends in a notch, overhung by a visor.

that carbonates pass into solution as bicarbonates is incorrect, for bicarbonate molecules do not exist in water that has dissolved limestone (Picknett et al. 1976).

Sea water off limestone coasts and ground water flowing from limestone cliffs is normally saturated with carbonates, and thus incapable of dissolving limestone on the shore, but limestone outcrops are certainly attacked by solution in the intertidal zone. Some cliff-base notches cut in limestone have been attributed to solution, rather than abrasion, because they are at least as well developed on the side sheltered from strong wave action (as on the lee side of islands and stacks). The problem of how sea water, already saturated with carbonates, can achieve further solution of limestone on the shore was analysed by Revelle and Emery (1957) on Bikini Atoll. They found strong diurnal alternations in the dissolving capacity of sea water, related partly to variations in the temperature and carbon dioxide content of water in the shore zone. Nocturnal cooling of sea water increases its capacity to take up carbon dioxide (which is more soluble in water at lower temperatures), permitting it to dissolve more limestone, and the biochemical activities of marine organisms lead to the production of carbon dioxide (from plant and animal respiration) which is used by the plants (chiefly algae) in photosynthesis by day but which accumulates at night when the absence of sunlight halts photosynthesis. This nocturnal increase in the acidity of coastal water permits limestone to be taken into solution, and calcium carbonate dissolved during the night may be precipitated by day, when temperature rises, photosynthesis revives and the capacity of sea water to retain dissolved carbonate diminishes. Much of the precipitated carbonate is quickly carried away from the limestone shore by waves and currents. As nocturnal solution processes operate mainly in the intertidal zone, they remove calcium carbonate in solution down to a level (close to low tide) at which permanently saturated and submerged limestone is being dissolved more slowly, if at all, by sea water.

Solution of limestone shores may be achieved by surf in breaking waves and by sea spray, thrown into the air as the waves break, particularly during storms and when strong winds blow onshore. Surf and sea spray are aerated and dispersed forms of sea water, and associated atmospheric carbon dioxide may render them corrosive. Limestone surfaces subjected frequently to the impact of sea spray become pitted and irregular in a manner suggesting that solution is in progress. This was observed at Diamond Bay, Australia, after a rock pinnacle was broken off just above high tide level during a 1968 storm on a cliff of dark, intricately weathered dune calcarenite. This formed a pale, freshly-exposed smooth surface of unweathered sandstone, but within a decade the exposed surface had become colonized by algae and was as dark as the surrounding rock, and after 25 years it was as irregular and intricately pitted as the surrounding weathered rock surface (Bird 1993a).

Bioerosion

Marine organisms certainly contribute to the etching of limestone on the shore, and Emery (1960) indicated that in favourable conditions biochemical processes can consume rock at least as quickly as physical and purely chemical erosion. The various organisms that inhabit the intertidal zone (especially shelly fauna) contribute to erosion by drilling, scraping, plucking and grazing, and solution by exuded fluids. Algae can excavate notches and perforate limestone, grazing molluscs (limpets, mussels and winkles) scrape the surface, dislodging rock fragments and excavating cavities, and marine worms, sponges and sea urchins all drill or quarry the rock (Figure 3.1). Where shells occupy hollows on a limestone surface it is tempting to deduce that they have excavated these hollows, but this cannot apply where they are found occupying similar hollows on hard flint surfaces on chalk coasts.

Bioerosion occurs when algal mats that have formed on rock surfaces dry out and peel off with adhering rock particles, especially in dry, windy weather during periods when the zone above high neap tides is not submerged. Some would go as far as to suggest that a notch can be almost literally eaten out by the shore fauna that occupy the high tide zone (Healy 1968). Shore organisms occur across the intertidal zone and into subtidal environments, but are most active and vigorous between low tide and the splashed zone above high tide. Biogenic effects are most obvious on soft limestones and calcareous sandstones and siltstones on low energy shores, for on stormy coasts the organisms are inhibited, and their effects obliterated by wave abrasion.

Rates of bioerosion have been summarized by Trenhaile (1987, Table 4.3).

Frost shattering

On coasts in cold regions repeated freezing and thawing result in physical weathering of rock outcrops on the shore, and the formation and melting of sea ice lead to plucking and dislocation of rock fragments. The disintegrated rock is later removed by wave sluicing, or mobilized by waves to achieve abrasion of the rocky shore. Freezing and thawing also occur during cold winters in temperate regions, and on the chalk shore platforms of the Sussex coast Robinson and Jerwood (1987) found that frost shattering during the cold months of January 1985 and February 1986 contributed to erosion and spalling on chalk boulders, the effects diminishing seaward to the low tide line. Shore ice abrasion has contributed to the cutting of low tide cliffs at the outer edge of shore platforms cut in shale on the stormy Gaspé coast in Quebec (Trenhaile 1978).

Induration

Induration of shore rocks is a form of weathering that increases resistance to erosion, and has contributed to the persistence of rock outcrops in cliffs, on stacks and islets, and on shore platforms and rocky shore topography. Indurated rock outcrops are less affected by other weathering processes or by wave abrasion. Various kinds of rock have become indurated by processes at work on the shore, the outcrops becoming much harder than those seen in quarries or encountered in boreholes in the same geological formation a short distance inland. Case hardening may result from superficial enrichment of shore outcrops by the precipitation of carbonates, or of ferruginous or siliceous cementing compounds. The process is well known to quarrymen: in south Devon the Beer Stone is a fine-grained soft homogeneous chalk, moist when it is cut in underground caverns: it is carved and sculpted before being allowed to dry and harden under exposure to the atmosphere.

Induration has contributed to the evolution of shore platforms where rock outcrops have been hardened by the precipitation of cementing materials, notably calcium carbonate or iron oxides.

Seaward-sloping shore platforms

Seaward-sloping shore platforms (Figure 4.29A) – Type A as defined by Sunamura (1992) – extend from the cliff base at about high spring tide level and slope gently, but not always uniformly, to pass beneath low spring tide level, where there may be a sharp drop (low tide cliff) or a gradual decline to the sea floor. They have been shaped primarily by wave abrasion, with only minor contributions from weathering, bioerosion or solution processes. The width of such platforms at low tide is determined by the relative rates of retreat of the cliff base and the seaward edge, and is thus influenced by exposure to wave action, tide range and transverse shore gradient, rock resistance and the length of time that processes have operated at present sea level (Trenhaile 1987). Some shore platforms diminish in gradient as the cliff recedes, others retain their slope as the sea floor is worn down at a rate that matches cliff recession. Shore platform width may also be influenced by the dip of outcropping rocks. Wide abrasion platforms have been cut in front of chalk cliffs where stratified chalk dips gently seaward on the Kent coast in Thanet (So 1965), and along the Sussex coast east of Brighton (Figure 4.2). Their width is also related to tide range, here about four metres. Shore platforms are poorly developed on the chalk of the Dorset coast, where the chalk outcrop restricted by a steep dip and tide range is small, but a seaward-sloping shore platform has been cut across steeply-dipping chalk at Whitecliff Point at the eastern end of the Isle of Wight: it is not clear why there is no such platform on similar steeply-dipping chalk at Ballard Point in Dorset.

On soft rock outcrops the seaward-sloping shore platform has a slightly concave shore profile descending from the cliff base and declining in gradient seaward below the low tide line (Zenkovich 1967). The profile is similar to the longitudinal profile of a river valley or the transverse profile of a beach (p. 122). There is often a beach at the base of a cliff, and an intertidal veneer of sand and gravel. Such a profile is seen fronting cliffs of glacial drift on the east coast of England, notably in Holderness and on clay outcrops on the Dorset coast, as on the Weald Clay in Worbarrow Bay. A similar shore profile fronts the soft receding sand cliffs of Alum Bay and Whitecliff Bay, on the Isle of Wight. It is probable that this is the profile that would develop in front of cliffed coasts generally, were it not for the complications of structure, resistant rock outcrops, weathering effects and induration.

As a cliff recedes wave abrasion commonly cuts a basal notch, the lower part of which is a seaward sloping abrasion ramp which declines into the usually gentler gradient of the shore platform and its undersea continuations. The abrasion ramp typically shows freshly scoured rock, without a weathered surface or plant growth or marine organisms. On the Liassic rocks of the Yorkshire coast Robinson (1977) found

that the cliff-base ramp was being lowered by abrasion at up to three centimetres per year. Cliff recession is generally accompanied by downwearing of the shore platform in such a way as to maintain the transverse profile and the seaward slope. Persistence of small mesas, capped by flint layers, rising above the general level of the shore platform on the coast near Birling Gap in Sussex are an indication of this downwearing. The adjacent sea floor is also worn down by abrasion, so that the whole coastal profile migrates landward.

There are variations in the upper limits of shore platforms around the coasts of England and Wales. The junction between the cliff base and shore platform is often obscured by beach deposits, but where it is exposed it is generally a sharp break of slope close to high tide level. In detail the level varies from about mean high neap tide and just above mean high spring tide, and can show lateral variation of up to four metres. It rises with tide range because high tide reaches a higher level above Ordnance datum. On the chalk coast of Thanet the junction between cliff base and shore platform is up to 3.6 metres higher at the heads of bays than on intervening headlands (Wright 1970). However, there are sites where the junction is higher on headlands than in adjacent bays, possibly because of storm wave action operating to higher levels (So 1965), or because the water table is higher on the headland, and weathering processes (such as wetting and drying) are shaping platforms at a higher level.

More often, wave abrasion has cut a gently sloping shore platform across the structure of coastal formations, so that the surface shows a pattern of truncated rock outcrops. Seaward-sloping shore platforms have been cut across steeply dipping Old Red Sandstone striking east to west along the south coast of Wales west of Tenby, notably at Manorbier, where the planed-off strata run parallel to the coastline. On the west-facing Atlantic coast of Cornwall and Devon between Widemouth Bay and Hartland Point seaward-sloping shore platforms have been cut across intricately folded sandstones and shales, truncating numerous anticlines and synclines that trend east–west, at right angles to the coastline. Similar platforms occur near San Sebastian on the north coast of Spain, and on Aoshima Island, Kyushu, Japan (Sunamura 1992), while swirling patterns of planed-off strata are seen on the abrasion platform cut in strongly folded Lias limestones on the south Wales coast near Nash Point.

The chalk cliffs and shore platforms on the English and French coasts bordering the English Channel are largely the product of wave abrasion, the waves being armed with flint nodules eroded out of the chalk (Prêcheur 1960, So 1965). Chalk is a relatively homogeneous limestone formation, apart from the flint nodules, which are usually in layers along the bedding-planes. As they weather out of the chalk they are mobilized by breaking waves and used as tools of abrasion as the tide rises and falls. They cut grooves at right-angles to the coastline on the chalk shore platforms of the Kent and Sussex coast. Fresh white chalk exposed on the shore platform and on the cliff-base ramp after stormy periods is evidence of wearing and scouring by waves armed with flints, but other processes (such as weathering and solution) are also at work, and their effects can be detected wherever wave abrasion diminishes.

Examination of fallen boulders shows that the chalk surface has been pitted by solution due to rainfall, sea spray and aerated surf, and that the rock has been modified by the physical and biochemical effects of shore flora (chiefly marine algae) and fauna (limpets, mussels, and winkles). These processes play an important part in the reduction and eventual disappearance of chalk, releasing further flint nodules which are then used in abrasion.

Seaward-sloping shore platforms are sometimes found where structureless sandstone or shale formations have been eroded by wave action, but their development requires a delicate balance between rock resistance and the intensity of wave abrasion. Relatively resistant formations may be eroded into steep cliffs and seaward-sloping shore platforms on a high wave energy coast but not where wave energy is low. Shore platforms are rarely found on massive igneous rock outcrops, but there are shore ramps on sectors of the granite coast of the Land's End peninsula and at Cape Woolamai in south-eastern Australia where the granite is more closely and intricately jointed, and more readily quarried by strong wave action. Similar features are seen where igneous rock formations are deeply weathered, and the underlying solid rock is exposed on the coast by marine erosion as a shore platform.

Seaward-sloping shore platforms may be cut in relatively weak formations on a low wave energy coast, but higher wave energy may destroy them. There are vertical cliffs on sandstones with little or no development of fronting shore platforms on the Bridport Sands at West Bay in Dorset and the Hastings Beds on the Sussex coast. In each case the rock outcrops are sufficiently coherent (and slightly indurated) to sustain a vertical cliff, but not sufficiently resistant to form a shore platform. Similar features are seen on cliffs cut into

volcanic ash and agglomerate, as on Anak Krakatau in Indonesia (Figure 4.18).

Under such a balance of conditions the morphology of the receding coast can be maintained during recession of cliffs and downwearing of intertidal platforms and the adjacent sea floor. It is possible that the Holderness coast, eastern England, cut into soft glacial drift deposits, is maintaining its whole profile as the coastline recedes. The cliffs here have been receding at rates of one to two metres per year, and if this has been maintained since the sea reached its present level the coastline has retreated six to twelve kilometres over 6000 years, and the sea floor at the initial coastline has been lowered at least 10 metres. Elsewhere the evolution of a receding coast is usually complicated by the effects of structure, lithology and weathering processes, but the balance of these may sometimes yield a deceptively simple cliff-and-platform profile, as on certain chalk coasts (Figure 4.2), where solution and bioerosion have accompanied wave abrasion. More often, variations in structure and lithology in the shore zone persist in irregularities of profile with seaward sloping segments of shore platform developed locally as planation corridors between ridges of harder rock, and channels where less resistant outcrops have been excavated (Figure 4.33). Increasing wave energy exploits structural and lithological contrasts, forming an irregular rocky shore with ribs and reefs of harder rock between deeper clefts and channels, as on the southern coast of the Lizard peninsula, where the cliffs are fronted by reefs and islets of hard Palaeozoic and Pre-Cambrian rock. There are rocky reefs and skerries instead of shore platforms in the shore and nearshore area. Similar features are seen on the granitic coast of the Bullers of Buchan in eastern Scotland.

On some coasts the present seaward-sloping shore platform is developing as the result of dissection of a similar higher shore platform cut in Late Pleistocene times when the sea stood a little above present high tide level. An example has been documented from Caamaño, Galicia, by Trenhaile et al. (1999). As dissection proceeds the shore becomes very rugged and irregular, with remnants of the emerged platform separated by transverse gullies and corridors along which a smooth slope, cut by abrasion, ascends landward. The modern shore platform is being cut at a lower level, and where the emerged platform has been completely removed the modern seaward-sloping shore platform runs back to a sharp junction at the base of the cliff, sometimes with an undercut notch. Evolution of a seaward-sloping shore platform by the dissection and reduction of an emerged shore platform is seen on

Figure 4.33   Corridor planation between ridges of more resistant rock on the New South Wales coast.

parts of the coast of south-west England, particularly in Gerrans Bay and Falmouth Bay in south Cornwall. At Whitsand Bay segments of the emerged platform are preserved on ribs of Dartmouth slate which run out from the cliff base between sandy coves and segments of the modern seaward-sloping shore platform.

As has been noted (p. 33) emerged shore platforms of Pleistocene and Holocene age occur on many coasts, some related to a Late Pleistocene interglacial high sea level, others the outcome of land uplift. The Pleistocene emerged shore platform at about three metres above present sea level in southern Britain is found at various other levels, ranging up to six metres in northern Britain where isostatic rebound has followed deglaciation (p. 34) (McCann 1964)

Structural shore platforms

Structural shore platforms are found where waves have exposed the surface of a flat or gently dipping resistant rock formation, usually a bedding-plane, as on Broad Ledge near Kimmeridge (Figure 4.28). The structure and lithology of coastal rock formations influence the development of shore platform profiles, notably where sandstones or limestones are interbedded with softer shales, as in the Lias of the Lyme Regis coast. Wave action has excavated bedding planes to produce struc-

tural ledges on the upper surface of each resistant layer, terminating in small scarps. Some are horizontal, some slope seaward or landward, and others alongshore. Segments with intricate scarps and dip slopes on Liassic limestones and shales are interspersed with seaward-sloping shore platforms at Robin Hood's Bay in Yorkshire. Limestone strata also form a stepped shore topography at Kimmeridge Ledges on the Dorset coast, each ledge disintegrating into large angular blocks as the underlying shales are etched out by wave scour.

Where soft formations contain minor resistant layers these may persist as segments of structural shore platform. This is well illustrated on the south-west coast of the Isle of Wight, where the soft Wealden shales contain thin sandstone horizons that rise and fall gently across the Brighstone Anticline and form structural shore platforms exposed at low tide, as at the Pine Raft at Hanover Point (Bird 1997a). Structural shore platforms are also found locally on horizontal or gently dipping Triassic sandstones of the Devon coast between Torquay and Sidmouth and on the carboniferous limestone of Northumbria, where undulating ledges follow shore outcrops of thick layered strata.

In Australia stepped cliff profiles have developed on outcrops of Triassic sandstones in the Sydney district, and there are structural shore platforms that coincide with the upper surfaces of resistant strata: they are horizontal where the bedding is flat, and inclined where the rocks are dipping. It is possible to follow a particular ledge along the shore, down the dip of a resistant sandstone layer, until it passes below sea level, when the ledge on the next higher resistant rock outcrop begins to dominate the shore profile as it declines into the sea. This kind of coastal topography results from storm wave abrasion along bedding-planes and joints and the removal of dislocated rock masses to lay bare a structural shore platform. Where the rock formation is horizontal the shore platform is similar to the Type B shore platforms discussed below, but where the rock formation dips gently seaward the profile may resemble a seaward-sloping (Type A) shore platform, passing below low tide level, differing in that abrasion has not truncated shore outcrops. Where ledges or platforms have developed on flat or gently dipping formations above high tide level they could be emerged features formed during an earlier phase of higher relative sea level, but on high wave energy coasts storm waves can cut structural ledges at various levels (Figure 4.6).

Structural platforms are often seen on coasts cut into volcanic rocks, where the platforms coincide with the upper surfaces of lava flows, which may be dissected along joint planes. There are good examples on the northern and western coasts of Skye in Scotland and on Phillip Island in Australia. Abrasion, weathering and bioerosion have contributed to the shaping of structural shore platforms, and wave sluicing (and occasionally wind action) is necessary to sweep away weak, weathered or unconsolidated material to lay bare the upper surface of a resistant rock formation.

High tide shore platforms

The term high tide shore platform (Figure 4.29B) is a brief way of describing a platform that is horizontal, or nearly horizontal (with a very gradual seaward slope, usually $<1°$), developed at, or slightly above, mean high tide level (Type B of Sunamura 1992). Such subhorizontal platforms are exposed to subaerial processes for much of the tidal cycle, but overwashed by storm waves and submerged by high spring tides (Figure 4.34). They typically end abruptly seaward in a steep drop (low tide cliff), below which the sea floor declines, usually with a broad and gentle concave profile. A sandy sea floor is indicated by the clouds of sand that rise in suspension with the up-current in incoming wave fronts (Figure 2.1). These high tide shore platforms are well developed in microtidal, low wave energy environments, particularly on sandstones, mudrocks and other

Figure 4.34   High tide shore platform on the Otways coast in south-eastern Australia, showing recession of the steep outer edge.

permeable fine-grained rock formations including basalt and consolidated volcanic ash. Originally observed and studied in New Zealand, Hawaii, and Australia, they occur widely on the islands and shores of the Pacific and Indian Oceans, and have also been noted locally on the Atlantic coast. They are quite distinct from seaward-sloping shore platforms, and although some are structural (coinciding with the upper surface of an outcropping rock formation) most truncate local geological structures and cannot be explained in terms of lithological control.

High tide shore platforms are produced by weathering processes, especially wetting and drying, rather than wave abrasion, although wave sluicing removes the weathered material down to a particular level (the upper level of rock saturation, no longer subject to recurrent drying) to lay bare the horizontal platform. Wetting and drying has contributed to the lowering and flattening of shore platforms on basalt and sandstone in south-eastern Australia (Hills 1971) and on mudrocks on the Kaikoura Peninsula in New Zealand (Kirk 1977). Wetting and drying is usually accompanied by salt crystallization, and the two processes cause pitting and honeycombing on the cliff face and on upstanding rock outcrops on high tide shore platforms. Such weathering produces rock fragments, mainly of silt and sand size, which are sluiced away by waves at high tide. Pools and channels on the platform surface become enlarged and integrated as their overhanging rims recede, and gradually the rock surface is reduced to a high tide shore platform. The process has also been described as water-layer weathering, and as the level of rock saturation need not coincide with bedding-planes it offers a mechanism by which shore planation may form platforms which transgress local geological structures (Hills 1949). It also accounts for the fact that high tide shore platforms are almost horizontal, often with a raised rim or rampart at the outer edge where the rock is permanently saturated to a higher level by breaking waves. This is often more pronounced where the rocks are dipping seaward (landward scarp) or landward (seaward scarp) than where they are horizontal. In places the seaward rampart has been removed by erosion, but where it persists there is an implication that the outer edge of the shore platform has been cut back little since the sea attained its present level in Holocene times, and that cliff recession has been accompanied by the lowering and flattening of the shore platform behind the rampart as the result of weathering processes and the sluicing away of weathered material by wave action.

Shore platforms of this kind are poorly represented around the British Isles, probably because of the relatively high tide ranges, stormy seas, and the rarity of suitable lithologies (Wright 1967). Type B shore platforms occur in the sub-Antarctic South Shetland Islands on sheltered parts of the coast where there is intertidal shore ice plucking and bulldozing of rock outcrops, and abrasive wave action being inhibited by the freezing of the sea for several months each year. On more exposed parts of the coast stronger wave action forms sloping intertidal Type A platforms (Hansom 1983).

In places high tide (Type B) shore platforms pass laterally into seaward-sloping (Type A) shore platforms, notably where there is an increase in rock fragments that can be moved to and fro by waves to achieve abrasion, where there is exposure to stronger wave action, or where the shore is dominated by a structural feature such as a seaward-sloping hard rock formation. On the New South Wales coast high tide shore platforms give place to seaward-sloping shore platforms on the more exposed sectors, and where wave abrasion is facilitated by the presence of locally-derived shingle (Bird and Dent 1966).

Rates of lowering of these subhorizontal shore platforms can be measured using micro-erosion meters. On the Yorkshire coast, for example, Robinson (1977) found that a subhorizontal platform up to 200 metres wide had been downwasting at one to two millimetres per year. Stephenson and Kirk (1998) used a micro-erosion meter to measure rates of erosion on shore platforms cut in mudstone at Kaikoura Peninsula, New Zealand, and found that subhorizontal platforms were being lowered at 0.733 millimetres per year, whereas seaward-sloping shore platforms had a lowering rate of 1.983 millimetres per year.

It has been suggested that high tide shore platforms are essentially storm wave abrasion platforms, produced by waves driven across them during storms when the cliff at the rear is cut back; in calmer weather, wave action is limited to the outer edge, which gradually recedes (Cotton 1963). It is difficult to accept this as an explanation for high tide shore platforms, except in the special case where a structural platform coincides with the upper surface of a horizontal rock formation at high tide level. Storm waves are unlikely to concentrate energy at any particular level because they are occasional, of varying duration, and come in a variety of dimensions, operating over a height range related to the rise and fall of tides. Horizontal or gently inclined platforms that truncate local geological structures (Figure 4.35) cannot be explained in terms of

Figure 4.35 Truncation of dipping Devonian sandstones and mudstones on the shore platform near Cape Liptrap in south-eastern Australia. Reproduced by permission of Neville Rosengren.

storm wave attack, which has a destructive influence on these features.

Evolution of high tide shore platforms as the result of weathering processes is indicated where they are as well, or better, developed on sectors of the coast that are sheltered from strong storm wave activity. On the New South Wales coast the strongest storm waves arrive from the south-east, but the high tide shore platforms are at least as broad and often better developed on the northern sides of headlands and offshore islands, as on Broulee Island. On the more exposed southern sides the platforms show evidence of dissection and destruction by wave abrasion at their outer margin and quarrying by waves along joints and bedding-planes, and there is evidence that this recession results in the extension of the concave abrasion platform at a

lower level, beyond the seaward drop, which is extending landward as the result of this cutting back of the high tide shore platform (Bird and Dent 1966). Wave abrasion, operating alone, tends to develop the simple profile of a steep cliff bordered by a seaward-sloping intertidal shore platform, within constraints imposed by the structure and lithology of coastal outcrops.

It is possible that subhorizontal shore platforms at or slightly above mean high tide level were originally formed as seaward-sloping shore platforms by wave abrasion at an earlier phase when sea level was higher, and have subsequently been flattened by weathering with the sea at its present level and diminished wave energy. This could have occurred where the Late Quaternary marine transgression attained a maximum slightly above present sea level before dropping back during a phase of Holocene emergence, and where high tide shore platforms are backed by degraded cliffs which have not been kept fresh by marine attack. Dissection of the outer edge of the platforms can then be interpreted as the result of the cutting of a new abrasion platform at a lower level following emergence.

Some high tide shore platforms may have been formed as the result of dissection and downwearing of similar subhorizontal shore platforms that developed when the sea stood at a higher level in late Pleistocene times (p. 33), in which case fragments of older and higher shore platforms may persist locally as emerged terraces. However, many high tide shore platforms are backed by actively receding cliffs, often with a small basal abrasion ramp, and it is not necessary to invoke an episode of higher sea level in Holocene times to account for their development.

High tide shore platforms are not found on soft rock outcrops such as clay or sand, where wave abrasion proceeds too rapidly for weathering effects to persist, or on massive hard rock outcrops such as granite or quartzite, on which weathering is very slow.

The width of a high tide shore platform is determined by the relative rates of recession of the cliff at the rear (by removal of weathered material and occasional storm wave abrasion) and along the seaward margin (by more continuous wave action on permanently saturated and unweathered rock). It is also related to the height of the cliff (i.e. the volume of rock to be removed as the platform is cut: small where the shore platform has developed as the result of dissection, and downwasting of an earlier shore platform cut at a higher relative sea level) and the length of time since the sea reached its present level (Trenhaile 1999).

Variations in high tide shore platforms are related to the lithology and structure of coastal rock outcrops, the effectiveness of weathering processes, nearshore topography, wave regime and tide range, as well as the availability of abrasive debris. These factors vary intricately on a coast of irregular configuration, with local variations in aspect resulting in changes in the width and transverse gradient of the shore platform, the persistence of upstanding rock outcrops, and the degree of dissection by wave abrasion and quarrying, particularly along the seaward margin. Broad, flat high tide shore platforms cut in basalt on the ocean coast near Flinders in Australia, become more irregular, with stronger structural features, on the adjacent more sheltered shore of Westernport Bay.

Development of shore platforms by weathering and the washing away of disintegrated material down to the level of rock saturation is seen on Old Hat islands, where residual central hills are encircled by flat rock ledges. Old Hat islands have formed on impermeable, homogeneous rock formations in the low wave energy landlocked Bay of Islands in New Zealand (Cotton 1974). They have formed where the land had previously been deeply weathered: wave sluicing has removed fine-grained weathered material to expose a shore platform on the upper limit of unweathered rock. This implies that the depth of prior weathering coincided with present day high tide level, but there is a possibility that some shore platforms at high tide level have formed where weathered rock has been re-indurated by secondary cementation (p. 80).

Little attention has been given to shore platforms on emerging coasts. On Cape Flattery, on the coast of Washington State, there are seaward-sloping shore platforms and weathering subhorizontal shore platforms in the present intertidal zone on a coast that has been rising tectonically at about 1.6 millimetres per year. Bird and Schwartz (2000) concluded that these platforms had been lowered by abrasion and weathering at about the same rate as the falling relative sea level.

Some shore platforms cut in ferruginous sandstone owe their persistence to induration, resulting from intertidal precipitation of iron compounds derived from glauconite, which originally forms a coating on quartz grains. These compounds dissolve in percolating groundwater and as the rocks dry out they are drawn to the rock surface and precipitated. The ferrous iron compounds then oxidize to form less soluble ferric minerals, such as brown goethite and reddish haematite, which bind the sand grains more firmly and produce a hardened sandstone. On the coast of Victoria, Australia outcrops of Tertiary ferruginous sandstones are grey or yellow and unconsolidated when first exposed to the atmosphere, but become darker and harder as oxidized iron compounds, mainly goethite, are precipitated near the rock surface. The transition in colour and hardness can often be traced from pale sandstones in cliffs behind bays to dark brown sandrock on headlands and rocky shores (Figure 4.36). Fallen boulders also harden and darken on the shore as the result of iron oxides precipitation. The process is aided by the recurrent wetting of rocks in the base of the cliff and on the shore by sea water and sea spray, and subsequent drying. As a consequence, wave abrasion is reduced, and the indurated outcrops persist on the shore to form structural shore ledges with upstanding ridges and reefs in the intertidal zone (Bird and Green 1992). Similar surface hardening is seen on the

Figure 4.36   Cliff and shore platform cut in Pliocene sandstone at Black Rock Point, Port Phillip Bay, Australia, showing induration increasing seaward from soft sandstone at the cliff base (A) to harder rock on the shore platform (B) and strongly indurated rock at the seaward edge (C).

cliff of ferruginous sandstone at Redend Point, near Studland in Dorset.

Low tide shore platforms

Low tide shore platforms are horizontal or almost horizontal platforms (with a very gradual seaward slope), exposed only for a relatively brief period when the sea falls to low tide level (Figure 4.29C). They are also Type B platforms, and are called low tide shore platforms because they are subaerially exposed only when the tide is low. They are best developed where the tide range is small (microtidal coasts) on limestone coasts. Low tide shore platforms may be broad and almost flat to the cliff base, which frequently has a notch overhung by a visor (Figure 4.32). As on high tide shore platforms there is sometimes a slightly higher rampart at the outer edge, in this case formed by an encrustation of algae in a zone that is kept wet by wave splashing even at low tide, and there is usually a sharp drop at the seaward edge, down to a sea floor with a broad gently concave profile, possibly a developing seaward-sloping subtidal platform (Figure 4.30).

Subhorizontal shore platforms of this kind have developed by the removal of limestone in solution (Wentworth 1939). As has been noted (p. 79), corrosion of shore limestones can be achieved by rain water (at low tide), by aerated surf and sea spray, and by sea water when its carbon dioxide content increases, notably during nocturnal cooling. Percolating groundwater and daytime sea water are usually saturated with dissolved carbonates, and unable to corrode limestone outcrops; they may instead precipitate carbonates. Limestone shore platforms are developed by marine solution processes on the coasts of arid regions, so solution by rain water is not essential, but it has undoubtedly contributed to shore limestone corrosion in humid environments.

Limestone cliffs and rocky shore outcrops show surface pitting and irregular dissection into networks of ridges and pinnacles (lapies) as the result of corrosion, especially by sea spray (Figure 4.32). Contrasts in the features of limestone coasts in tropical and temperate environments were identified by Guilcher (1958). On limestone coasts in cool temperate regions (where sea water is colder and less saturated with carbonates), as on the Burren coast in western Ireland, the shore rocks are intricately corroded, with numerous pinnacles and small pools. On warm temperate coasts, as in Portugal and Morocco, solution of shore limestone produces pitted rock outcrops and flat-floored pools (at various levels) with overhanging rims which may be residual limestone with or without algal or vermetid encrustations. As the pitted rock outcrops are consumed the flat-floored pools grow larger, and coalesce, eventually forming a broad shore platform at about mid tide level. On tropical limestone coasts (including emerged coral reefs) notches produced by solution are prominent, with overhanging visors bearing pinnacles formed by solution in sea spray and rain. They are conspicuous on the limestone

cliffs of Phang-nga Bay in Thailand (Figure 3.2), on the great limestone cliffs of northern Palawan in the Philippines, and on emerged coral reef islands in the Pacific Ocean.

A flat-floored notch with an overhanging visor at the cliff base and around rock outcrops rising above the shore platform can be produced by solution processes, which are most rapid (removing up to a millimetre of rock annually) just above mid tide level (Hodgkin 1964). Notches of this kind are well developed on tropical coasts, particularly behind shore platforms cut into emerged coral reefs. The importance of solution processes is indicated where such notches extend around stacks and islets (forming mushroom rocks), and are at least as well developed on the side sheltered from incoming waves (Figure 4.37). As the notch is cut back, the shore platform is extended. The level of the shore platform is determined by the downward limit to which solution processes are effective, and at this level the precipitation of carbonates begins. It is

Figure 4.37 Notch and visor around a mushroom rock at Sorrento, south-eastern Australia.

noteworthy that coral reefs are built up, and emerged coral reefs planed off, to this same level, as seen on the shores of Mbudya Island in Tanzania.

The level of planation (mainly by solution) on limestone coasts is thus slightly lower than planation (mainly by weathering) on fine-grained sedimentary and volcanic rocks. At Cape Schanck in Australia the rising tide submerges subhorizontal low tide platforms cut in Pleistocene dune calcarenite before it reaches similar high tide platforms cut in Tertiary basalt.

Shore platforms on limestone coasts often coincide with a horizon of induration which develops where the precipitation of carbonates and downward-percolating groundwater meets carbonate-saturated sea water and solidifies and hardens the rock outcrop. This is most obvious on Pleistocene dune calcarenites, as on the Nepean ocean coast in Victoria, Australia (Figure 4.38), where the shore platform coincides with the surface of a horizon where the calcareous sandstone has been indurated by carbonate precipitation (Hills 1949). The calcarenite is less resistant both above and below

this horizon, which thus becomes a structural influence on shore platform evolution. Induration also occurs on blocks and boulders of dune calcarenite that fall to the shore. Locally, an unusual form of shore pothole (p. 77) develops where soil pipes that had been excavated into old land surfaces by solution below plant root systems (the surrounding calcarenite becoming indurated by associated carbonate precipitation) are exhumed and washed out as circular rimmed vase-like depressions (Bird 1993a).

Apart from solution, other weathering processes are at work on limestone coasts. Wetting and drying contributes to the disintegration of limestone surfaces, particularly on dune calcarenites, but ceases at high tide level, above the limit of solution processes. Salt crystallization also contributes to the erosion of sea-splashed surfaces that dry out sufficiently for crystals to form. Bioerosion is certainly active, with numerous shelly organisms attacking the rock surface in the intertidal zone, where many species are present, and shelly organisms certainly contribute to pitting and the excavation

Figure 4.38   Cliff and shore platform cut in Pleistocene dune calcarenite at Jubilee Point, near Sorrento in south-eastern Australia.

of notches (Spencer 1988), but they do not play a major role in shore planation because they are also present, though in reduced numbers, below mid tide level. Instead, they may form bioconstructional features such as the algal (mainly *Lithophyllum*) and vermetid ledges projecting from the base on notches on some Mediterranean limestone coasts, notably in Crete (Kelletat and Zimmerman 1991).

Wave sluicing at high tide washes away weathered, eroded and precipitated fine-grained sediment from limestone shore platforms. Where sand and gravel are present wave abrasion becomes effective, and waves armed with rock debris cut an inclined abrasion ramp towards the rear of the platform, rising to the base of the receding cliff. As cliff recession proceeds the inclined abrasion ramp is lowered and flattened by solution and the shore platform is extended landward. While cliff-base and stack-base notches on limestone coasts are largely the outcome of solution processes, wave abrasion can modify them, producing abrasion notches with scoured sloping ramps, particularly on sectors exposed to strong wave attack. Quarrying by waves can dissect shore platforms, especially along joints or faults, and by undermining the cliff at the seaward edge. As on high tide shore platforms, strong wave action armed with abrasive debris can disrupt limestone shore platforms and replace them by seaward-sloping shore platforms of the kind seen on chalk coasts, where waves achieve abrasion by mobilizing flint nodules. Such sloping platforms are more likely to develop on stormy high wave energy coasts and on mesotidal and macrotidal coasts than where the tide range is small.

Rocky and bouldery shores

Some coasts have irregular rocky shores, either where shore platforms have been greatly dissected, or where they have failed to form because of an intricate geological structure and much variation in rock type and hardness. The surface is rugged, with many ridges, promontories and stacks, and complex patterns of grooves and channels. There are often intervening small bays, some of which may contain beaches of sand or gravel, and usually there are many dislodged blocks and boulders, so that some shores can be described as boulder-strewn. The boulders may have been produced by disintegration of shore rock outcrops or imported as glacial erratics, relinquished from a former ice sheet, as on parts of the Baltic coast.

STRANDFLATS

Strandflats are extensive coastal platforms of problematical origin found in front of mountainous slopes on the fiord coasts of Norway, Spitzbergen, Iceland, and Greenland. They are up to 50 kilometres wide, and have a relative relief of ± 200 metres. They are not strictly shore platforms, although they may grade into seaward-sloping shore platforms that extend beneath shallow coastal waters. They may be partly emerged shore platforms, for numerous rocky hills (interpreted as former islands) rise above them. They occur only on cold coasts, and must be related to past glacial or periglacial processes. It is possible that they developed in Pleistocene times as the result of plucking and disintegration of coastal rock outcrops by sea ice that rose and fell with the tide, followed by the sweeping away of debris by wave action when the ice melted (Tietze 1962). Alternatively, they may be the outcome of prolonged coastal periglaciation, repeated freezing and thawing producing a frost-shattered rock mantle which was sluiced away by the sea, exposing the underlying unweathered rock. It is not clear why strandflats have formed on some glaciated coasts but not on others: they are not seen in Alaska or British Columbia, in Patagonia, or in South Island, New Zealand.

PLUNGING CLIFFS

Reference has been made to plunging cliffs (Figure 4.29D), which pass steeply beneath low tide level without any shore platforms or rocky shore outcrops. It is necessary to explain why the processes that have produced the various kinds of shore platform have not succeeded in forming one where there are plunging cliffs. There are several possible explanations. Plunging cliffs can be produced by recent faulting, the cliff face being the exposed plane of the fault on the up-throw side, the down-thrown block having subsided beneath the sea, there having been insufficient time for marine erosion to cut a shore platform at present sea level. Plunging cliffs along the line of the Wellington fault, on the western shore of Port Nicholson in New Zealand, formed in this way. They slope at about 55°, show little evidence of marine modification at the intertidal level, and descend to the 12 fathom (about 20 metres) submarine contour, close inshore and parallel to the coastline.

Guilcher (1965) suggested that powerful waves breaking against a soft rock coast (such as the cliffed drumlins in western Ireland) produce plunging cliffs by cutting a platform below low tide level. This may explain why the Nullarbor cliffs in southern Australia, cut in soft Miocene limestone, have no shore platforms, but descend to a submarine platform a few metres below low tide level, which declines gently seaward. The Portland stone cliffs of the south coast of Purbeck in Dorset plunge into deep nearshore water where the underlying soft formations (Portland sand and part of the Kimmeridge clay) have been swept away by strong waves and currents to leave submerged limestone ledges (Bird 1995). Where stacks rise from the nearshore sea floor, rather than from a shore platform exposed at low tide (Figure 4.16), the implication is that cliff recession has been accompanied by down-wearing of the shore to produce a platform below low tide level.

Plunging cliffs may be formed by tectonic subsidence of coastal regions partially submerging coastal escarpments or former cliffs, possibly with shore platforms that are now below low tide level. The plunging cliffs of Lyttleton Harbour and Banks Peninsula in New Zealand originated as the result of subsidence of this area during and since the Late Quaternary marine transgression. The explosive eruption of Krakatau, a volcano in Sunda Strait, Indonesia, in 1883 left steep plunging cliffs on residual islands facing into the caldera, and similar features are seen on Santorini, north of Crete, which exploded about 3500 years ago. On the Scottish island of Skye vertical inland lava cliffs pass laterally into plunging sea cliffs formed by marine submergence.

Coasts built up recently by volcanic activity, as on the island of Hawaii, show plunging cliffs on sectors where there has not yet been time for shore platforms to develop. The absence of shore platforms on very sheltered sectors of coast bordering rias and fiords, and on sectors where coastal rock outcrops are extremely resistant, may be due to the fact that the period of up to 6000 years since the sea attained its present level has been too brief for marine processes to develop platforms.

Figure 4.39   Plunging cliff at Wilsons Promontory, south-eastern Australia.

Plunging cliffs are seen on resistant formations, such as the granites on the Land's End peninsula and the Bullers of Buchan on the east coast of Scotland, which form steep rocky coastal slopes descending beneath the sea, and similar features are seen on the various igneous and metamorphic rocks that form the high and rugged cliffs on the Lizard peninsula in Cornwall, exposed to strong wave action from the Atlantic Ocean. The high cliffs of Moher in County Clare, Ireland, also plunge into the sea.

In Australia there are plunging cliffs on massive granite on Wilson's Promontory (Figure 4.39), the rock being too resistant for shore platforms to have developed within the past 6000 years: the cliffs are the partly-submerged slopes of granite hills and mountains.

Where plunging cliffs descend to submerged terraces, as on parts of the Scottish coast, it may be that the terraces were cut under colder conditions, when freeze–thaw and sea ice processes contributed to rapid shore planation at a lower sea level (similar to strandflats). Alternatively, the cutting of a shore platform at the present sea level could have been inhibited by continuing land subsidence or frequent sea level changes.

5

Beaches

INTRODUCTION

Beaches fringe about 40 per cent of the world's coastline, and generally consist of unconsolidated deposits of sand and gravel on the shore. Some are long and almost straight or gently curved; others are shorter, and include sharply curved pocket beaches in bays or coves between rocky headlands. Many are exposed to the open ocean or stormy seas, but others are sheltered in bays or behind islands or reefs. Beaches form the seaward fringes of barriers (Chapter 6), which are banks of beach material deposited across inlets and embayments to enclose lagoons and swamps. Some beaches are bordered by deep water close inshore; others face shallow or shoaly water, often with bars, which are intertidal or subtidal ridges of sand or gravel running parallel, or at an angle to, the high tide shoreline. There are beaches that have been fairly stable over periods of years or decades, but most show rapid changes, especially in stormy weather. Some beaches are obviously gaining or losing sediment; others consist of sediment in transit (migrating along the coast), and others remain in position and may be relict, without any present-day sediment supply. Many beaches change in plan (i.e. the shape seen on a map or vertical air photograph) and profile (transverse to the shore), either rapidly over a few hours or days, or slowly over several decades or centuries. Some changes are cyclic (the beach returning to the same plan and profile over varying periods); others are prograding (advancing seaward by accretion) or receding as the result of continuing erosion.

Beach profiles are surveyed using graduated poles and a level, a theodolite, an electronic distance measurer or a wheeled vehicle designed to register rise and fall along a beach transect. Profile changes can be measured with reference to inserted stakes, or surveys along a pier or groyne. Beach morphology can be mapped using conventional survey methods, or remote sensing.

Beaches are essentially similar on coasts in various climatic environments, except in cold regions where wave action ceases (at least in winter) because of the freezing of the sea. When the ice fringe melts, wave action shapes characteristic beach forms, including beach ridges, spits, and tombolos, as on the northern coasts of Alaska and in Antarctica.

BEACH SEDIMENTS

Beach sediments consist of sand or gravel particles of various sizes (Table 5.1), the proportions of which can be determined by grain size (granulometric) analysis, as shown and explained in Table 5.2 and Figure 5.1. Some are coarse, dominated by cobbles and pebbles; others finer, with various grades of sandy sediment, granules being relatively rare. Some are uniform (i.e. well-sorted), with granulometric analysis showing a high proportion of a particular size grade; others are more varied in texture, sometimes with contrasting zones of coarser and finer material (p. 123) along the beach face. Beaches are better sorted on high wave energy coasts, particularly if they are swash-dominated (p. 120).

Many of the world's beaches are sandy, but coarser particles (gravels, which comprise granules, pebbles or cobbles, generally of stone but sometimes shelly) are often present, and may be scattered across a sandy beach or arranged in patterns such as cusps or ridges running parallel to the shore. Sand and gravel particles may be angular or subangular in shape, but usually become rounded by abrasion and diminished by attrition as they are worn by the action of waves agitating the beach. Beaches composed entirely of well-rounded pebbles and cobbles are known as shingle beaches, especially in Britain, and there are also boulder beaches, with heaps of stones ranging in diameter up to more than a metre, as on parts of the Dorset coast where they are derived from disintegrating limestone and sandstone strata.

Table 5.1 Beach Grain Size Categories.

The Wentworth scale of particle diameters. The ø scale is based on the negative logarithm (to base 2) of the particle diameter in millimetres [ø = log 2d], so that coarser particles have negative values.

Wentworth scale category	Particle diameter	ø scale
Boulders	>256 mm	below −8ø
Cobbles	64 mm – 256 mm	−6ø to −8ø
Pebbles	4 mm – 64 mm	−2ø to −6ø
Granules	2 mm – 4 mm	−1ø to −2ø
Very coarse sand	1 mm – 2 mm	0ø to −1ø
Coarse sand	$\frac{1}{2}$ mm – 1 mm	1ø to 0ø
Medium sand	$\frac{1}{4}$ mm – $\frac{1}{2}$ mm	2ø to 1ø
Fine sand	$\frac{1}{8}$ mm – $\frac{1}{4}$ mm	3ø to 2ø
Very fine sand	$\frac{1}{16}$ mm – $\frac{1}{8}$ mm	4ø to 3ø
Silt	$\frac{1}{256}$ mm – $\frac{1}{16}$ mm	8ø to 4ø
Clay	$<\frac{1}{256}$ mm	above 8

An alternative is a decimal scale centred on 2 mm, in which sand ranges from 0.2 mm – 2.0 mm, but this does not correspond well with generally perceived grain size categories. The sand range excludes sediment that would be generally classified as fine to very fine sand, and the coarser (>2.0 mm) and finer (<0.2 mm) divisions do not match widely accepted categories of pebbles and cobbles or of silt and clay.

Table 5.2 Granulometric Analysis of Beach Sediment.

A sample of beach sediment, washed and dried, is passed through a set of sieves of diminishing mesh diameter to divide it into size grades which are weighed separately. The results are presented as a graph of grain size distribution (Figure 5.1), the steepness of which increases with the degree of sorting. Beach sediments are generally well sorted, so that the bulk of a sample falls within a particular size grade (fine sand in Figure 5.1) and the central part of the graph rises steeply. The grain size distribution of a beach sediment is usually asymmetrical and negatively-skewed (the mean grain size being coarser than the median), as the result of the removal of fine particles by wave (and on sandy beaches wind) action.

The shape of the curve can be characterized numerically by using the median diameter (P_{50}, the 50th percentile) and selected higher and lower values, such as the 16th and 84th percentiles, as an indication of the relative proportions of coarser and finer particles. In Figure 5.1 the median (P_{50}) is 1.4, P_{16} is 0.7 and P_{84} 1.9. Values for the mean, sorting (standard deviation) and skewness can be calculated from the following formulae:

Mean $= \frac{1}{2}(P_{16} + P_{84})$ - in this case $= \frac{1}{2}(0.7 + 1.9) = 1.3$

Sorting $= \frac{1}{2}(P_{84} - P_{16})$ – in this case $= \frac{1}{2}(1.9 - 0.7) = 0.6$

Skewness $= \dfrac{\text{Mean} - \text{Median}}{\text{Sorting}}$ – in this case $= \dfrac{(1.3 - 1.4)}{0.6} = -0.17$

There are descriptive categories for sorting and skewness. A sediment with sorting <0.35 is very well sorted, 0.35–0.50 well sorted, 0.50–1.00 moderately well sorted, 1.00–2.00 poorly sorted, 2.00–4.00 very poorly sorted and >4.00 extremely poorly sorted. A sediment with skewness −1.00 to −0.30 is strongly negatively skewed, −0.30 to −0.10 negatively skewed, −0.10 to 0.10 nearly symmetrical, 0.10 to 0.30 positively skewed, and 0.30 to 1.00 strongly positively skewed. The sample in Figure 5.1 is therefore moderately well sorted (0.60) and negatively skewed (−0.17). These parameters can be calculated with the aid of a computer program, but this example indicates the principles involved.

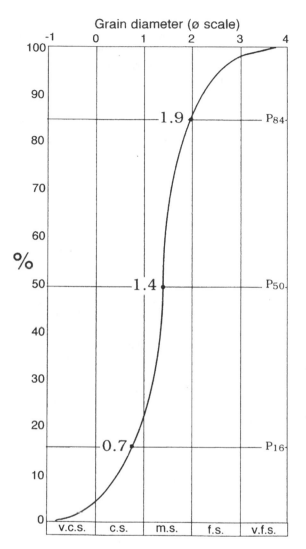

Figure 5.1 Grain size analysis graph (see Table 5.2).

Descriptions of beaches sometimes consider the shape of beach sediment particles (usually sampled from the modal size range). An index of roundness can be obtained by dividing the minimum diameter by the maximum length (a spherical pebble = 1.0) and an index of flatness by dividing the sum of the length and width by twice the thickness (minimum flatness of a spherical pebble = 1.0). Some beaches consisting of well-sorted and well-rounded, highly polished sand grains emit a squeaking noise when walked on, as at Whistling Sands on the Lleyn Peninsula in north Wales, and Squeaky Beach on the west coast of Wilson's Promontory, Australia.

Beaches dominated by granules (grain diameter 2–4 millimetres) are comparatively rare, but they do occur locally where there is a source of material of this calibre (e.g. quartz and felspar grains in decomposed granite) or where a vein of quartz or basalt has disintegrated into gravel that has been reduced to this size range by attrition. Vault Beach, near Dodman Point in Cornwall, is dominated by quartz granules. Conventionally, deposits of sediment finer than sand (i.e. silt and clay) are regarded as muddy shores rather than beaches.

EVOLUTION OF BEACHES

In Chapter 3 it was shown that the sea has risen and fallen several times around the world's coastline during the past few million years, so that former coastlines, some with beaches, are found at various levels above and below present mean sea level. Existing beaches are geologically of recent origin, having formed as the Late Quaternary marine transgression slackened, or gave place to a Holocene sea level stillstand: on most coasts about 6000 years ago. Sand and gravel from various sources have been delivered to beaches along steep coasts, except where there are cliffs plunging into deep water, or where the shore is too rocky and rugged to retain a beach. There are steep coasts without beaches, for example, around the Land's End peninsula in Cornwall, around high islands as in the Seychelles, and on steep promontories such as Rocky Cape in northern Tasmania, where plunging cliffs pass below low tide level, but there is abundant sand on the sea floor. The nearshore profile is evidently too steep and reflective for waves to be able to carry sand shoreward to form beaches. Beaches have also formed along the fringes of coastal lowlands, except where wave energy is too low, and the shore becomes muddy and marshy. Beaches may be coarser on high wave energy coasts, but there are many sandy beaches on high wave energy coasts where there is (or has been) a sandy supply but coarser sediment is unavailable, while gravelly beaches occur on low wave energy coasts where there is a source of coarse material.

Some beaches remain narrow, fringing cliffs and steep coastal slopes or bordering alluvial plains and wetlands, while others have widened with the addition of successively-formed backshore beach ridges (p. 141), which may bear dunes built of sand winnowed from the shore.

There are relationships between patterns of refracted waves approaching the shore and the sediment characteristics of beaches. Where convergence of wave ortho-gonals (p. 11) indicates augmented wave energy (i.e. larger waves breaking on the beach) beaches become generally steeper, higher and better sorted; erosion is more severe and divergence of longshore currents causes sediment dispersal. Divergence of orthogonals indicates a low wave energy coast, with the reverse of these conditions: lower beaches with gentler gradients, generally finer and less well sorted beach sediment, reduced erosion or perhaps accretion, and convergent longshore currents bringing in beach sediment. These relationships are complicated, however, by other factors, such as the nature of available sediment.

PROVENANCE OF BEACH SEDIMENTS

The origin of the various kinds of beach sediment can be determined with reference to petrological and miner-alogical characteristics, and to patterns of sediment flow produced by waves and currents on the coast and in nearshore areas. Beaches have received their sediments from various sources (Figure 5.2). Some have been supplied with sand and gravel washed down to the coast by rivers: either large rivers draining a catchment which yields an abundance of such sediment (e.g. the De Grey River in north-western Australia) or where rivers drain steep hinterlands (e.g. on the west coast of South Island, New Zealand). Others consist of material derived from the erosion of nearby cliff and foreshore outcrops, particularly where these include weathering sandstones and conglomer-ates. Sand and gravel have been washed in from the sea floor by waves and currents to form beaches, parti-cularly on oceanic coasts, and in a few places beaches have been supplied with sand delivered by winds blow-ing from the hinterland. In recent decades many beaches have been augmented by the arrival of sedi-ment produced as the result of human activities, such as agriculture or mining on the coast or in the hinter-land. Some beaches have been artificially nourished or replenished, especially at seaside resorts. While many beaches are still receiving sediment from one or more of these sources, some have become relict, and now consist of deposits that accumulated in the past, but are no longer arriving. The following sections exemplify these several categories.

Beaches supplied with fluvial sediment

Fluvial supply of sediment to beaches occurs where sand and/or gravel have been washed down to the

mouth of a river and are carried along the coast by waves that arrive at an angle to the shore (Figure 5.3). Alternatively rivers may sweep sediment out to the sea floor, where it is reworked by waves and currents, and some of it (mainly the coarser sand and gravel) delivered to the nearby coast. Sand and gravel may accumulate on the shores of a symmetrically grow-ing delta, or be distributed alongshore in either direc-tion by waves and currents to form beaches and spits that can extend for several kilometres along the coast. The coarser sand and gravel often remain on beaches near the river mouth while the finer sediment is carried further along the shore. Rivers that flow into inlets, estuaries or lagoons may deposit their loads of sand or gravel before reaching the sea.

The nature of sediment supplied to beaches by rivers depends on the types of rock that outcrop along the river channel and within the catchment basin, where runoff delivers surface material formed as rock out-crops decompose or disintegrate by weathering. The volume and calibre of fluvial sediment loads are also influenced by the steepness of the hinterland, the vigour of runoff produced by rainfall or the melting of snow or ice, the effects of earthquakes and volcanic eruptions and the extent and luxuriance of the vegetation cover, which has often been modified by agriculture, forestry, mining or urban development. Examples of each of these are given below.

Catchments including weathered granite or sand-stone produce sandy sediment, often dominated by quartz, the most durable of common rock minerals and so the most widespread, dominating most sandy beaches. Felspars and micas are less durable, but may persist in some beach sediments, especially those receiving abundant material from steep hinterlands and those where felspathic sandstones (arkoses) out-crop in river catchments, as on the Otway and Gippsland coasts in Victoria, Australia. There are vary-ing proportions of heavy (ferromagnesian) minerals in material derived from such rocks as granite, and these may become concentrated in beach sediments as layers of mineral sand, such as the rutile and ilmenite found on parts of the Australian coast, magnetite in southern Japan, and gold in Alaska. Sand brought down from volcanic hinterlands also reflects the composition of the source rocks: weathered basalt, for example, yields grey or black sand dominated by the dark mineral olivine, as on the beaches of Hawaii, Indonesia and Kamchatka.

Some rivers have been delivering sand or gravel to beaches near their mouths, but some of the sand in estuaries may be of marine origin, washed in from the sea by waves and inflowing tides. It is necessary to seek

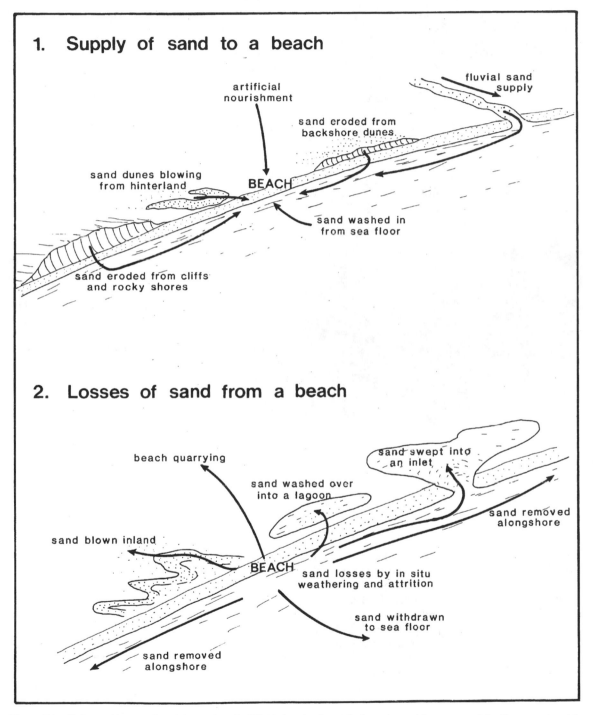

Figure 5.2 Gains and losses of sand from a beach. Shingle beaches have similar gains and losses, except for those produced by wind action.

Figure 5.3 Fluvial sediment supply to beaches, illustrated by the River Don in north-eastern Australia. Sandy material in the channel is washed down to the river mouth during periods of strong fluvial discharge, and is then reworked and distributed along the coast by wave action.

Mediterranean and on the Pacific coast of the United States, notably those that have been supplied by the Columbia River, extending southward into Oregon and northward into Washington State. Sandy beaches supplied with fluvial sand derived from sandstone catchments are extensive in southern California, and in eastern Australia, particularly in Queensland, where beaches include substantial quantities of sand and some gravel brought down by rivers from steep catchments east of the Great Dividing Range. Beaches near the mouths of the Shoalhaven, Moruya, Bega and Towamba rivers in New South Wales consist largely of fluvially-supplied quartz and felspar sand originating from weathered granite. Beaches composed primarily of fluvially-supplied sand are also seen on coasts adjacent to the mouths of the Gascoyne, Ashburton and De Grey Rivers in north-western Australia, each of which has a large catchment of sandy desert from which substantial quantities of sand are carried downstream occasionally, when rapid runoff is generated by the heavy rainfall that accompanies tropical cyclones (Bird 1978b). Occasional sudden downpours flood wadis in eastern Egypt and Sinai, washing terrigenous sand down to beaches on the shores of the Red Sea. In Kenya sand supplied by the Galana River has drifted southward to prograde the beach at the resort of Malindi (Figure 5.4).

Shingle beaches are found where rivers drain catchments with fissile rock outcrops or extensive gravel deposits and flow strongly enough to deliver this stony material to the shore. In the North Island of New Zealand Quaternary volcanic deposits include much coarse gravelly material, some of which has been carried downstream by such rivers as the Mohaka, and deposited to form pebbly beaches along the north coast of Hawke Bay (Bird 1996b). Where the hinterland is steep and mountainous, as on the west coast of South Island, New Zealand, or the Caucasian Black Sea coast, strongly-flowing rivers carry gravel as well as sand down to the coast, to be incorporated in beaches.

Shingle beaches are rare in the humid tropics because coastal and hinterland rock outcrops are generally deeply weathered and contain little gravel. There are exceptions on coasts bordering steep hinterlands, as along the north coast of New Guinea below the Torricelli Mountains, and at Lae, which has a cobble beach supplied with gravels from the Markham River.

Beach gravels are common in areas where past or present glaciation or periglaciation has produced

confirmation that sand or gravel have been derived from source areas upstream and carried down to the river mouth before deciding that beaches near the river mouth have been fluvially nourished. In Britain there are few fluvially nourished sandy beaches. In north-east England the Tyne and the Tees have carried sand down to the coast, but the sand in the Tweed estuary has been washed in from the North Sea, where it is likely that eskers similar to those seen in Tweedsdale have been submerged, reworked, and swept shoreward.

Sand from the Loire and the Gironde rivers has been delivered to beaches on the west coast of France, the Tagus brings sand down to the Portuguese coast, and there are fluvially-fed sandy beaches around the

Figure 5.4   Prograded beach at Malindi, Kenya. The vegetation has advanced on to the backshore as the beach was widened by sand accretion.

large quantities of stony morainic debris which can be washed down to the coast, as in southern Alaska where streams swollen by spring meltwater carry large quantities of sand and gravel derived from moraines and glacial outwash down to the coast to form beaches on the shores of Icy Bay (Molnia 1985). Shingle beaches on the north coast of Scotland include gravelly material derived from Pleistocene glacifluvial deposits and brought down by the rivers Spey and Findhorn. In Peru pebble beaches have been derived from Pleistocene conglomerates deposited by Andean rivers during phases of wetter climate and stronger runoff.

Beaches receive accessions of fluvial sediment during episodes of flooding after heavy rainfall or the sudden melting of snow or ice in the river catchment. Floodwaters carry sand and gravel out to form shoals off river mouths (Figure 4.1), and when these shoals are reworked by waves and currents after the floods abate some of the sand or gravel is washed onshore to be added to the beach. On steep coasts, such as those in southern Brazil, beaches are periodically renourished with sand brought down by river floods from a steep hinterland where soil erosion has generated an abundance of fluvial sediment. In 1967 torrential rainfall in the Serra do Mar mountains south west of Rio de Janeiro caused extensive erosion, catastrophic river flooding, and the delivery of large quantities of sand, silt and clay to the sea, notably in Caraguatatuba Bay, whence wave action subsequently sorted this material and washed the sandy fraction onshore, to be added to the beaches (Cruz et al. 1985). Antarctic beaches receive ice-rafted erratic rock fragments and the debris yield of melting coastal glaciers in summer.

Earthquakes in the hinterland can increase fluvial sediment yields. The severe tremors which shook the Torricelli Mountains in northern New Guinea in 1907 and 1935 caused massive landslides which delivered large quantities of sand and gravel to rivers draining northward to the coast, where beaches were widened, and beach material subsequently drifted alongshore from river mouths, eastward to Cape Wom (Bird 1981).

Volcanic eruptions within a river basin can also generate downstream movement of large quantities of sand and gravel. In the North Island of New Zealand river loads have been occasionally increased as the result of episodes of volcanic activity. The Tarawera volcanic eruption in 1886 produced sands and gravels that moved down rivers such as the Rangitaiki, and beaches on the shores of the Bay of Plenty (notably on Matakana Island) prograded when this material reached the coast (McLean 1978). In a similar way successive eruptions of Merapi volcano, north of Jogjakarta in Java, have resulted in a downstream flow of sand and gravel in the Opak River, the arrival of this sediment at the river mouth being followed by the progradation of beaches of grey volcanic sand at and west of Parangtritis (Bird and Ongkosongo 1980). In south-east Iceland glacifluvial streams have had their sand and gravel loads increased by the eruptions of hinterland volcanoes, which also cause catastrophic ice melting and river flooding, leading to the formation of wide outwash plains, known as sandur, the seaward fringes of which are subsequently reworked by wave action to form beaches (Bodéré 1979). However, although the Mount St Helens eruptions in 1980 delivered vast quantities of sediment to the Columbia and other rivers flowing down to the Pacific coast, much of it was silt and clay, and there has been little ensuing progradation of the sandy beaches in Oregon and Washington.

Runoff from unvegetated slopes is much more rapid than from slopes that carry a plant cover which intercepts, retards and recycles rainfall and thus diminishes fluvial sediment yields. Rivers draining catchments where the vegetation cover has been reduced or removed deliver larger and coarser sediment loads to the coast. Around the Mediterranean, for example, deforestation, overgrazing and excessive cultivation of hinterlands over the past 2000 years has led to increased runoff, soil erosion and river flooding, so that larger quantities of fluvial sediment have reached the mouths of rivers, and beaches have prograded. In Greece and Turkey rapid deposition at and around river mouths led, for example, to the historical infilling of the Maliakos Gulf (Bird 1985a). Such augmentation of fluvial sediment supply depends on the continued presence of weathered material on the hinterland slopes, and where this is removed entirely the sediment yield diminishes. The Argentina River, on the Ligurian coast of Italy, had a phase of rapid delta growth and beach progradation when sediment yield from the steep, eroding catchment was high, but in recent decades the supply has declined as bedrock became widely exposed upstream (Bird and Fabbri 1993).

Some beaches have been enlarged by the accumulation of sand or gravel brought down by rivers from mining areas in the hinterland. On the island of Bougainville, north of New Guinea, tailings from copper mines increased the load of the Kawerong River, and have been carried downstream to be deposited on beaches at and around the river mouth at Jaba, on the shores of Empress Augusta Bay (Brown 1974). Beaches have been augmented on the Pahang delta on the east coast of peninsular Malaysia by the arrival of large quantities of sand generated by alluvial tin mining upstream. In New Caledonia extensive hill-top quarrying of nickel and chromium ores has resulted in massive spillage of rocky waste down steep slopes into river valleys and the delivery of increased and coarsened fluvial loads to prograde beaches at and near river mouths, as at Karembé on the western coast and Thio and Houailou on the eastern coast (Bird et al. 1984). In Chile the arrival of sandy tailings washed downstream from copper mines resulted in progradation of the beaches in Chañaral Bay (Paskoff and Petiot 1990), and in south-west England tin and copper mining and the quarrying of china clay from Hensbarrow Downs, a granite upland north of St Austell, produced large quantities of sand and gravel waste with much quartz and felspar, some of which was carried down rivers to prograde beaches at Par and Pentewan (Everard 1962).

The contribution of river sediment to beaches has thus been substantial in many parts of the world, but few beaches are entirely of fluvial origin and many have been nourished primarily from other sources now to be considered.

Beaches supplied from eroding cliffs and foreshores

Beach sediment derived from the erosion of cliffs and foreshores has characteristics related partly to the lithology of the outcropping formations and partly to the energy of the waves and currents which erode these outcrops and carry sediment along the shore. Sand eroded from cliffs cut in soft sandstone has nourished beaches at Point Reyes on the coast of southern California, and where cliffs of Tertiary sand line the shores of Bournemouth Bay in Dorset.

In general, gravel beaches are found where the coastal rock formations have yielded material of suitable size, such as fragments broken from thin resistant layers in sedimentary rocks, or intricately fissured formations, or pebbly conglomerates. Weathered granite yields quartz and felspar sand, forming cobbles and pebbles only where it disintegrates along intricate closely-spaced joint planes into blocks and angular fragments that become rounded as cobbles and pebbles. At Kimmeridge in Dorset limestone ledges on the shore have broken up along joint planes to form a blocky beach. Closely-jointed basalt also yields blocks that become cobbles and pebbles, but otherwise weathers by superficial flaking to black sand. Shingle beaches are not found where the coastal rock outcrops are homogeneous, as on massive granites, or where they are soft and fine-grained.

Extensive shingle beaches have been derived from the nodules of hard flint eroded from chalk cliffs on the south coast of England and the north coast of France. As the cliffs are cut back the flint nodules released by weathering and erosion form gravelly beaches, which are agitated by wave action and soon become well-rounded shingle. Flint nodules exposed in chalk cliffs are black, often with a white rind, and recently derived flint cobbles and pebbles on adjacent beaches are black or blue. The shingle train that has drifted eastward along the south coast of England to accumulations such as the cuspate foreland at Dungeness contains large quantities of brown flints that have not come directly from the chalk. They have had a much longer and more complicated history, having been weathered during residence in various Tertiary or Quaternary gravel deposits on land, or on the emerged sea floor during low sea level phases: the change from black to brown results from oxidation of iron compounds, and there has often also been some leaching of silica. The rate of change is not known, but it seems unlikely that many brown flints originated from Holocene erosion of chalk outcrops.

As well as flint, the shingle beaches of south-east England include cherts from the Upper and Lower Greensands and quartzites, sandstones and limestones from various Jurassic, Cretaceous and Tertiary rock outcrops along the coast. Each of these rock types is present in beach material in bays along the cliffy coasts of Dorset, and shingle has been supplied from conglomerates or breccias in coastal rock outcrops, as at Budleigh Salterton in Devon, where a cobble beach has been derived from coarse Triassic (desert wadi) gravels exposed in the adjacent cliffs.

Cliffs cut in glacial deposits (rubble drift, gravelly moraines or drumlins) have produced sand and gravel delivered to beaches around Puget Sound in Washington state (Shepard and Wanless 1971) and on other similar coasts, as in New England, the Danish archipelago, and the southern shores of the Baltic, notably in Poland. In the British Isles sand and gravel from cliffs of glacial drift have supplied beaches on the North Sea coast (Clayton 1989) and on the west coast, particularly around the Irish Sea. Beaches occur where glacial moraines intersect the coastline and cliffs cut into these deposits have provided a source of sand and gravel which has been spread along the coast. Examples have been noted in southern Norway along the Ra moraine, and in Poland and Estonia (Orviku et al. 1995). The redistribution of material from a morainic zone has nourished beaches such as Dingle Bay and Tralee Bay on the west coast of Ireland and spits such as Blakeney Point in Norfolk and Whiteness Head in Scotland. Sand, gravel and boulders eroded from cliffs cut in Pleistocene glacial drift on the east coast of England at Holderness has been supplied to local beaches and drifted south to Spurn Head. In north Wales the bouldery beaches of Barmouth Bay and the Harlech coast have been eroded from cliffs of glacial boulder clay, and the beach of sand backed by pebbles and cobbles in Hell's Mouth (Porth Neigwl) in north Wales, has been derived from erosion of cliffs of glacial drift. Erratic boulders (i.e. not of local derivation) are found on beaches and in the nearshore zone on the formerly glaciated shores of the Baltic Sea and Puget Sound.

Beaches on a cliffed coast cut in massive rock formations, homogeneous limestone or fine-grained sediment may be derived from an outcrop of a rock formation alongshore yielding sand or gravel. On the Port Campbell coast in Australia soft Miocene calcareous sediments form vertical cliffs, and there are few beaches except where cliff recession has intersected a dune calcarenite ridge, from which sandy sediment has been derived.

Erratic gravel and boulders of unusual origin are seen on the shores of Petrifaction Bay, on Flinders Island, Australia, where granite blocks that had been transported in a Tertiary lava flow are being eroded out of clay cliffs in the now deeply-weathered volcanic rock. On the north coast of France near Port d'Ailly, west of Dieppe the beach below chalk cliffs contains scattered boulders of quartz sandstone (known as sarsens) that have fallen from a capping of Tertiary deposits.

In south-west England beaches include sand and gravel derived from cliffs cut in the periglacial Head deposits (frost-shattered earthy rubble) that mantle coastal slopes (p. 54), and quartzite pebbles have come from disintegrating outcrops of vein quartz from cliffs and shore outcrops in the Devonian rocks. Many beaches in Devon and Cornwall incorporate sand and gravel from Pleistocene emerged beach deposits which stand a few metres above high tide level, were buried by Late Pleistocene periglacial Head deposits and subsequently exposed and cut back as cliffs by marine erosion, as on the shores of Falmouth Bay in Cornwall. On the similar coasts of Brittany, sand and gravel beaches formed during the Last Interglacial (Eemian) period were on a larger scale than their modern counterparts, which are partly derived from them (Guilcher 1958).

On granite coasts the contrast between the upper, leached pale layers of weathered sandy material (including slope wash) and the underlying darker layers, containing illuvial organic matter and iron oxides, can yield contrasts in the colour of derived beaches, as at White Beach and Yellow Beach near Lady Barron, on Flinders Island, Tasmania.

Gravel beaches are rare on humid tropical coasts, but they occur where there is a suitable source, such as the lateritic ironstone crusts that outcrop in coastal cliffs and rocky shores in the Darwin district in Australia and at Port Dickson in Malaysia. Sand and gravel beaches on the shores of the young volcano at Anak Krakatau, Indonesia have been derived from the erosion of cliffs cut in unconsolidated volcanic ash and agglomerate (Bird and Rosengren 1984), and there are similar beaches around the modern volcanic island of Surtsey, which began to form off the south coast of Iceland in 1963 (Norrman 1980).

Some beaches have incorporated sand or shingle derived from Pleistocene barrier formations (p. 172) truncated by marine erosion, as in the sandy cliffs near Seal Rocks in northern New South Wales. Beaches have been supplied with sand, gravel and boulders by landslides which form protruding lobes, usually with much silt and clay, as on the shores of Lyme Bay and on the north Norfolk coast in England. As these are consumed by wave action the finer sediment is dispersed, and the sand, gravel and boulders are released to nearby beaches.

Beaches supplied with sediment from the sea floor

Shoreward drift of sediment from the sea floor has contributed to beach deposits, and is most obvious on coasts where there is no supply from rivers or cliff and shore erosion. Sediment washed in to beaches from the sea floor includes sand or gravel eroded from submerged geological outcrops or collected from unconsolidated bottom deposits. The latter include sediment that originated from the hinterland, and was spread across the continental shelf by river outflow, glaciers or wind action during phases of lower sea level in

Pleistocene times. Just as there are river catchments on land from which coastal sediment has been derived and delivered to the coast at river mouths, so are there sea floor catchments on the continental shelf from which beach material has been, and may still be, carried shoreward, and delivered to parts of the coast. The contrast between the wide calcareous beaches and dunes on the west coast of the Penthièvre isthmus on the Presqu'ile de Quibéron in Brittany, and the narrow quartzose beaches on the east coast is related to distinct sea floor catchments. On King Island, Tasmania, the contrast between calcareous beaches and dunes on the west coast and the quartzose beaches and dunes on the east coast is related to contrasting sea floor catchments, the continental shelf to the west being an area of calcareous biogenic sediment production (cold upwelling water from the Southern Ocean and a meagre terrigenous input from a low-lying, arid continent) and the floor of Bass Strait to the east (strewn with quartzose sediments derived from weathered granites and fluvial deposits at low sea level stages).

The depth from which material can drift shoreward varies in relation to sea floor topography, wave and current movements on the sea floor and the size, shape and specific gravity of the available sediment, as well as the presence of sea floor vegetation, notably seagrasses such as *Zostera*, *Posidonia* and *Cymodocea*, which can inhibit sediment flow (p. 203). These grow luxuriantly in the Mediterranean and off the South Australian coast. In the Mediterranean, Van Straaten (1959) found that waves move sand on to beaches on the French coast from a depth of 9 metres, and on coasts receiving long ocean swell the process may be effective from greater depths than this. Off southern California there is only limited and occasional movement of sea-floor sand by ocean swell in depths exceeding 18 metres, indicating the probable maximum limit. Sand dumped in water 12 metres deep off the New Jersey coast was not delivered to the beach (Harris 1955), and was therefore beyond the limit for shoreward drifting.

The importance of shoreward drift of sediment to beaches was deduced in the Isles of Scilly by Barrow (1906) because of the lack of eroding cliffs and sediment-yielding rivers, sand and gravel (including shelly debris) was swept in from the surrounding sea floor by wave action. Shoreward drift has occurred where the nearshore waters are shallow, or are becoming shallower because of land uplift or a falling sea level. This can be seen in the Gulf of Bothnia, where isostatic uplift is causing emergence, so that sand and gravel from submerged but shallowing glacial drift deposits are being carried shoreward on to beaches, as at

Storsand in Sweden, Brusand in Norway and Kalajoki in Finland. Similar emergence has led to the formation of a wide sandy beach on the south-western shore of the island of Laesø in the Kattegat (Møller 1985). During the phase of lowering of the Caspian Sea between 1930 and 1977 shoreward movement of sea floor sand took place, and beaches were widened by accretion as the shores emerged, and around the Great Lakes in North America similar beach progradation has occurred as the result of shoreward drift during each of several phases of falling lake level (Olson 1958, Dubois 1977). On the coasts of Scotland, where post-glacial isostatic recovery is in progress, emerged Holocene shingle beaches and beach ridges are often bordered seaward by younger sand deposits in beaches and dunes, indicating a diminution in the calibre of sediment carried shoreward from the shallowing sea floor.

Widespread shoreward drift took place on many coasts during the Late Quaternary marine transgression, when the sea advanced across shoaly topography and waves washed sediment on to the shore, forming beaches and dunes (Figure 5.5A–C). As the sea rose some of the beach material was washed or blown landward, and it is probable that minor oscillations of sea level occurred, facilitating shoreward sweeping of sand and gravel on to beaches. Chesil Beach in Dorset is an example of a shingle barrier (p. 173) that formed and was driven shoreward as the sea rose, with a lagoon (The Fleet) on its landward side backed by a submerging, indented hinterland coast that was never exposed to wave action from the English Channel to the south-west (Steers 1953). The Loe Bar in Cornwall consists of flint shingle swept up on to a coast which has no cliff or shore sources of flint (Figure 5.6).

Glacial drift deposits submerged on the floor of the North Sea by the Late Quaternary marine transgression have been reworked by wave action, with sand and gravel carried shoreward to beaches on the east coasts of Scotland and England, and on the Dutch, German and Danish North Sea coasts. Along the Atlantic coasts of Britain and Europe aprons of peri-glacial Head deposits that extended out on to the sea floor during the Last Glacial low sea level phase (p. 54) have been submerged and reworked by wave action to yield sandy material that in some places (Corrubedo and Laxe in Galicia, for example, and Figuera do Foz in Portugal) is still moving onshore to beaches.

Shoreward drift of sand and gravel has produced beaches in suitable niches, such as shallow bays, coves, and inlets on steep and rocky coasts. On oceanic coasts swell has shaped long gently-curving beaches, often backed by wide sandy plains with multiple beach ridges and dunes that formed after the Late Quaternary marine transgression came to an end. The bulk of shoreward sand drift on the Australian coast took place during this transgression, when the near-shore zone migrated shoreward over unconsolidated deposits stranded during the previous sea regression. Beach progradation occurred in the ensuing stillstand as ocean waves swept sand in from nearshore shoals (Bird 1978b). Relics of such nearshore shoals persist in sheltered areas to the east of Wilson's Promontory and in the shallows between Robbins Island and Hunter Island off north-west Tasmania, and in both cases sand is still moving onshore to beaches.

On the Gippsland coast in south-eastern Australia, where most of the rivers deposit their loads in estuaries or lagoons (such as the Gippsland Lakes, p. 243) behind wide dune-capped Holocene coastal barriers, and where eroding cliffs are very limited, the Ninety Mile Beach (p. 176) has been formed almost entirely of sand swept in from the sea floor during the Late Quaternary marine transgression and the Holocene stillstand. Successively-formed beach or dune ridges backing the Ninety Mile Beach include evidence of Holocene progradation, which in places is still continuing. At the south-western end of the Ninety Mile Beach sand is still moving shoreward on to the beach, and there is similar shoreward movement of sand from shoals in Streaky Bay in South Australia (Bird 1978b) (p. 139).

Other examples of coasts where sand from nearshore shoals is being carried onshore to be added to beaches include south-western Denmark, where the sandy islands of Fanø, Mandø and Romø have beaches prograding in this way, parts of the southern coast of Florida, and near Montevideo in Uruguay. Sand is also being washed in from a shallow sea floor to prograde beaches on the shores of Carmarthen Bay in south Wales, the wide sandy beaches on either side of Holy Island in Northumberland, and Studland Bay, Dorset. Progradation has taken place along several kilometres of beach at Tentsmuir in Scotland, and as the beach does not widen towards the mouth of the River Tay, it is deduced that shoreward drift from the sea floor has been more important than fluvial sand supply.

Sand supplied from the sea floor may earlier have been of terrigenous origin, laid down during phases of low sea level and subsequently reworked by wave action during marine transgressions. On the New South Wales coast, sand delivered from the sea floor has smaller proportions of the less resistant felspar and mica grains than freshly supplied fluvial sand, the

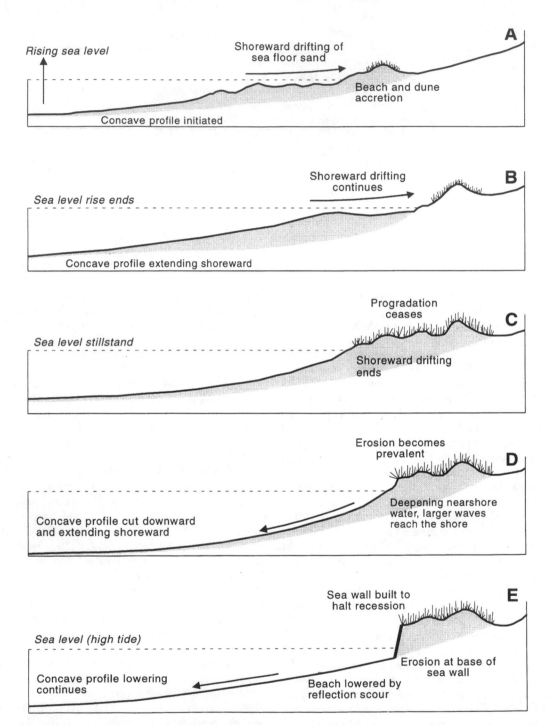

Figure 5.5   On many coasts sand drifted shoreward from sea floor shoals during the world-wide Late Quaternary marine transgression (A), and for a time after that transgression came to an end (B), so that beaches prograded. With the attainment of a smooth concave sea floor profile progradation ceased (C) and subsequently the landward migration of that profile has been accompanied by beach erosion and coastline retreat (D). Where a sea wall has been built along the eroded shore to halt coastline recession there has been further lowering of the beach (E).

Figure 5.6 The Loe Bar, a shingle barrier enclosing Loe Pool on the south coast of Cornwall.

beaches becoming predominantly quartzose away from river mouths.

Shoreward drift is impeded by fringing coral reefs or shore platforms, sediment accumulating against the outer margin of such structures being dispersed along-shore until some of it drifts up through gaps to form pocket beaches on the backing coastline. At Singatoka in Fiji a beach of black volcanic sand has been washed in from the sea floor through a gap in the fringing coral reef. Here, as on many oceanic islands of volcanic origin, there are alternations of black sand beaches with pale calcareous sand derived from corals or shells.

The conclusion that sandy beaches in southern Australia have been supplied largely by shoreward drift from the sea floor, with only minor contributions from cliff and rocky shore erosion or from fluvial sources (Bird 1978b) was contested by Davis (1989), who found up to 15 per cent felspar in a beach on the Otways coast, where felspathic sandstones and mud-stones are being derived locally from eroding cliffs and shore platforms. However the beaches on that part of the coast consist largely (usually >90 per

cent) of light brown quartz and carbonate sand washed from the sea floor into small coves and inlets along the rocky shore. Weathering of the felspathic rocks on the Otways coast yields mainly fine-grained sediment (silt and clay), and there are no beaches of cliff-derived sand along the cliff base behind the shore platforms. In general, beaches on the coast of south-eastern Australia contain small amounts of quartz and felspar sand derived from cliff and rocky shore erosion, and varying proportions of biogenic carbonate sand. Their similar-ity along cliffed coasts of granite, basalt, sandstone and limestone suggests that they are largely of sea floor origin (Bird 1993a).

Shelly and calcareous beaches

Beach deposits originating from the sea floor include sand or gravel derived from marine organisms, notably shells, which may be intact or broken, or comminuted to calcareous sand, by the time they arrive on the beach. On tropical coasts there are beaches of coralline and algal gravel derived from the disintegration of fringing and nearshore reefs, and shelly beaches where shells or organisms living in nearshore mud have been sorted by wave action and washed ashore, as on the Malaysian coast at Sembilang, north of the Klang delta.

Some shelly beaches are composed of rock-dwelling species while others have come from shallow sandy and muddy environments in sheltered bays and estuaries (Gell 1978). On the Texas coast Watson (1971) found a shelly beach left by deflation of a sandy matrix by onshore winds. Other examples of shelly beaches are seen on Herm in the Channel Islands, at Barricane Beach near Woolacombe in north Devon, and on the shores of the Firth of Forth near Edinburgh. There are mussel shell beaches on the rocky coast of Maine and bordering the Sea of Azov, sandy beaches composed of oolites formed in the clear warm seas of the Bahama Banks in the Atlantic, and shell grit beaches on the shores of the Hebrides in western Scotland. Cockle shells are abundant on beaches and intertidal sands on Traigh Mhôr, on the Hebridean island of Barra (Figure 5.7). Shelly beaches are also common on the

shores of estuaries and lagoons, sometimes bordering salt marshes or emplaced on them as cheniers (p. 145), as on the Essex coast. Coarse calcareous sand, known locally as calcified seaweed, is extensive on the sea floor off south Cornwall, particularly in Falmouth Bay. It is biogenic sand, formed by the alga *Lithothamnium calcareum*, but little of it drifts on to the beaches along the present coast, which are dominated by quartzose sand and gravel and material eroded from coastal rock outcrops. Beach accumulations of *Lithophyllum* form the mearls of Brittany, and shelly *Crepidula* species introduced from the United States (possibly during the Normandy landings in 1944) have become so common that they now form wide shelly beaches on the shores of the Bay of Mont Saint Michel in north-western France.

In southern and western Australia there are calcareous sandy beaches consisting partly of shelly debris, and partly of biogenic material from sand-sized organisms such as foraminifera and bryozoa carried in from the continental shelf by wave action (Bird

Figure 5.7   A shelly beach and intertidal zone (Traigh Mhôr) on the Island of Barra in the Hebrides, Scotland. The intertidal sand flat is firm enough to be used as an airfield at low tide.

1978b). In Pleistocene times similar calcareous beaches supplied dune sands that were lithified to form the dune calcarenites mentioned previously, a rock sufficiently coherent and resistant to have been eroded into cliffs, shore platforms and nearshore reefs (p. 90). Disintegrating calcarenite provides another source of beach sand, and is a prominent constituent of bay and cove beaches interspersed with cliffs along calcarenite coasts, particularly in Western Australia.

Shelly beaches often contain varying proportions of inorganic sand or gravel, but beaches bordering low-lying (i.e. uncliffed) desert coasts may be strongly calcareous because the meagre runoff has yielded so little fluvial terrigenous sediment supply to the shore, beaches being dominated by material washed in from the sea floor. The Eighty Mile Beach, in north-western Australia, exemplifies this, and there are shelly beaches bordering the dry northern shores of Spencer Gulf near Whyalla in South Australia. As sources of natural beach nourishment diminish it is likely that the proportion of biogenic sediment on beaches will increase on coasts where nearshore waters provide a rich environment for shelly organisms.

An important consequence of the prolonged sea level stillstand of the last 6000 years has been the development of a transverse sea floor concave profile on which the shoreward drift of sand and shingle that occurred during the later stages of the Late Quaternary marine transgression is no longer possible. This has resulted in erosion of beaches that had previously been built up (Figure 5.5D). Movement of sea floor sediment is now predominantly offshore, and the transverse profile is migrating landward.

Beaches supplied with wind-blown sand

Beaches have been supplied with sand blown from the hinterland where there is a suitable source of unconsolidated sand with little or no retaining vegetation, and winds blow from the land to the sea. This happens on arid coasts, as in Angola, where barchans spilling on to the shores of Tiger Bay have added sand to local beaches (Guilcher 1985b), on the desert coasts of Namibia between Sandwich Harbour and Conception Bay (Bremner 1985), in Mauritania (Vermeer 1985), the Bahia de Paracas in Peru (Craig and Psuty 1968), and Qatar in south-eastern Arabia (Sanlaville 1985).

Where dunes have been built by onshore winds, sand may occasionally be swept back to the beach, and into the sea by winds that blow from the land. An example

of this is seen on the north-facing shores of the Slowinski National Park (Figure 5.8), on the Polish Baltic coast, where a wide beach has been partly nourished with sand blown by southerly or south-westerly winds from poorly vegetated backshore dunes (Borowca and Rotnicki 1994). There is similar eastward movement of dunes from the backshore to beaches on Prince Edward Island in Canada, and westerly winds occasionally sweep sand from backshore dunes to beaches north and south of Holy Island on the north-east coast of England, and on the Sands of Forvie in eastern Scotland. Sand blown from backshore dunes tends to flatten beach and nearshore profiles, thereby diminishing incoming waves and reducing beach erosion.

Where the prevailing winds blow more or less alongshore, dunes may drift across promontories and headlands to nourish beaches on the lee coast. There are examples of this on the south-facing Cape Coast of South Africa, notably at Port Elizabeth, where sand driven by westerly winds across Cape Recife has been spilling on to the eastern beaches (Heydorn and Tinley 1980). There are similar situations in Uruguay, where dunes spilling on to the shore are nourishing the beach near Castillos (Jackson 1985), in Paracas Bay, Peru, where wind-blown sand has spilled across the Paracas Peninsula (Craig and Psuty 1968) and on the Victorian coast in south-eastern Australia, notably at Cape Woolamai and on the Yanakie Isthmus, Wilson's Promontory (Bird 1993a). On Cape Otway Pleistocene dune calcarenites extend across the headland, indicating that this process was active at earlier stages.

Beaches made or modified by human activities

Many beaches contain small proportions of sand or gravel formed from fragments of glass, concrete, brick, and earthenware produced by human activities. These are prominent on coasts with a long history of human settlement, as around the Mediterranean, or close to large urban or industrial centres. At Workington, on the Cumbrian coast, north-west England, the beach is dominated by basic slag waste from a former steel works. While the steel works was active, dumping of this material prograded the coastline, comparison of 1884 maps with 1981 air photographs showing an advance of up to 200 metres. The slag tip has since been cliffed by marine erosion, and as the cliffs retreat, sand and gravel derived from the waste drifts northward alongshore, augmenting the natural

Figure 5.8   Westerly winds blow sand along the beach in the Slowinski National Park on the Baltic coast, Poland.

beach as far as the Pell Mell breakwater at the mouth of the Derwent River (Empsall 1989). Beaches near ports have sometimes been augmented by sand or gravel from ballast brought in by ships and dumped before taking on a cargo, as at the former coal port of Saundersfoot in south-west Wales and the china clay port at Charlestown in Cornwall. The beach in Oriental Bay near Wellington, New Zealand, has received sediment from ships' ballast.

Reference has been made to beaches supplied with fluvial sediment increased by mining activities in the hinterland, but some beaches have been directly supplied with material from quarry waste spilling on to the shore. At Hoed in Denmark wave action has distributed gravelly flint waste dumped from a coastal quarry cut in chalky glacial drift, and built it into a series of low parallel beach ridges fronted by a shingle beach extending north and south from the quarry (Figure 5.9) (Bird and Christiansen 1982). Gravelly waste from coastal quarries tipped over cliffs into the sea on the east coast of the Lizard peninsula, in south-west England has accumulated in adjacent coves at

Porthallow and Porthoustock as widened gravelly beaches (Figure 5.10) (Bird 1987). The beach at Rapid Bay in South Australia widened by up to 230 metres as the result of spillage of quarry waste (Bourman 1990). On the west coast of Corsica the arrival of waste debris from an asbestos mine produced a beach in a cove where none existed previously (Paskoff 1994). Other examples include the Nganga Negara tin mine site on the west coast of peninsular Malaysia, where the dredging of tin from a depositional apron and coastal plain fronting the granitic Segari Hills generated large quantities of quartzose sand, gravel and boulders, which were heaped as a high tailings bank along the coastal fringe. Waves have reworked this material to form a gently shelving sandy beach, finer and better sorted than the tailings sediment. Waste material from a large granite and diorite quarry at Ronez, on the north coast of Jersey, has spilled down to the shore and formed a beach downdrift to the east.

Waste dumped on the shore from coal mines has developed into a beach of black pebbles and sand at

Figure 5.9   Shingle beach ridges on the coast at Hoed in Denmark, formed by accretion of gravel waste dumped from a coastal quarry in chalky glacial drift. The house (arrowed) stands on the eighteenth century coastline behind the prograded beach ridge plain.

Lynemouth in Northumberland, and similar material from coastal collieries has been dumped on several parts of the Durham coast, notably between Seaham and Easington and at Horden (Carter 1988). On Brownsea Island, in Poole Harbour, Dorset, wave action has distributed debris from a former pipeworks to build a beach of broken subangular earthenware.

Beaches can be formed from sediment obtained from the sea floor (notably in the approaches to ports) and dumped on or near the coast, as at Odessa on the Ukrainian Black Sea coast (Shuisky 1994).

Driftwood, including sawn timber, is extensive on beaches bordering high forested hinterlands, as in British Columbia and the Washington and Oregon

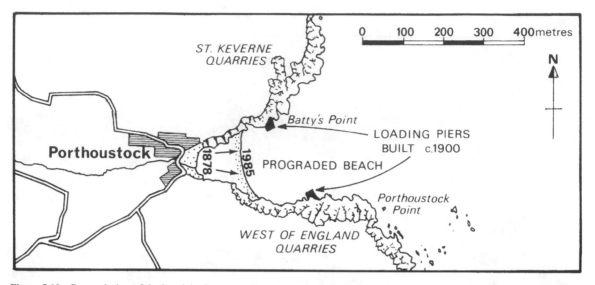

Figure 5.10   Progradation of the beach in the cove at Porthoustock, Cornwall, was due to the arrival of gravelly material derived from waste tipped into the sea from nearby coastal quarries.

coasts. On the beaches of the Westland coast, New Zealand, driftwood consists mainly of trees washed down to river mouths, but coasts adjacent to lumbering areas in western Canada and northern Russia are heaped with sawn timber washed up on the shore. Many beaches contain material derived from garbage of various kinds, including fragments of broken glass, metal, brick and plastic, derived from bottles, cans, containers and other litter dumped on the shore, carried down by rivers, or washed in from the sea.

Artificially nourished (or renourished) beaches, formed partly or wholly by the dumping of sand or gravel, brought to the shore from inland quarries, alongshore sources or dredged from the sea floor, are now extensive on the shore, particularly in the United States, Europe, and Australia.

Beaches of mixed origin

Most beaches contain sand or gravel from more than one of these sources, the proportions received from rivers, eroding cliffs and shores, the sea floor or aeolian inputs varying considerably along coastlines. In southern California beaches have been largely fed with fluvial sand from such rivers as the Santa Clara and Ventura, but they also include material eroded from coastal cliff outcrops, as well as sediment, including shelly debris, swept in from the sea floor (Emery

1960). On the Cape Coast of South Africa beaches include sand of aeolian origin where dunes are spilling along the shore, as well as fluvial, marine and cliff-derived sediment (Heydorn and Tinley 1980). Fluvially-supplied beaches downdrift, from a river mouth, may gradually become mixed with sediment from other sources, such as eroding cliffs or the sea floor, as on the shores of Hawke Bay in New Zealand (Bird 1996b). The fluvial contribution may remain identifiable by mineralogical evidence, as on the Ninety Mile Beach in south-eastern Australia, which consists largely of well-rounded quartz sand and shell fragments washed in from the sea floor, but in the vicinity of the mouth of the Snowy River also contains fluvial sand, supplied by occasional river floods, with distinctive minerals, including augite (Bird 1993a). North Queensland beach sediments are varying mixtures of sand from rivers, eroding cliffs and nearshore reefs. Samples from Garners Beach, near Bingil Bay, contain quartz and felspar sand supplied by nearby rivers, fragments from local rock outcrops, particles of disintegrated beach rock, ferruginous sandrock from an eroded Pleistocene sandy beach formation, and coralline and shelly material from an adjacent reef (Bird 1971a). On the western shores of the Arabian Gulf the beaches consist largely of sand blown from the desert, but also include shelly and in-organic sediment washed in from the sea floor (Sanlaville 1985). Near Broome in north-western

Australia there are beaches derived from the red pindan sands (quartzose sand with an iron oxide stain) in elongated desert dune ridges that have been submerged and reworked by wave action to produce clean white quartz where the iron oxide stain has been removed by abrasion. Along the Eighty Mile Beach this quartzose sand is mixed with pale grey calcareous biogenic sand washed up from the sea floor. Pumice derived from submarine volcanic eruptions floats on sea water, and is a common, and often far-travelled, constituent of many oceanic beaches, particularly around the Pacific.

Relict beaches

Some beaches have become relict, the sediment sources that originally supplied them being no longer available. This may be due to the natural or artificial diversion of a river, the construction of a dam, or the implementation of successful anti-erosion works in a catchment, so that the former fluvial supply of sediment to the beach has ceased. Alternatively, it can be the result of the halting of cliff erosion by sea wall construction, or emergence due to land uplift or sea level lowering, resulting in withdrawal of wave attack from the cliff base. On many oceanic beaches there is evidence that sediment is no longer being washed in from the sea floor because the supply from unconsolidated shoaly deposits has run out (p. 106). Most British shingle beaches are relict in the sense of being coarse material that was collected and delivered to the coast during the later stages of the Late Quaternary marine transgression and soon after it came to an end, but in the vicinity of eroding chalk cliffs or Tertiary or Quaternary gravel deposits, beaches are still receiving shingle.

Relict beaches, continually reworked by wave action, can become very well sorted, or develop lateral grading in grain size in relation to incident wave regimes. Chesil Beach, on the shores of Lyme Bay, in Dorset, is essentially a relict beach that is well sorted, with lateral grading from small pebbles in the west to large cobbles in the south-east (Figure 5.11). Such adjustments are impeded where there is a continuing supply of fresh sediment. There is also a tendency for the grain size of beach sediment to diminish as the result of gradual attrition by wave action (more rapidly in the intertidal to high tide swash zone). Pebbles on the beach south of Bridgwater Bay in Somerset diminish in size as the result of attrition as they drift along the shore. Slapton Beach in Devon is an example of a well-sorted relict beach of coarse sand and fine shingle, the calibre of which has been reduced by prolonged

attrition as the result of agitation by wave action. Flint cobbles have been gradually reduced by attrition to pebbles and eventually flint sand on the beach on the south-west coast of the Isle of Wight, while the diminution of basalt cobbles to pebbles, granules and sand can be traced along beaches at Flinders on the coast of the Mornington Peninsula in Australia.

WEATHERING OF BEACH MATERIAL

The rounding and reduction of beach sand and gravel particles by abrasion is a form of physical weathering. As rounding and attrition proceed, sand grains become smooth and highly polished, while pebbles and cobbles tend to become slightly flattened, and thinner at right-angles to their longest axis. The rate of such weathering depends on the hardness and structure of the grains, quartz sand and flint gravel being relatively durable, whereas sandstone fragments may disintegrate and disperse. Guilcher (1958) observed in Brittany that some beaches had attained smooth, gently-curved outlines even though their component gravels remained poorly rounded and poorly sorted.

Angular gravel dumped on the shore from coastal quarries becomes rounded pebbles on the beach face washed regularly by wave action, up to just above high tide level, angular or subangular material persisting at the top and bottom of the beach. This has been observed at Delec in Brittany (Berthois 1951) and below the quarry at Grassy on King Island, Tasmania. Beach gravels remain angular, however, on arctic coasts where rounding by wave action is halted during prolonged winter sea freezing, and frost shattering occurs: the addition of frost-shattered and solifluction gravel outweighs the rounding and smoothing effects of summer wave action. Arctic beaches are also less well sorted and often include muddy zones. Freeze-and-thaw processes form cracks and mounds and produce stone polygons on the poorly sorted beaches (Nichols 1961).

Attrition of pebbles may produce fine powdery material, as on chalk coasts where the sea is often milky with chalk in suspension. On beaches no longer receiving sediment physical weathering gradually reduces the beach volume, lowering and flattening the beach profile.

Beach volume is also modified by chemical weathering, notably the dissolving and removal of carbonates (shells, coral, or limestone fragments) by percolating rain water or corrosive groundwater. As has been noted, sea water is usually saturated with dissolved

Figure 5.11   Lateral grading on Chesil Beach, Dorset. The scale in both photographs is a one foot (approximately 30 centimetre) ruler. On the left is the beach near the western end (*A* in Figure 5.21), where the mean pebble diameter is between one and two centimetres, and on the right the beach towards the south-eastern end (*B* in Figure 5.21), where the mean pebble diameter is about five centimetres.

carbonates, but at lower temperatures (at night, during winter, or in cold climates), and in aerated sea spray, the carbon dioxide content rises, causing acidification, which leads to further solution of carbonates (p. 79).

BEACH ROCK

Precipitation of carbonates in the zone of fluctuating water table within a beach (related to the rise and fall of tides and alternations of wet and dry weather) can cement beach sand into hard sandstone layers known as beach rock, which may be exposed by subsequent erosion (Figure 5.12). Beach rock has been found on the shores of the Caribbean Sea, around the Mediterranean, the Red Sea and the Arabian Gulf,

and on the coasts of Brazil, South Africa, and Australia. It has been quarried for use as building stone from the beaches of Kuwait.

Formation of beach rock is assisted by high evaporation, which causes upward movement of water and dissolved carbonates in the beach sand. Beach rock forms as a layer of beach sand becomes consolidated by secondary deposition of calcium carbonate (as calcite or aragonite), precipitated from ground water in the zone between high and low tide level. Precipitation of calcium carbonate may be aided or caused by the activity of micro-organisms, such as bacteria, which inhabit the beach close to the water table.

Beach rock is frequently found on tropical beaches, especially on coral cays (p. 274). Cementation of beach

Figure 5.12   Beach rock exposed on the shore of a sandy cay on Gili Bidara, Lombok, Indonesia.

sediment occurs in warm environments where the interstitial water has a temperature exceeding 20°C for at least half the year, but occasionally beach rock is found on temperate coasts where the cementing calcite has been supplied by seepage of carbonate-rich water from the hinterland. An example is seen on the beach in Harlyn Bay, on the north coast of Cornwall, where slabs of beach rock have been exposed by beach erosion near high tide (Bird 1998). On the dune calcarenite coasts of southern Australia beach rock occurs in association with carbonates deposited by seepage from the cliffs.

Cementation can proceed rapidly, for artefacts such as bottles have been found incorporated in beach rock. Where the cemented beach material includes angular gravel (often coralline) it is termed a beach breccia, and where rounded pebbles are enclosed, a beach conglomerate. Exposed by beach erosion as seaward-sloping layers of calcareous sandstone (possibly further hardened on subaerial exposure), beach rock can be undercut by wave action and broken into flagstones, which may be thrown up beach by tsunamis or storm surges, as at Port Hedland (Figure 2.6). Patterns of eroded beach rock can be used to trace changes in beach outline, particularly on cays. Beach rock found above the present intertidal level is an indication that coastal emergence has taken place (p. 33). A horizontal variant of beach rock, known as cay sandstone, occurs where carbonates leached by rainwater from the upper layers of a beach are precipitated below, as at Belize in the Caribbean and Diego Garcia in the Indian Ocean.

Another form of induration is seen where the beach surface develops a coherent (biscuit-like) crust as the result of interstitial deposition of fine-grained sediment or precipitated salt which binds the sand or gravel. Where such a crust has developed, wave action on the beach face may cut a small cliff, capped by the indurated layer: a process known as scarping. The profile of beaches formed by deposition of coal-mining waste on the Durham coast in north-east England includes a small scarp cut into a surficial layer of sand bound by clayey downwash. Beach surface crusts formed by the precipitation of salt in dry weather may inhibit movement of sand by wind action, but these crusts are usually too soft and friable to impede wave action.

NEARSHORE PROCESSES

Sediment that has been supplied to a beach is subject to various processes that change its calibre and composition, and result in the shaping of beach morphology. Waves that break parallel to the coastline move sand or gravel either shoreward, when the swash is stronger than the backwash, or seaward, when the backwash is stronger, thereby producing alternations of onshore and offshore drift.

Waves that arrive at an angle to the coastline produce a transverse swash, running diagonally up the beach, followed by a backwash that retreats directly down into the sea. This results in the movement of beach material alongshore, the zigzag beach drift being accompanied by sediment movement along the nearshore zone, generated by the longshore currents that accompany obliquely-arriving waves. The combined effect of these processes is longshore drift (also known as littoral drift, but the term longshore drift is more accurate because littoral drift could be in any direction across the shore) of sediment to beaches and spits downdrift. Longshore drift is most rapid when wave crests approach the shore at an angle of between 40° and 50°, where the coastline is straight or gently curved and unbroken by headlands, inlets or estuaries, and where the nearshore sea floor profile is smooth. It increases with wave energy and is aided by a small tide range, which results in more continuous and concentrated wave action than where the zone of breaking waves rises and falls over a substantial tide zone.

Beach sediments may move first one way, then the other, according to the direction from which the waves approach. If waves arrive as frequently from one direction as the other, the resultant (or net) drift over a period will be negligible, but usually one direction predominates and there is long-term drift indicated by the longshore growth of spits, the deflection of river mouths, or the accumulation of beach sediment alongside headlands or breakwaters (Jacobsen and Schwartz 1981). The predominance of southward longshore drift on the east coast of England is indicated by the southward growth of spits at Spurn Head and Orfordness, the southward deflection of the River Yare at Yarmouth and the accumulation of beach material on the northern sides of harbour breakwaters at Lowestoft and Southwold. It is interrupted by predominant westward longshore drift on the north Norfolk coast, shown by the westward growth of spits at Blakeney Point and Scolt Head Island. This divergence of longshore drift could be related to the impact of north-easterly waves from the North Sea arriving on the large salient of East Anglia and generating westward and southward longshore drift from the vicinity of Happisburgh. Divergent longshore drift is also indicated on the Fylde coast in Lancashire by the growth of spits at Rossall Point, Fleetwood to the north and Lytham to the south.

The beaches of southern England are trains of shingle that drift mainly eastward. Drifting shingle accumulates alongside headlands, landslide lobes and breakwaters, and is interrupted by rocky or bouldery shores, at least until sufficient sand or gravel has arrived to fill in crevices and form a smooth enough surface for drifting to proceed. At Seven Rock Point, near Lyme Regis in Dorset shingle drifting eastward from Pinhay Bay was delayed for several years by an irregular bouldery shore below a landslide, but resumed after pebbles had filled the gaps between the boulders.

On the north Queensland coast in Australia the dominant waves generated by the prevailing south-easterly trade winds have drifted sand northward from river mouths to form a succession of beaches and spits that deflect the river mouths northward. The Burdekin delta, for example, has sandy beaches extending northward from the mouths of each of its distributary channels, culminating in the long recurved Bowling Green spit (p. 254). On the Adelaide coast in South Australia south-westerly waves move sand northward to accumulate on the prograding beach at Largs Bay, where longshore drift is intercepted on the southern side of the Outer Harbour breakwater.

Beaches showing alternations of longshore drift as waves arrive from each direction occur on the Pacific coast of North America because of seasonal changes in the direction of incident swell (p. 9). In Half Moon Bay and on Boomer Beach in California waves from the south-west move sand northward in summer and waves from the north-west drive it back south in winter (Emery 1960). Further north on the Pacific coast the Columbia River has delivered large quantities of sand which have drifted northward in summer when the waves arrive mainly from the south-west, and southward in winter when north-westerly waves prevail. The stronger northward drift has carried sand up the Washington coast to supply beaches as far as Cape Shoalwater, and southward drift has built the beach that extends down to Tillamook Head.

On the north-east coast of Port Phillip Bay, Australia, beaches are diminished by erosion at their southern ends and increased by accretion at their

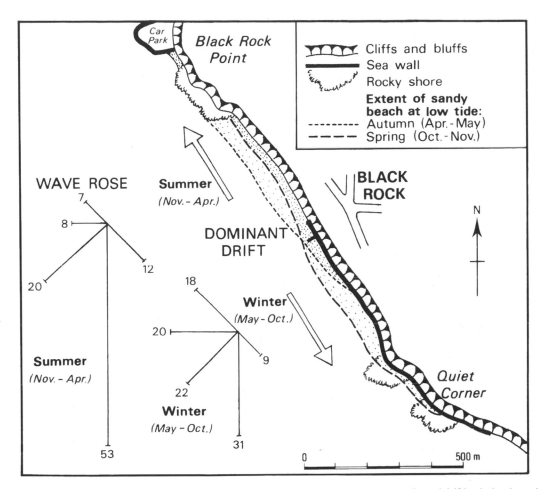

Figure 5.13 Seasonal variations on Black Rock Beach, Port Phillip Bay, Australia, where northward drifting is dominant in the summer (November to April) and southward drifting in winter (May to October). The wave roses are based on the percentage frequency of onshore winds > Beaufort Scale 3 (Figure 5.14), which determine wave action on this part of the coast.

northern ends during summer, and this pattern is reversed in winter in response to seasonal variations in the direction of the onshore winds that generate wave action (Figure 5.13) (Bird 1993a). The breakwater at Black Rock shows accretion on its southern side in summer, and on its northern side in winter as longshore drift alternates.

BEACH MORPHOLOGY

The shape of a beach changes as beach sediments are moved by waves and currents from one sector to another. In addition to the longshore drift of sand and

gravel when waves arrive at an angle to the shore, sediments are moved to and fro between the beach and the nearshore zone (onshore–offshore drift) when waves break parallel to the coastline. Currents generated by winds, waves or tides move sediment in the nearshore and intertidal zones as the tide rises and falls. As they move sand or gravel offshore, onshore or alongshore, they promote erosion or accretion, influencing the form of the waves that subsequently break upon the beach, and so contributing indirectly to the shaping of beach morphology.

Beach morphology also changes frequently in response to rising and falling tide and to wind action. As the tide rises stronger waves arrive through deepening water, winnowing fine sediment from the beach face, and as it falls a veneer of fine sediment may be deposited. Wind action can deliver hinterland sand to the shore (p. 180), and beach morphology is modified as onshore winds sweep beach sand to backshore dunes, longshore winds carry it along the beach or offshore winds blow it into the sea.

The beach face is often diversified by ephemeral zones of finer and coarser material arranged in various patterns and by minor features such as ridges, terraces and cusps. In general the coarser the beach material the steeper the beach face, but after a stormy episode the beach face often shows a steep or vertical scarp, particularly when it is cut into damp, coherent sand.

Nearshore processes shape beach morphology (Hardisty 1990), but there is feedback, whereby beach morphology influences the processes at work in the nearshore zone (Krumbein 1963, Komar 1976). This adjustment between nearshore processes and beach morphology as these interactions proceed was illustrated by Short (1992) in his analysis of beach systems on the central Netherlands coast.

Beach morphology is three-dimensional, but is usually analyzed in terms of beach outlines in plan and beach outlines in profile.

BEACH OUTLINES IN PLAN

The orientations of beaches are related partly to wave patterns and partly to the general trend of the coastline, notably the position of prominent headlands. Beaches often have smoothly-curved outlines in plan, concave seaward (Figure 2.1), shaped by the pattern of approaching waves that have been refracted so that they anticipate, and on breaking, fit the plan of the beach. Such swash-aligned beaches are well developed on oceanic coasts, receiving refracted ocean swell. Examples include the Loe Bar in Cornwall, the beach on the west coast of South Uist in the Hebrides, the Ninety Mile Beach and Encounter Bay in Australia. South-easterly swell entering Jervis Bay in eastern Australia is refracted into patterns that fit the curved outlines of bordering beaches that face in various directions. Once established, these gently-curving beaches maintain their outlines even when beach material is withdrawn seaward from them by storm waves.

Changes in configuration occur on beaches that have not yet become adjusted to the prevailing wave regime. Where the beach outline is more sharply curved than the approaching waves the breakers produce a convergence of longshore drift of beach sediment towards the

centre of the bay, where the beach prograades until it fits the outline of the arriving swell. Convergent longshore drift in Byobugaura Bay, near Tokyo, has prograded the central part at Katakai in this way, with sediment derived from the erosion of the northern and southern parts, and a similar evolution has been taking place at Guilianova on the east coast of Italy. Where the beach outline is less sharply curved than the approaching waves it sets up a divergent longshore drift from the centre until the beach outline fits the wave pattern. This has occurred on the shores of the Andalusian Bight, in southern Spain, where erosion of the central part has been balanced by progradation at Matalascanas to the south and Mazagon to the north. Beaches have been reorientated where sand or shingle eroded from one coastal sector has drifted alongshore to accrete on another sector downdrift, particularly where it has been intercepted by headlands and breakwaters.

Beaches shaped by waves generated by local winds (i.e. not receiving ocean swell) have outlines related partly to the direction, strength and frequency of onshore winds, partly to variations in the length of fetch (i.e. open water across which waves are generated by those winds), and partly to wave refraction. The orientation of such beaches is the outcome of the long-term effects of waves arriving from various directions (Figure 5.14). These can be expressed as a resultant of wind-generated waves, calculated from

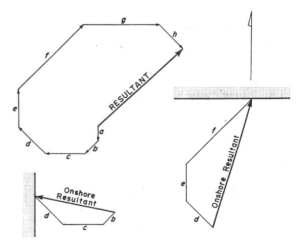

Figure 5.14 Wind resultants formed by drawing vectors (a–h) obtained by multiplying the frequency of winds in each Beaufort Scale (>3) category by the cube of the mean velocity of that category in a directional diagram. Onshore resultants are obtained by taking only the vectors of winds to which a coastline is exposed.

records of the frequency and strength of onshore winds, (Beaufort Scale >3, that is >20 kilometres per hour), which produce only small waves that have little effect on beaches. Typically such beaches are modified by erosion and accretion until they become orientated at right angles to this onshore wind resultant. Where there are marked contrasts in fetch these must be taken into account, for strong winds blowing over a short fetch may be less effective in generating wave action than gentler winds blowing over a long fetch. If the direction of longest fetch coincides with the onshore wind resultant, the beach becomes orientated at right-angles to this coincident line, but where the two differ the orientation becomes perpendicular to a line that lies between them. On coasts of intricate configuration (as in the Danish archipelago or Puget Sound, Washington state) wave action is determined more by the fetch

than the onshore wind resultant, and many beaches have become aligned at right-angles to the maximum fetch (Schou 1952). The relationship is well illustrated by the shingle beaches (known as ayres) of the Orkney and Shetland Islands. Similar patterns in relation to wind regime and fetch are seen on beaches on the shores of landlocked embayments or coastal lagoons.

Where waves arriving at an angle to the coastline are refracted round headlands they shape asymmetrical (crenulate, half-heart or zeta-form) beaches in the adjacent embayment, as in Venus Bay and Waratah Bay, Victoria, Australia (Figure 5.15). On the New South Wales coast a succession of such beaches shows sharply-curved southern ends, with finer sand, a gentler beach face and diminished berm height, and sand coarseness increasing on the straighter northern shores. Although this pattern is related to refracted

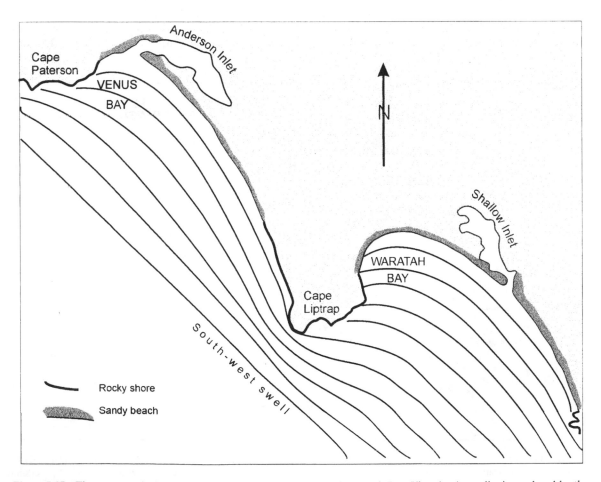

Figure 5.15 The asymmetrical curvature of beaches in Venus Bay and Waratah Bay, Victoria, Australia, is produced by the refraction of ocean swell arriving from the south-west.

south-easterly ocean swell there is usually also longshore drift of sand northward round headlands from one zeta-form beach to the next. A series of asymmetrical beaches on the south coast of the Lleyn Peninsula in Wales occupies embayments cut out by refracted south-westerly waves in glacial drift between bedrock head-lands, the most prominent of which is Pen-y-chan, east of Pwllheli, which has intercepted sand and shingle drifting eastward. On the east coast of Sri Lanka similar successions of asymmetrical beaches have been shaped by refracted south-easterly ocean swell between deltaic salients formed at river outlets.

Where waves are refracted by nearshore islands, off-shore reefs or shoals on the sea floor, they move into the shore and build cuspate beach outlines in their lee. If these offshore features remain stable the cusps and bays persist, but changes in sea-floor configuration, when currents scour out hollows or build up shoals, or when banks or bars migrate shoreward, seaward or along the coast, modify patterns of wave refraction and so change the outlines of the beach. As a shoal moves

along the coast the cuspate beach in its lee is eroded on one side and accreted on the other, so that the cusp also migrates. Longshore migration of sand and shingle forelands at Benacre Ness and Winterton Ness, on the East Anglian coast, is the outcome of changing wave and current patterns related to movement of shoals off-shore (Robinson 1966).

A distinction is made between swash-dominated beaches built parallel to incoming wave crests (particu-larly refracted ocean swell) with little or no longshore drift but alternations of onshore–offshore drift, and drift-dominated beaches with alignments parallel to the line of maximum longshore sediment flow, gener-ated by obliquely-incident (typically 40°–50°) waves. Beaches exposed to refracted ocean swell have swash alignments, as do bay-head beaches, but waves moving into embayments build beaches on drift alignments, often terminating in spits. An example of both swash-dominated and drift-dominated beaches in the south-west of Westernport Bay, Australia is given in Figure 5.16. In general, swash-dominated beaches are

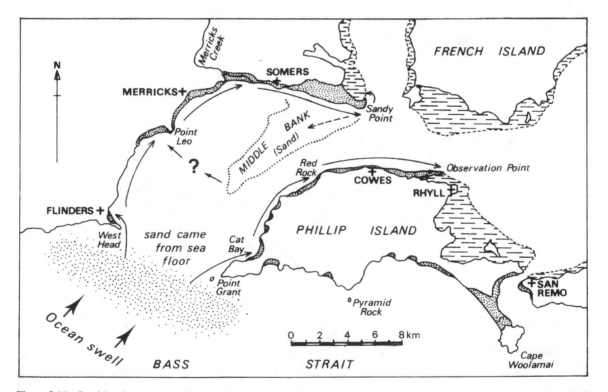

Figure 5.16   Sand has been washed in from the sea floor to beaches in the south-west of Westernport Bay, Australia. Drift-dominated beaches have formed as longshore drift carried sand along the coasts from Cat Bay to Observation Point and past Point Leo to Sandy Point. Ebb currents have swept sand out south-westward from Sandy Point to form Middle Bank, and it is possible that sand moves intermittently back on to the coast near Point Leo, completing a sediment circulation.

smoother in outline than drift-dominated beaches, which are typically sinuous, with intermittent migrating lobes and slightly divergent longshore spits. On the coast of Wales the beach in Porth Neigwl on the Lleyn Peninsula is swash-dominated, whereas the beaches on either side of the southern entrance to the Menai Strait are drift-dominated, ending in spits, and similar features are seen in Portmadoc Bay, to the south. With downdrift progradation such convergent spits can grow to form a curving swash-dominated beach.

Few beaches are entirely swash or drift-dominated. Longshore drift is alternating and balanced on largely swash-dominated beaches such as Chesil Beach in Dorset and Encounter Bay in South Australia, and onshore–offshore (swash–backwash) sequences occur on largely drift-dominated beaches such as Sandy Hook in New Jersey or the beaches north of the Columbia River in Washington state. The curving beach behind Disaster Bay, New South Wales, a narrow embayment between high parallel sandstone cliffs, is almost entirely swash-dominated, with an outline shaped by incoming refracted ocean swell but too limited in aspect for oblique waves to generate longshore drift. Numerous parallel dune ridges indicate that progradation has maintained this curved plan.

Most British beaches are drift-dominated with frequent alternations of longshore drift, swash-dominated beaches being largely confined to the Atlantic coasts receiving ocean swell, such as Rhossili Bay in south Wales. Orfordness is an elongated drift-dominated beach on the East Anglian coast, but it includes multiple parallel beach ridges that indicate occasional swash domination.

Shaping of beach outlines also depends on patterns of sediment supply, with a tendency for beaches to become narrower downdrift from a sediment source such as a river mouth or eroding cliff. Beach accretion is influenced by the pattern of incoming waves in relation to the source of the material. On deltaic coasts where sand or gravel are being delivered to the mouth of a river, waves arriving parallel to the coastline diverge along both sides of the delta and distribute the fluvially supplied sediment to produce symmetrical beaches, as on the Tagliamento delta on the northern shore of the Adriatic Sea, or trailing lateral spits, as on the shores of the Ebro delta in Spain (p. 255).

Equilibrium beach plans

Where beach outlines have become adjusted to the prevailing pattern of refracted waves they are more stable than those that have not yet achieved such an adjustment, and may attain an equilibrium with processes at work on them. The asymmetrical swash-aligned beaches produced where incoming waves are refracted round a headland, as on the New South Wales coast in Australia, have been cited as an example (Silvester 1974), and the inference is that artificial beaches shaped in this bay will be stable. However, where such zeta-form beaches have been established, on the south-east coast of Singapore Island and on the north-east coast of Brunei, beach erosion has simply continued on the zeta-form alignments.

Equilibrium is defined in the Oxford English Dictionary as a condition of balance between opposing forces, the forces being so arranged that their resultant is zero. Because of the variability of beach features in response to changing processes it is necessary to state a time scale over which such an equilibrium can exist. A cyclic equilibrium is one that returns to its original condition after being disturbed, and a dynamic or shifting equilibrium is one that changes while remaining in balance with driving forces. A beach may show cyclic equilibrium when losses during phases of erosion are balanced by gains during phases of accretion (p. 122) so that it is neither gaining nor losing sediment, or as long as losses are compensated by gains. Dynamic equilibrium may be achieved when a beach is either prograding (advancing seaward by accretion) or eroding (receding landward) but maintaining its outline in plan, but the coastline is not stable under such conditions. Stability can only be attained where there is a sufficient input of sediment to balance episodic losses, and so maintain the outline of a beach in plan and in position.

Beaches that show alternations of longshore drift, one end being prograded as it receives sediment from the other, in response to changing directions of wave approach, may achieve cyclic equilibrium if they return regularly to earlier configurations, there being no net gain or loss of beach material. On drift-dominated beaches a cyclic equilibrium may be attained where the beach attains a curvature adjusted in such a way that waves impinging on the shore provide sufficient energy to transport the sediment arriving at one end of the beach through to the other, the configuration being maintained. There is an analogy with the graded stream concept, adapted for beaches with longshore sediment flow, but there are complications where sediment is coming in from the sea floor or where there are losses of sediment offshore.

Equilibrium concepts have been applied to beach outlines in profile as well as in plan. Beach outlines in plan may persist even when gains or losses steepen or flatten beach profiles. Alternatively beach profiles may be maintained even though there are gains or losses in beach volume resulting in the advance or retreat of the coastline.

BEACH OUTLINES IN PROFILE

Beach profiles are related to the nature of beach sediment and to wave conditions during the preceding few days or weeks: the effects of a severe storm may still be visible several months later. The profiles are shaped largely by wave action, notably the swash generated as waves break upon the shore and the ensuing backwash (p. 13), which generate onshore and offshore movements of sediment. There is often an upper beach above normal high tide level, only occasionally submerged by large breaking waves or exceptionally high tides, a middle section, often more steeply sloping, and a lower gently sloping or concave section which extends down to and below low tide.

Swash and backwash velocities can be measured with a dynamometer in waves close to the shore, and correlated with onshore and offshore movements of beach sediment, as indicated by profile changes or accumulations in trays placed to intercept beach material drifting shoreward or seaward. As has been noted (p. 13) waves which have steepness ($H_0 : L_0$) in deep water of less than 0.025 produce constructive (spilling) breakers which move sediment shoreward on to the beach, while those with higher ratios are destructive (plunging), withdrawing sediment to the sea floor. Spilling breakers occur where the ratio of breaker height to deep-water length $H_b : L_0$ is greater than 2.4 $(\tan \beta)^{1.8}$, β being the nearshore slope gradient (Sunamura 1992). However, grain size and beach permeability also influence swash–backwash conditions, and breaking waves are more likely to build up the profile on a shingle beach, where backwash energy is diminished by percolation, than on a sandy beach, where the backwash is more effective in moving sediment back into the sea. It is not uncommon for plunging breakers to pile up and steepen an upper beach of shingle or coarse sand while combing down a lower, flatter beach of finer material, a form of onshore–offshore sorting that leaves a sharp boundary between the two (Figure 5.17).

Beach profiles also vary with tide range. On microtidal coasts, as around the Baltic and Mediterranean, wave action is concentrated within a relatively narrow vertical zone and the beach profile is often steeper than on the same size beach material where tide range is larger. On sheltered sandy microtidal beaches in Western Australia (within bays or behind reefs) profiles varied from gentle concave, through stepped, to moderately steep (Hegge et al. 1996).

Cut and fill

Beach profiles are lowered and cut back by strong wave action, especially during stormy periods, when plunging and surging waves, with limited swash and stronger backwash, withdraw sand and shingle to the nearshore zone, leaving a concave profile (Figure 5.18). In calmer weather spilling waves (p. 13), with a strong swash and lesser backwash, move sand and shingle up the beach and rebuild a convex profile. Alternations of beach erosion by plunging or surging waves and beach accretion by spilling waves are known as cut and fill, and are associated with alternations of seaward and shoreward drift of sediment.

Cut and fill occurs over time-scales varying from a few days to several years, but as storms are usually more frequent in winter many beaches show a seasonal sequence of winter cut and summer fill. On the Pacific coast of the United States sandy beaches are scoured by storm waves in winter to expose rocky and bouldery shores, but in spring gentler wave action in calmer weather restores the sand cover. This was demonstrated by profile monitoring along the beach and nearshore zone alongside the Scripps Institution Pier at La Jolla (Shepard 1973). On monsoonal coasts beach erosion occurs during storms in the wet season, with recovery in calmer dry season weather. The Darwin beaches in northern Australia are cut back and lowered during the wet summer and rebuilt in the drier winter (Gowlland-Lewis et al. 1996), and similar seasonal cut and fill is seen on the east coast beaches of peninsular Malaysia, where erosion prevails during the north-east monsoon (Teh Tiong Sa 1985). In Marsden Bay, north-east England, King (1972) found that sand moved shoreward on to beaches, when winds blowing offshore flattened incoming waves, whereas seaward movement occurred when steeper waves were generated by winds blowing onshore.

Whether waves of a particular form will push sediment up on to a beach or withdraw it from the beach face depends also on the pre-existing transverse profile of the shore and the grain size of shore sediments. Shoreward movement is more likely where the transverse profile is gentle, and the sediment predominantly fine-grained.

Figure 5.17 An upper beach of shingle and a lower beach of sand on the east coast of Dungeness in south-eastern England, separated by swash–backwash sorting. The arrow indicates the high spring tide line.

Alternations of cut and fill can take place over longer periods. The fortnightly sequence of increasing high tides from high neap to high spring tides is, in effect, a short-term marine transgression, and is often accompanied by the erosion of beaches. It is followed by a sequence of diminishing high tides from springs to neaps, effectively a regression of the sea, which is often accompanied by beach accretion. Seasonal alternations of beach erosion and accretion occur where there is a sharp contrast between a wet and a dry season, between stronger winter and gentler summer wind and wave action, or between the direction of approach of dominant winds in different seasons.

Sweep zone

As a result of cut and fill alternations over varying periods of time the beach profile shows changes in form, being higher and wider after phases of accretion and lower and flatter (with a backing rise or cliff) after phases of erosion. The vertical section of a beach within these alternations (subject to removal and replacement) is known as the sweep zone. It can be measured by making successive surveys across the beach to determine the cross-sectional area between the accreted and eroded profiles, and when this is multiplied by the length of the beach the volume of sand gained or lost can be calculated. Measurements on the Ninety Mile Beach have shown that the seasonal sweep zone has a mean cross-sectional area of 42 square metres, indicating a loss of 6 million square metres of sand from this beach during stormy phases, which are most frequent in winter, and equivalent replacements during fine weather. On such a beach profile variations are cyclically restored by natural processes over periods of time.

Beach gradient and sediment size

The transverse slope of a beach profile is also related to the grain size of beach sediments. Gravels tend to assume steeper gradients than those on sand (usually

Figure 5.18 Concave profile produced on a shingle beach at Ringstead in Dorset, cut back by preceding storm wave action. Beach cusps have been formed on the concave slope. View eastward to White Nothe (W).

<5°), chiefly because they are more permeable, and beach face slope increases with pebble size. On Chesil Beach the beach face slope increases from about 5° on fine shingle at the western end to more than 20° on the large cobbles near the south-eastern end.

As waves break, the swash sweeps sediment up on to the beach. Backwash tends to carry it back, but the greater permeability of gravel and coarse sand beaches absorbs and diminishes its effectiveness, so that swash-piled sediment is left at relatively steep gradients. Fine sand beaches are more affected by backwash, and develop gentler slopes as more sediment is withdrawn seaward.

Storm waves thus steepen the profiles of shingle (gravel) beaches and flatten those of sandy beaches. On the north Norfolk coast the 1953 storm surge lowered and cut back sandy beaches on Scolt Head Island, but piled up the crests of shingle beaches, parts of which were rolled landward, as at Blakeney Point (Steers 1964). Both sectors of coastline retreated, but the shingle beaches were left with steeper profiles than the sandy beaches, although the latter were generally backed by cliffed dunes. On the sandy beaches of the Atlantic coast of Britain, as at Braunton Burrows in north Devon, storm waves are almost entirely destructive, lowering and cutting back sandy beaches, and similar effects have been observed on east coast sandy beaches, as at Goswick Sands and Gibraltar Point.

Nevertheless, there is great variation in sand and shingle beach gradients, depending on preceding wave conditions as well as grain size. Storm waves can cut vertical cliffs in sandy beaches and comb shingle down to a gentle slope (Kirk 1980).

Upper and lower beach

A beach with a mixture of sand and shingle (or fine and coarse sand) may be sorted by swash and backwash until the profile consists of a coarser, steeper upper beach and a finer and flatter lower beach, exposed as the tide ebbs. The upper beach is often of shingle, clearly demarcated from the lower beach of sand, a contrast resulting from the differing responses of shingle and sand to storm waves, which pile up swash-built shingle, that persists as backwash withdraws the sand (Figure 5.17). The shingle is permeable, and descends to a seepage zone and a gentler slope of wetter, firmly-packed sand across which it may be possible to drive a car. The backwashed sand is sometimes deposited in the form of a step at the foot of the beach, which may prograde seaward as the tide falls. On some coasts the lower beach passes seaward to intertidal bars and troughs, which are discussed in Chapter 8.

Beach stratification

During stormy phases the finer and lighter components of a beach sediment may be withdrawn by strong backwash, leaving the coarser and heavier material as a lag deposit on the beach face. In calmer weather gentle wave action moves fine sediment back on to the beach face. Where overall beach accretion has taken place such alternations form stratified (or laminated) deposits, with coarser seams between layers of finer sediment, seen when a section is cut across a beach. Within each stratified layer the distinctive sediment is well sorted, being related to a particular phase of deposition or reworking by waves. If there is a phase of erosion these may outcrop as contrasted zones along the face of the beach.

Beach berms

Swash deposition can build up and prograde a flat berm (beach terrace) at about high tide level or form a swash berm, a subdued ridge parallel to the coastline along the length of the beach, sometimes backed by a shallow lagoon that drains out as the tide falls. On sandy beaches the force of breaking waves may compact the sand so firmly that it is possible to ride a bicycle, drive a car or land an aircraft on a berm. On shingle beaches breaking storm waves throw some pebbles forward and up the beach, to build a berm at the swash limit.

Coarse sand can be built into berms by storm waves which erode beaches of finer sand. Storm waves wash coarse sand from the beach to the backshore and accompanying onshore gales blow sand landward. The gradient and morphology of the nearshore sea floor is also relevant, for where it is gentle and shallow, storm waves lose some of their energy and break as constructive surf with a strong swash, instead of the plunging breakers seen on steeper shores. Under these conditions even medium to fine sand can be built up as berms along the shore.

On microtidal coasts there is typically a single swash-built berm, whereas on mesotidal coasts (as in East Anglia) there are often two berms, one just above high tide level, the other just above low tide level, swash action being more prolonged at these two levels than when the tide is rising or falling. On macrotidal coasts, especially where the shore is sandy, there are multiple berms, grading into foreshore ridge and runnel topography (p. 199). Grain size also influences berm development, so that a shingle beach passing laterally into a sandy beach may have one shingle berm that divides alongshore into two sandy berms.

Beach cusps

The beach face is often divided into zones of shingle, shelly gravel and coarse and fine sand that run parallel to the coastline, sorted and shaped by swash and backwash processes. On some sand and shingle beaches the pattern takes the form of beach cusps, consisting of regular successions of half saucer shaped depressions containing finer sediment, up to a metre deep between cuspate points of slightly coarser material. They are generally found where waves arrive parallel to the coastline, rather than obliquely, and are ephemeral features (persisting for a few hours to several days), but once formed they influence the patterns of swash as the waves break. They are often found on beaches exposed to ocean swell.

Beach cusps are shaped by breaking waves that generate swash salients with currents that divide at points where coarser sediment is deposited, and coalesce in intervening areas to form backwash that scours out depressions (Bagnold 1940). Their regular spacing results from the interaction of the swash salients with orthogonal edge waves (Guza and Inman 1975). The spacing of cuspate points (two to five times breaker height) increases as wave height is augmented, and is also related to wave length. Miniature cusps about 30 centimetres apart have been seen on a low energy lagoon shore beach, but generally they are spaced at intervals of 10 to 20 metres, attaining 50 metres on beaches receiving ocean swell. In some beach cusps the erosional depressions are backed by miniature cliffs cut into the beach face (Allen et al. 1996).

Beach cusps are most clearly developed where the beach sediment has a bimodal grain size distribution. They are frequently formed on shingle beaches in Britain where there is finer material occupying the depressions (Figure 5.18). They are less clearly developed where there is a lateral transition to more uniformly sorted beach material: if the proportion of pebbles to sand diminishes the cusps are broken into less regular shingle patches and the depressions become fainter on a well-sorted sandy beach. Cusps of coarser sand and depressions of finer sand occur on the sandy beaches on the Gulf and Atlantic coasts of the United States. In Australia they frequently form on steep coarse sandy beaches in embayments along the Sydney coast, and on the sandy shores of landlocked embayments such as Jervis Bay and Botany Bay. Beach cusps on the Perth coast fade when wave height increases during sea breeze episodes and revive in calmer weather when longer ocean swell dominates (Masselink and Pattiaratchi 1998).

Less regular cuspate re-entrants are excavated in sandy beaches behind the heads of rip currents, where larger waves move in through water deepened by outflow scour. Such local scour has been described on sandy beaches along the Atlantic coast of the United States (Dolan 1971) and on Siletz Spit in Oregon (Komar 1983).

Wind action on beaches

Beach profiles are also modified by wind action, when sand is blown along or across the beach, lowering some parts and building up others. Anemometers and intercepting trays can be used to measure quantities of sand moved across or along a beach in relation to the direction, strength and duration of wind action. Sand blown to the backshore can be banked up against cliffs or bluffs, or form coastal dunes (p. 179). Dunes may later be cliffed by marine erosion, the beach regaining some of the sand previously lost to the backshore, so that over time there are beach–dune interactions (p. 180).

Offshore winds sweep sand back across the beach and into the sea, to be reworked and transported by waves and currents in the nearshore zone. Winds blowing along the beach can carry substantial amounts of sand, especially when the beach face is dry (thereby augmenting the effects of longshore drift by waves and nearshore currents). This was demonstrated by So (1982) on the sandy beach at Portsea, in Victoria, Australia, and is a common occurrence on windy North

Sea sandy beaches at low tide, as at Bray in north-eastern France.

Fine sand is more readily mobilized and transported by wind action than coarse (or heavy mineral) sand, and pebbles (lag gravel) are left behind when beaches of sand and gravel are winnowed, in much the same way as when wave backwash withdraws finer sediment from the beach face. A strong wind can produce a sheet of drifting sand, with sand grains airborne (in suspension), bouncing (saltation), and rolling along the beach. Under these conditions barchans (p. 188) may form on the shore, and migrate downwind. Movements of wind-blown sand to backshore dunes are discussed in Chapter 7.

Downwashing

Runoff and seepage during heavy rain or as the result of melting snow or ice can lower the beach profile by washing sand or gravel into the sea. On coasts where rainfall is high, seeping groundwater may take the form of springs which wash out irregular pits and form stream-cut furrows on the sandy foreshore at low tide, as on Three Mile Beach in north-west Tasmania. Antarctic beaches are covered by snow in winter, but receive silty solifluction downwash during the summer thaw.

Response to tidal oscillations

Beach profiles change as the tide rises or falls. In general the sea level rise as the tide comes leads to recession of the beach profile and withdrawal of sediment seaward, while emergence as the tide falls is usually accompanied by shoreward movement of sediment and beach progradation. Similar trends can be detected during the sequence of rising high tides between neaps and springs and of falling high tides between springs and neaps. There are, however, many complicating factors, including the effects of longshore drift, variability of wave action and rates of sediment input and output. A beach receiving abundant sediment (e.g. from a nearby river mouth or rapidly eroding cliffs) may prograde even when sea level is rising, while a beach that is losing sediment offshore or alongshore may be cut back even when the sea level is falling.

Equilibrium beach profiles

The question of beaches attaining an equilibrium in plan has been discussed (p. 121), and similar considerations apply to the evolution of equilibrium in beach profiles. Many coastal geomorphologists and engineers have accepted the idea that beaches can attain a profile of equilibrium (Silvester 1974), but there is some confusion over what this actually means (Pilkey et al. 1993).

After a phase of sustained and steady wave activity a beach profile becomes a smooth, concave upward curve, the gradient of which depends partly on the grain size of the beach sediment, gravel beaches generally having steeper profiles than sandy beaches. This can be considered an equilibrium beach profile, achieved by adjustment between the beach profile and the waves and currents at work on it. Dean (1991) noted that such a profile can be described as a concave upward curve:

$$h = Ax^{0.67},$$

where h is the still-water depth, x is the horizontal distance from the coastline, and A is a dimensional shape parameter, the shape of which depends on the grain size of the beach material. Bodge (1992) preferred an exponential expression:

$$h = B(1 - e^{-kx}),$$

in which B and k are depth and inverse distance. These equations fit many surveyed beach profiles, especially those without nearshore bars and shoals, which are complications that are difficult to match with mathematical curves. It should be noted that many beach profiles are surveyed when calm conditions have become established following a phase of stronger wave action.

The fact that mathematical curves fit surveyed profiles is not an indication of stability, for as beach profiles steepen or flatten in response to changes in wave regime or granulometric composition they continue to show profiles similar to those generated by the formulae of Dean (1991) or Bodge (1992). The smoothly-curved profiles may represent a condition that beaches attain when their granulometric composition has become adjusted to specific wave conditions. However, granulometric composition is modified by gradual attrition as the result of wave agitation, and by sorting, and this will lead to profile changes unless mean grain size is maintained by a continual inflow of coarser sediment as the finer fraction is removed.

The sequence of cut and fill observed on many beaches, with sediment removed during stormy phases and replaced during subsequent calmer weather, can be considered a cyclic equilibrium to the extent that the profiles are restored by natural processes. This can occur over periods ranging from a few days or weeks (a storm and its aftermath) to a year or more (seasonal alternations).

Geomorphologists have used the term dynamic equilibrium to describe the condition where landforms maintain their shape, even though changes (uplift, erosion, deposition or subsidence) have occurred. A dynamic equilibrium could exist on a beach when the profile remains the same (perhaps with cyclic alternations) while the beach as a whole is either advancing (prograding) or receding (retrograding), but many would argue that this advance (with land gained by accretion) or retreat (with land lost by erosion) of the coastline really indicates disequilibrium. As will be shown later, analyses of global patterns of beach change have indicated that over the past century most of the world's beaches have been retreating; some have been advancing, and a few have remained static or shown no net change. The supposition that a particular beach is (or has recently been) in some kind of equilibrium should therefore be treated with caution.

An equilibrium between the beach profile and nearshore processes may not have been attained before there is a change in processes as the sea becomes stormier or calmer. A new adjustment then begins. As nearshore processes are highly variable, most beach profiles are changing most of the time. This variability is compounded when the whole three-dimensional beach morphology is considered, for equilibrium in both plan and profile must be an unusual condition. When geomorphologists or engineers refer to beach equilibrium concepts, they should indicate which kind of equilibrium they envisage, the time scale they are using, and whether they are dealing with the beach in plan or in profile, or the whole three-dimensional beach system.

Eroding and prograding beach profiles

Beaches that are eroding typically show a gently inclined or concave slope backed by a cliff cut into beach sediments or backshore dunes. Sand and gravel removed from a beach by storm waves are spread across the nearshore sea floor as a terrace or deposited as bars and troughs running more or less parallel to the coastline. The intertidal shore profile may thus become wider and flatter after a phase of erosion.

Prograding beaches are typically convex, often with berms or beach terraces that are being built seaward and new foredunes developing above high tide level.

Sustained accretion increases beach volume, and the coastline progrades.

BEACH MORPHODYNAMICS AND BEACH STATES

Beach morphodynamics (the study of the mutual adjustment of beach morphology and shore processes, involving sediment transport) can be studied in terms of field measurements and analysed with the aid of computer simulations (Cowell and Thom 1994). Analyses of incident wave regimes on beaches has shown that wave energy is partially reflected by steep ($> 3°$) beaches (especially shingle beaches) whereas gently sloping (generally sandy) shores dissipate wave energy, the waves breaking and spilling across a wide surf zone. The outcome is beach states that correspond with surf scale categories based on the dimensions of breaking waves (Wright and Short 1984, Short 1993). Dean (1991) proposed a surf scaling parameter (Ω), based on breaker height (H_b), and obtained from the formula:

$$\Omega = H_b / wsT$$

in which *ws* is mean sediment fall velocity and T the wave period. This enables beaches to be described as reflective ($\Omega < 1$) where they receive surging breakers and have a high proportion of wave energy reflected from the beach face, or dissipative ($\Omega > 5$) where wave energy from spilling breakers is lost across a wide gently sloping beach. An intermediate category ($\Omega = 1$ to 5) is recognized (Masselink and Short 1993).

On the New South Wales coast, where fine-to-medium grained sandy beaches occupy asymmetrical embayments with microtidal moderate-to-high wave energy conditions, a distinction can be made between steep sandy beaches facing relatively deep water (without bars) close inshore, which are reflective because part of the incident wave energy (plunging waves) is reflected seaward, and flatter sandy beaches fronted by nearshore sand bars and wide surf zones, which are dissipative because much of the wave energy (spilling waves) is lost as waves arrive through the shallow water (Wiseman et al. 1979). The surf zone, where the waves have broken, is often less than 10 metres wide on reflective beaches, but may be at least 100 metres wide on dissipative beaches. Reflective beach profiles have upper slopes of between 6° and 12°, often with a distinct, steeper step at the base, then a gentler nearshore slope of 0.5° to 1°, and with well-formed ripples parallel to the coastline. Dissipative beach profiles are broader and flatter, with slopes typically less than 1°, and a nearshore zone diversified by multiple parallel sand bars.

On the Sydney coast many beaches are predominantly steep and reflective in late winter and spring, and predominantly gentle and dissipative, with wide surf zones and sand bars, in summer. The seasonal oscillations vary with degree of exposure (beaches in narrow bays between headlands being more often reflective), sand texture (coarser sand on reflective beaches) and wave height (larger waves develop wider and gentler dissipative shore profiles). Asymmetrical bay beaches receiving south-easterly wave action (particularly in winter) develop steeper reflective profiles in the southern part and wider dissipative profiles, with sand bars, on the more exposed central and northern parts.

Sandy beaches are more likely to be dissipative than shingle beaches, especially at low tide. They have gentler shore profiles than shingle beaches, usually with one or more sand bars, and they produce longer edge waves and more widely spaced rip currents and beach cusps. Shingle beaches are steeper, reflective, with little if any surf zone and no bars.

Studies of beaches in other parts of Australia and around the world have shown an association of high, steep waves and fine sand on dissipative beaches, while long, low waves and coarser sediment (including shingle) characterize reflective beaches. The intermediate category is found where the combination of moderate to high wave energy and fine to medium sand results in a transitional beach type. In Britain shingle beaches, like Chesil Beach, are reflective, and the wide sandy shores of Atlantic coast beaches usually dissipative. Some shingle-backed sandy shores are reflective at high tide and dissipative at low tide, as at Porth Neigwl in north Wales.

On beaches with sediment of a particular size wave conditions largely determine the beach state, which can change from reflective through intermediate to dissipative and back again with variations in weather. Reflective beaches can be cut back by moderate wave action, whereas dissipative beaches are scoured only when swash levels are augmented by storms, a heavy swell, or exceptionally high tides. Particular events, such as a severe storm, can at least temporarily convert a dissipative beach into a reflective one, while prolonged fine weather can restore a dissipative beach profile: essentially the sequence of cut and fill described previously (p. 122). The intermediate beach state may occur when a reflective beach is being modified to dissipative, with declining wave energy and the formation of bars, or when a dissipative beach is being modified to reflective, with increasing energy and bars losing sediment shoreward to the beach.

On the crenulate sandy beaches of the kind seen in embayments on the New South Wales coast, all three types exist. The relatively straight northern coasts, receiving relatively unrefracted ocean swell, are reflective, the slightly curved central sectors intermediate, with nearshore bars dissected by rip current channels, and the sharply curved southern shores dissipative, with waves much diminished by refraction over multiple sand bars, some of which are exposed at low tide. Within such an embayment it is possible for a moderate long swell to cut back reflective beaches towards the northern end at the same time as promoting accretion on the southern part. There is no simple relationship between the three beach states and long-term trends of erosion or accretion, for changes from one beach state to another occur during short-term cycles (e.g. cut-and-fill: p. 122) on beaches that may show long-term progradation, stability or erosion.

Classification into reflective, dissipative and intermediate beach states was originally devised on microtidal coasts. As tide range increases, wave action is dispersed over wider vertical and horizontal zones and tidal currents interact with incoming waves, reducing transverse gradients, smoothing bar topography and modifying rip currents. Beach profiles exposed at low tide show wider and flatter bars and troughs or low tide terraces.

Masselink and Short (1993) examined features of beach morphology associated with the interaction of wave height and tide range, and introduced the concept of relative tide range (RTR), where actual tide range (TR) is divided by breaker height (H_b):

$$RTR = TR/H_b$$

to distinguish categories of wave-dominated beach morphology as relative tide range increases. On swash-dominated beaches these range from reflective ($\Omega < 2$ and $RTR < 3$) to low tide terraces with rip channels ($\Omega < 2$, RTR 3 to 7) then low tide terraces without rip channels ($\Omega < 2$, $RTR > 7$); from intermediate ($\Omega = 2$ to 5, $RTR < 7$) to low tide bar and rip channels ($\Omega = 2$ to 5, $RTR > 7$); and from barred dissipative ($\Omega > 5$ and $RTR < 3$) to non-barred dissipative ($\Omega > 5$ and $RTR > 7$) and ultra-dissipative ($\Omega > 2$ and $RTR > 7$), with multiple lines of breakers moving in over a very wide low gradient profile, as on Sarina Beach in Queensland and Pendine Sands in Wales. There are further complications on drift-dominated beaches, where profiles vary laterally with such features as migrating cusps and lobes, and bars formed at varying angles to the coastlines moving alongshore.

Short (1991) found three categories of beach morphology on mesotidal and macrotidal coasts: high wave energy concave beaches with moderate gradient ($1°$–$3°$), moderate wave energy gentler beaches ($0.5°$) with multiple bars, and low wave energy backshore beaches behind wide tidal flats. Horn (1993) found another category on the Isle of Man: a steeper, reflective beach behind a shore with finer sediment grading seaward.

USE OF MODELS

Another approach to the study of beach morphodynamics is to set up scale models in tanks, in which waves and currents are generated and the rise and fall of tides simulated, in order to test their effects on beach morphology and sedimentation (Bagnold 1940). Such models can be used to demonstrate cut and fill sequences in response to changing wave conditions, the effects of wave refraction on beach outlines and longshore drift by waves arriving at an angle to the shore. Coastal and nearshore topography can be reduced to scale, as can waves and currents, but there are difficulties in scaling down natural sediment calibre and characteristics to fit model conditions. Sand grains can be used in a model to represent a shingle beach, but representation of a sandy beach to scale would require the use of silt, which has different physical properties and may not give the correct response of sand to wave and current action. Bearing in mind this limitation, scale models have been used to test harbour design and to test the effects of introducing artificial structures, such as breakwaters and marinas, to the coast.

An alternative has been to develop mathematical models, which provide computerized simulations of the response of coastal features to changing tide, wave and current conditions, using information obtained from surveys. Mathematical modelling is useful as a means of exploring process–response relationships, but coastal systems are complex and predictions can prove unreliable. Monitoring of coastal changes is needed to check predictions and obtain further data for refinement of models.

There is a trend towards field experiments, using temporary structures, as a prelude to more permanent coastal engineering works, and there are advocates of a trial-and-error approach in beach nourishment projects.

BEACH COMPARTMENTS

Many beaches occupy distinct compartments or cells on the coast, separated by rocky reefs or protruding headlands, particularly those that end in deep water or are defined by river mouths or the heads of submarine canyons close inshore. Some have been artificially delimited by structures such as breakwaters. Beach compartments have also been described as sediment cells or coastal sediment compartments (Davies 1974). Some contain relict beaches (no longer receiving sediment), and some show cyclic changes related to onshore and offshore movements of beach material. Others are less clearly defined, in the sense that beach material is moving in and out round the bordering rocky headlands, by-passing river mouths, or being lost into submarine canyons. There may be sediment cells between boundaries such as major promontories which prevent any longshore movement of sediment and sub-cells within embayments between minor promontories which can be by-passed by drifting sediment under unusual conditions such as severe storms. A sediment cell exists in the south-west of Westernport Bay, Australia, where sand is carried along the coast by longshore drift produced by southerly and south-westerly wave action from Point Leo past Somers to Sandy Point, where strong ebb currents sweep some of it out on to sand bars trailing south-eastward, whence south-easterly waves may move it back on to the downdrift shore (Figure 5.16).

Preliminary identification of beach compartments can be made from maps, charts and air photographs, but it is necessary to study patterns of coastal sediment flow to determine compartment boundaries. In Nova Scotia cliffed headlands cut in drumlins are sources of sand and gravel delivered to beaches in intervening sediment cells (Carter et al. 1990). There are also sediment cells with fine-grained material, silt and clay, usually supplied by a river and occupying the coast on either side of the river mouth. The term sector has been used in this book to describe a portion of coastline between any two points – it is not necessarily a cell, subcell or compartment.

There can be striking differences in the nature of beaches in adjacent bays, as on the steep coasts of south-west England, where the grain size or mineral composition of beaches varies from one cove to another, the beach compartments being separated by rocky promontories. Within each compartment wave action can move the confined beach material to and fro along the shore, such alternations of longshore drift building up first one side then the other, or change

the transverse profile by alternations of shoreward and seaward movements of sediment. Sand also moves from beaches to backshore dunes, and from dunes back to beaches (p. 180).

There are gains and losses from a beach compartment if the intervening headlands are small enough for sand and gravel to be carried past them on the sea floor by waves and currents. Such natural by-passing often takes place predominantly in one direction as a response to prevailing longshore wave and current action, particularly during storms. Beach sand may also be delivered to dunes that spill over headlands on to the next beach. Some beach compartments are delimited by river mouths or tidal entrances, where transverse ebb and flow currents impede longshore drift in much the same way as headlands or solid breakwaters. This can result in beach accretion on the updrift side, and erosion downdrift, but waves and associated currents may carry sand and gravel across the sea floor, by-passing the mouths of rivers and tidal entrances, sometimes with the formation of longshore bars.

On the south coast of England Lizard Point, Dodman Point, Start Point and Portland Bill are major compartment boundaries, but even these are by-passed by sand and gravel moving eastward along the sea floor. Bray et al. (1995) defined littoral cells on the central south coast of England, identifying Christchurch Bay as a sediment transport unit. Well-defined beach compartments occur within coves on the coasts of Devon and Cornwall, but Lyme Bay can be subdivided into several compartments, some (like Tor Bay) delimited by major headlands ending in deep water, other beaches (between Lyme Regis and West Bay) separated by rocky ebbs around which sediment may drift along the sea floor. Chesil Beach occupies a compartment that once extended eastward from Eype, but since 1866 it has been artificially delimited by the stone breakwaters built at the entrance to West Bay harbour.

There are beach compartments in southern California where sand delivered to the shore mainly by rivers (with some from eroding cliffs, and possibly some from the sea floor) drifts southward along the coast past headlands and through successive bays until it is lost into submarine canyons that run out from the southern ends of embayments (Emery 1960). A similar situation exists on fluvially-fed gravel beaches in Georgia, on the Caucasian Black Sea coast (Zenkovich 1973). These are unusual situations, however, for most beaches receive sediment from several sources, including rivers, eroding cliffs and the sea floor, and very few lose sediment into submarine canyons.

At Wewak on the north coast of New Guinea tectonic uplift led to emergence of fringing reefs as promontories which have segregated a beach-fringed coastline into a series of bays and cut off the former supply of eastward drifting sand delivered by rivers draining to the coast to the west. Thus deprived, the bay beaches have become relict, finer in texture, and are eroding (Bird 1981).

BEACH BUDGETS

Coastal sediment budgets deal with the volumes of sediment supplied to a particular sector by onshore and longshore drift and yields from the hinterland and the volumes of sediment lost offshore, alongshore, or landward over a specific period. Beach budgets show a net gain or a net loss in beach volume, which can be determined by making repeated surveys along and across a beach, using conventional methods with instruments such as a level or theodolite, to measure variations in the plan and profile of the beach. Beach profiles are generally surveyed at right angles to the coastline, from backshore datum points (marked so that they can be found for subsequent surveys) down to the low tide line, and some distance seaward in shallow water. They can be used to monitor the advance or retreat of the coastline, supplementing information from a series of dated air photographs. When they are linked by alongshore surveys, gains and losses of land area on the coast and in the intertidal zone can also be measured. Within defined beach compartments it is possible to compute beach budgets by multiplying the mean cross-sectional area of neighbouring profiles (using arbitrarily-defined basal and rear planes) by the intervening longshore distance to obtain volumetric changes. Pierce (1969) used these methods to assess a sediment budget for the beaches between Cape Hatteras and Cape Lookout on the Atlantic coast of the United States, identifying gains from shoreward and longshore drift and losses through tidal entrances and by wind and wave action over intervening barrier islands.

If the volumes of sediment gained and lost on each sector within a beach compartment over particular periods are calculated, these changes can be expressed in terms of a beach budget for that compartment. Beach budgets can also be established for a coastline that includes more than one beach compartment, or for any other defined sector, such as a seaside resort

waterfront. Figure 5.19 shows an example, based on regular surveys of profiles at intervals along a beach near Somers, in Westernport Bay, Australia. Volumes of sand in arbitrary beach compartments between successive pairs of profiles were calculated by multiplying their cross-sectional areas by the intervening distance (Bird 1985b). On the Natal coast in South Africa surveys showed that 380 000 cubic metres of sand moved northward past Durban each year, and that with inputs from sand-yielding rivers to the north this increased to 500 000 cubic metres per year at Richards Bay.

Alternatively, it is possible to make successive contour maps of the beach surface, and to determine patterns and volumes of gain or loss between each survey from these. Various tracking meters and vehicles have been developed to accelerate beach surveying, and use has been made of the satellite-based GPS (Global Positioning System), particularly on beaches firm enough for vehicle-based surveys, as on the Gulf Coast in Texas (Morton et al. 1993). On the Adelaide coast, in South Australia, GIS (Geographical Information Systems) have been used to generate serial contour maps showing the pattern of beach surface gains and losses as a basis for calculating beach budgets and planning beach nourishment (Fotheringham and Goodwins 1990).

TRACING BEACH SEDIMENT FLOW

Patterns of longshore drift on beaches can be deduced from accretion alongside headlands, groynes, breakwaters or landslides, migration of beach lobes, deflection of river mouths and lagoon outlets or growth of spits (Schwartz et al. 1985). Some of these indications of the direction of longshore drift may be misleading where patterns of beach accretion result partly from sediment movement in from the sea floor, rather than alongshore. Shoreward drift of sand and gravel by waves arriving parallel to the coastline can deposit beaches in patterns that result in river mouths or lagoon outlets migrating to sectors of lower wave energy, and accretion alongside headlands can be the outcome of shoreward drift of sediment from a nearby shoal. Paired spits (p. 167) may result from a convergence of longshore drift, but can also be formed by the breaching of a barrier (p. 172) or by wave refraction into a maintained river outlet or lagoon entrance. Usually at least some of the longshore drift of sand and gravel takes place in the nearshore zone, or by way of exchanges between the beach and the near-

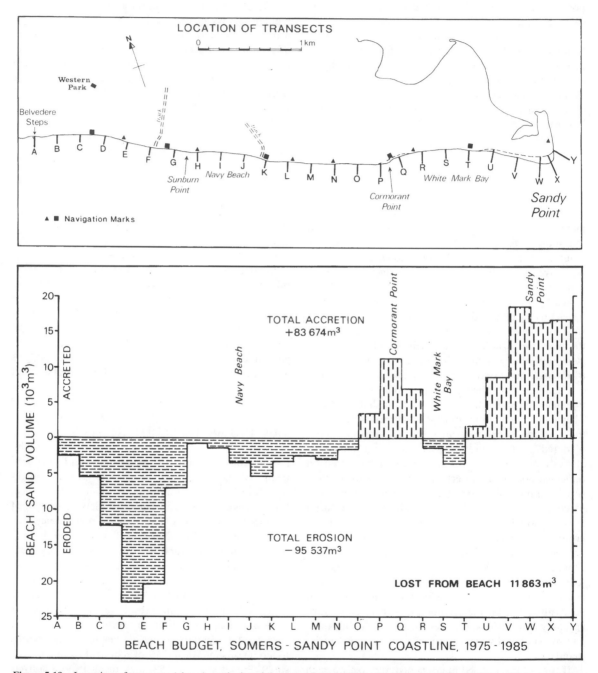

Figure 5.19   Location of transects (above) on the beach between Somers (Belvedere Steps) and Sandy Point, Westernport Bay, Australia. The pecked line indicates sectors of the 1975 coastline where substantial progradation occurred by 1985. The lower diagram gives a beach budget for 1975–1985, based on surveys of cross-sectional areas over the decade, multiplied by the intervening distances (generally 225 metres). The volume of sand gained on accreting sectors was less than that lost from eroding sectors.

shore zone, with subsequent delivery to the beach downdrift.

Pebbles of an unusual rock type or specific mineral sands may act as natural tracers indicating longshore drift from a source area such as a cliff outcrop or river mouth. Patterns of beach sediment movement can also be determined by introducing and following tracers. These are materials that move in the same way as the natural beach sediment, consisting of particles similar in size and shape, with similar hardness and the same specific gravity as the sediment already present on the beach. Tracer material is deposited on a beach or in the nearshore zone at a particular point or along a selected profile, and surveys are made subsequently to see where it has gone. It must be readily identifiable after it has moved along a beach or across the sea floor. Some may become buried within a beach or lost seaward from the nearshore zone, but usually a proportion remains on the beach face to indicate the direction in which sediment has moved.

Some tracer projects have used sand minerals that were not naturally present on the beach. This mineralogical method depends on a complete preliminary inventory of the minerals already present in the beach to be sure that the introduced material is indeed alien, and will be identifiable downdrift. Baker (1956) introduced alien minerals, such as iron pyrites (actually of higher specific gravity than the natural beach sand) to beaches of quartz and calcarenite sand on the west coast of Victoria, Australia, as a prelude to surveys of patterns of sand movement alongshore and across Portland Bay, where a harbour was to be constructed (Figure 5.20). The project indicated that most of the sand moving along the coast was passing out across the sea floor, and was unlikely to accumulate in the harbour, but there has been some harbour accretion as the result of inwashing of sand from the adjacent sea floor. There are risks in using a tracer of higher specific gravity than the natural beach sediment, because the introduced minerals may not reproduce the behaviour of the natural beach sediment. Lighter grains are moved more frequently and further than

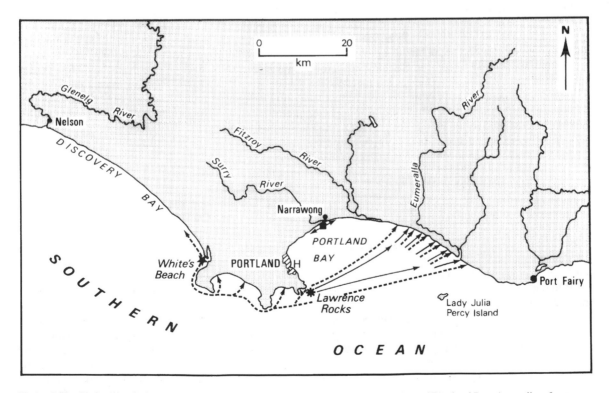

Figure 5.20 Pathways of mineral tracer sand movement on to beaches in Discovery Bay and Portland Bay, Australia, after tracer injection at White's Beach, Lawrence Rocks and Narrawong. The aim was to determine whether there would be significant sand movement into the harbour that was subsequently built at Portland (H). Accretion has occurred in this harbour, but it consists of fine sand washed up from the nearby sea floor rather than coarse beach sand moving along the coast.

heavy grains, and their net long-term drift may not be in the same direction.

Sand or gravel that is naturally or artificially coloured can be used as a tracer, but there are difficulties in observing coloured grains when they form only a small proportion of the sediment on a beach. More effective is the use of natural or artificial sand or gravel coated or imbued with a colloidal substance containing a fluorescent dye (Ingle 1966). Sand or gravel labelled in this way is dumped on a beach, and its subsequent movement traced by locating the dyed material at night, using an ultra-violet lamp, or sand samples taken from the beach can be examined under ultra-violet light in a darkroom, when the fluorescent stain stands out brightly. Fluorescent quartz has been used to trace beach sand movement along the shores of the Nile delta (Badr and Lotfy 1999), and Rink (1999) has explored possibilities of using quartz luminescence in tracing beach and nearshore sand movements.

Another method of tracing sediment flow depends on the use of radioactive materials, which can be followed by means of a detection meter (Geiger counter). Artificial sand can be made from soda glass containing scandium oxide which is ground down to the appropriate size and shape for use as a tracer. It is then placed in a nuclear reactor to acquire the radioactive isotope scandium 46, which has a half-life (i.e. time taken for the radioactivity to diminish to half its original strength) of 85 days, which means that it can still be detected by Geiger counter three or four months after it has been introduced. Gravels can be traced by using granules, pebbles or cobbles that have been taken from the beach and labelled with a radioactive substance, or artificial materials such as concrete in which radioactive material has been embedded. After the tracer has been deposited, surveys are made by carrying a detection meter over a beach, or dragging it across the sea floor mounted on a sledge. The location and intensity of radioactivity can thus be mapped, and paths of sediment flow deduced.

Radioactive tracers are expensive and a health hazard, but they provide a good means of tracing sediment movement, for the tracer can be detected even when it has been buried in a beach, where coloured or fluorescent tracer would not be visible. However, fluorescent tracers are cheaper and safer, and often more durable for long-term projects, whereas radioactive tracers permit only one project, and cannot be used again until all the tracer used in the first project has disappeared. Different coloured paints or fluorescent dyes can be used to trace sediment movement from

several places at the same time. An alternative on gravel beaches is to use radio transmitters embedded in artificial pebbles introduced to the beach to register their patterns of movement.

Tracers have been used to estimate rates and quantities of beach sediment movement. On Sandy Hook, New Jersey, Yasso (1965) introduced four grain size classes of fluorescent sand, and subsequent sampling on a profile 100 feet downdrift showed that the smallest particles (0.59–0.70 millimetres) arrived first, and the next class (0.70–0.84 millimetres) soon afterwards, the maximum rates of flow being 61 and 79 centimetres per minute respectively.

If the predominant direction of longshore drift is known it is possible to estimate the volume of sediment moving along the beach by introducing a standard quantity of tracer at regular intervals to a particular injection point, and taking samples at another point downdrift to measure the concentration of tracer passing by. As the concentration of tracer downdrift is proportional to the rate at which the sediment is moving, the quantities in transit can be measured. Jolliffe (1961) applied this tracer concentration method successfully to the measurement of rates of longshore drift on shingle beaches on the south coast of England, using pebbles coated with a fluorescent substance. It is more difficult to trace sand movement by this method because of the vast number of grains involved, but sand movement on the Caucasian Black Sea coast has been measured using fluorescent tracer sand down to a dilution of one grain in 10 million.

LATERAL GRADING

Some beaches show lateral grading from fine to coarse sediment along the shore. Grain size composition of beach material may vary in the vicinity of source areas such as eroding headlands, where the proportion of locally-derived coarse material is high, or river mouths, where a larger proportion of coarse fluvial sediment is likely to be present. There is no single, simple explanation for lateral grading of beaches, and several hypotheses have been put forward. One is that there has been longshore sorting of beach material by breaking waves and nearshore currents, so that a beach which initially had particles of various sizes shows selection in the course of longshore drift, the finer grains being moved further because they are more easily mobilized by waves and transported by associated currents. This is supported by Yasso's (1965) measurements at Sandy Hook, New Jersey.

However, the reverse is found on some beaches, the larger particles having been differentially moved further downdrift. McCave (1978) explained downdrift coarsening of sand to shingle beaches on the Norfolk coast as the result of preferential removal of sand to offshore bars as longshore drift progressed. At Hampton, on the shores of Port Phillip Bay, Australia, a beach renourished with sediment dredged from the sea floor, a mixture of sand and shelly gravel, was depleted by longshore drift, the shelly gravel moving downdrift farther and faster than the sand, and so that shells dominated the beach downdrift. Alternatively finer grains may be selectively removed from the beach, either because they are blown offshore or onshore by wind action, or because they are withdrawn seaward by backwash, the extent of such removal diminishing alongshore (Komar 1976).

Chesil Beach, a swash-dominated beach in Lyme Bay (Figure 5.21) in southern England, is subject to longshore drift when waves arrive obliquely to the shore. It consists mainly of flint pebbles, and grades from granules on the beach at the western end, near Bridport, through to pebbles and cobbles (mainly of locally-derived limestone) towards the south-eastern end, near Portland Bill (Figure 5.11). Grading is correlated with a lateral increase in wave energy, with larger waves coming in through deeper water to build the higher and coarser beach at the more exposed south-eastern end, but it is not clear how this contrast in wave energy has achieved lateral sorting. There may have been longshore edging of the finer particles from higher to lower wave energy sectors. There is no doubt that Chesil Beach used to receive shingle from the beaches of Lyme Bay to the west, but the dominant eastward longshore drift of shingle has been interrupted by landslide lobes and bouldery ebbs, and the harbour breakwaters at West Bay now form the western limit of a beach compartment in which the shingle deposits are now essentially relict (Bray 1997). The beach is still receiving small amounts of shingle from cliff erosion to the south-east, where Portland limestone yields cobbles and pebbles (mainly from quarries). These become mixed with the flint shingle, and it seems that pebbles of particular dimensions then drift quickly along the shore to their appropriate size sectors, and are there retained, while pebbles that are larger or smaller move away. It has also been suggested that Chesil Beach may originally have been a beach formation of uniform grain size, and that as it was rolled obliquely landward by storm surges it became modified to a laterally graded beach by attrition, the farther-travelled western sector being

reduced to finer particles than the less-travelled eastern sector.

Bascom (1951) related similar grading in beach sand particle size on the shores of Half Moon Bay, California, to a variation in degree of exposure to refracted ocean swell coming in from the north-west, coarse sand on the exposed sector grading to fine sand in the lee of a headland at the northern end. A similar pattern is seen in Wineglass Bay, on the east coast of Tasmania, where a curving sandy beach is steeper, narrower and coarser on the more exposed northern sector and becomes flatter, wider and finer towards the more sheltered southern end. In both cases there is an implication that lateral sorting developed as sand was edged from the higher to the lower wave energy sector.

Another possible explanation of lateral grading on beaches is where longshore drift is stronger in one direction than the other, as the result of contrasts in the height or frequency of incident waves. Larger waves carry the whole range of available coarse to fine particles in one direction, but smaller waves take back only the finer material. Alternations of longshore drift can thus result in separation of sand from shingle. Chesil Beach could have become laterally sorted in this way (Jolliffe 1964), and the other beaches of Lyme Bay are also finer to the west and coarser eastward in response to similar alternations of stronger south-westerly wave action and weaker south-easterly wave action (Bird 1989). Farther east, between Weymouth and Osmington (in the lee of Portland Bill), the beaches are graded in the opposite direction because south-westerly wave action is much reduced, and south-easterly wave action becomes dominant.

On drift-dominated beaches lateral grading may be the outcome of wearing and attrition of sediment derived to a particular sector of the coast as longshore drift carries it away from the source. This may explain the graded beach on the coast of Hawke Bay, New Zealand (Figure 5.22), which is supplied with gravelly material largely from the Mohaka River, and shows a gradual reduction of mean particle size from pebbles to granules, then coarse to fine sand, along the shore eastward from the mouth of the river (Bird 1996b). In the Bay of Anges, between Nice and Antibes in southern France, pebbles delivered by the River Var became reduced in size (and more rounded) as they moved along the 15 kilometre shore. The mean diameter of pebbles was 9.7 centimetres at the river mouth, 5.8 centimetres in the middle of the bay, and 4.3 centimetres at Antibes. The pebbles were large

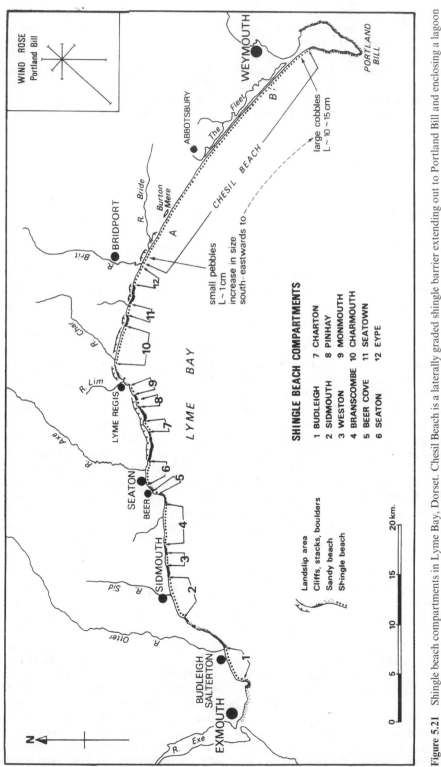

Figure 5.21   Shingle beach compartments in Lyme Bay, Dorset. Chesil Beach is a laterally graded shingle barrier extending out to Portland Bill and enclosing a lagoon (The Fleet). *A* and *B* indicate locations of photographs shown in Figure 5.11.

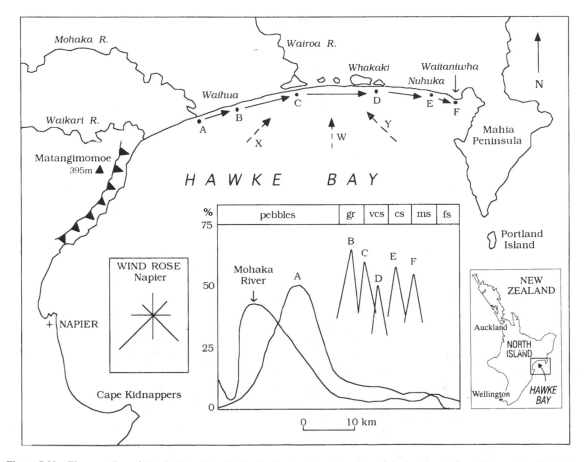

Figure 5.22   The coastline of Hawke Bay, New Zealand, showing lateral grading of the beach east from the mouth of Mohaka River (A) to Waitaniwha (F), as indicated by modal grain size graphs. The wind rose shows the prevalence of south-westerly waves (X) and south-easterly waves (Y), but there is also a southerly swell (W). Predominant longshore drift is eastward, and grading has been attributed to attrition of pebbles as they move alongshore.

and ovoid at the Var mouth while at Antibes, smaller, flatter, and rounded pebbles predominated. However, it is possible that longshore sorting has contributed to this grading, the initially smaller pebbles travelling farther (Cailleux 1948). Landon (1930) described how angular gravel from a cliff on the west coast of Lake Michigan became rounded and reduced in size as it drifted southward along the shore, and similar longshore attrition may explain the reduction of cobble size that occurs northward along the spit at Westward Ho! in Devon. Flint and chalk fragments eroded from the chalk cliffs south of Studland in Dorset become rounded as they drift northward along the beach, the chalk particles being reduced much more quickly than the harder flints.

Where lateral grading is the outcome of more rapid attrition on a beach sector subject to strong wave action, the finer sediment will be found in the high energy sector. Where there is a correlation between higher wave energy and coarser sediment passing laterally to lower wave energy and finer sediment the implication is that there has been little or no attrition. It is difficult to explain lateral grading on a swash-dominated beach unless there has been sufficient longshore drift to achieve sorting.

Lateral variation may also occur in the mineral composition of drifting sediment, but it may also reflect the pattern of sources. In Encounter Bay on the South Australian coast the south-eastern end of the beach, near Kingston, consists almost entirely of calcareous sand (almost 90 per cent calcium carbonate) washed up from the sea floor, but the proportion of quartz sand increases north-westwards to Goolwa, at the mouth of the Murray (calcium carbonate content less than 10 per cent). This lateral variation is the outcome of a pattern of earlier deposition of fluvial quartzose material on the sea floor off the mouth of the Murray during Pleistocene low sea level phases, so that the sand swept onshore is there dominated by quartz rather than the calcareous sand of marine origin further along the bay shore (Sprigg 1959).

As some beaches show lateral grading it is worth asking why others do not. More intricate patterns of sorting are found on beaches behind intermittent reefs or shoals, as in the shingle beach in Ringstead Bay, Dorset, where cobbles are concentrated in the lee of limestone reefs and smaller pebbles in intervening small bays. The swash-dominated cobble beach at Newgale in Pembrokeshire is composed of gravel that has been washed in from the floor of St Brides Bay, and is not laterally graded.

King (1972) suggested that coarser beach material is found on sectors where wave energy is higher, but there are many exceptions to this. Grain size of beach sediments is not necessarily well correlated with wave energy sectors, for although storm-piled cobble and shingle beaches are more likely to be found on high wave energy coasts, many oceanic coasts have sandy beaches. On the other hand beaches on low wave energy coasts may be gravelly if there is a local source of coarse material, even if wave action is rarely strong enough to mobilize it.

PROGRADING BEACHES

Beaches that show net accretion, receiving more sediment from various sources than they lose onshore, offshore or alongshore become higher and wider, prograding seaward. Their transverse profiles may be convex, or in the form of a terrace ending in a seaward slope. The dimensions of a prograding beach are determined by the balance of gains from fluvial sources, cliff and rocky foreshore yields, the sea floor, or dunes blown from the hinterland against losses alongshore and offshore, the removal of sand by wind to build landward dunes, and the washing of sediment into estuaries and tidal inlets (Figure 5.2). Beach progradation may be induced by human activities, notably where the sediment supply has increased as the result of coastal quarrying or where fluvial sediment yields

have been augmented by hinterland deforestation or mining (p. 102).

Progradation of a beach is often indicated by the formation of successive beach ridges of sand or shingle, built above high tide level (p. 141). Successively-formed foredunes mark stages in the seaward advance of a sandy coastline, as on Winterton Ness in Norfolk and on the shores of Carmarthen Bay in south Wales, where sand is being washed and blown onshore from extensive shoals exposed at low tide and the coastal dunes at Pembrey are extending seaward.

Some beaches are prograding as the result of shoreward drift of sand or shingle from the sea floor, notably where there are bars or shoals in the nearshore area. On the west coast of Wales there is northward longshore drift from Cardigan Bay and eastward longshore drift along the Lleyn Peninsula, converging in Tremadoc Bay, where there are extensive sandy shoals and intertidal flats washed into the Afon Glaslyn estuary and built up as bordering beaches, spits and dunes. In south-eastern Australia beaches are receiving sediment washed in from shoals in Streaky Bay, South Australia (Figure 5.23), off Hunter Island in north-west Tasmania and off Corner Inlet, east of Wilson's Promontory, Victoria. The sand shoals east of Vansittart Strait, between Flinders Island and Lady Barron Island off north-east Tasmania are a graphic example of sea floor sand drift which in this case has passed through a strait instead of accumulating on west-facing beaches on these islands.

Beach progradation on the Danish island of Kyholm resulted from the disappearance of seagrass meadows in the surrounding waters and the ensuing shoreward movement of fine sand that had previously been retained on the sea floor by this vegetation (Christiansen et al. 1981). Progradation is more likely to take place on coasts where emergence is in progress, stimulating shoreward drift of nearshore sediment, as at Kalajoki on the Finnish coast of the Gulf of Bothnia. In northern Britain uplift is continuing as the result of postglacial isostatic recovery, and progradation is taking place on the north-east coast of England, where wide sandy beaches have formed on either side of Holy Island in north-east England and at Tentsmuir on the east coast of Scotland.

Progradation continues on the ends of spits, as on Hurst Castle spit in Hampshire and at the western end of Blakeney Point in Norfolk (p. 165), and on the shores of cuspate forelands, as on the eastern shore of Dungeness in Kent and the north shore of Benacre Ness in Suffolk (p. 169). Beaches have also prograded on coasts where longshore drift has been intercepted by

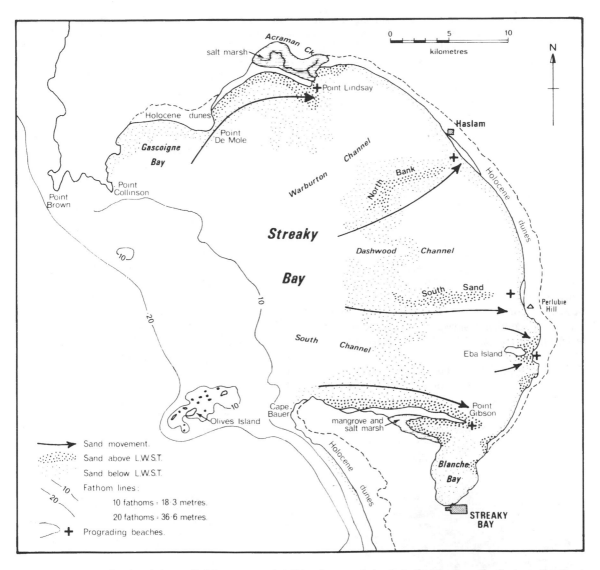

Figure 5.23 Prograding beaches supplied by shoreward drifting from sand shoals in Streaky Bay, South Australia. Beach accretion has been shaped by convergence of south-westerly waves refracted by the linear shoals, and in the lee of Eba Island.

a protruding headland, so that there is accretion on the updrift side, as at Point Dume in California, Apam in Ghana, and Cape Wom on the north coast of New Guinea. Such beaches will continue to prograde until sand or gravel by-pass the promontory to reach the downdrift shore.

Beach sediment drifting alongshore is intercepted where jetties or breakwaters have been constructed, usually to stabilize a river mouth or lagoon outlet, or shelter a harbour from winds and waves. This results in progradation updrift of the intercepting structure. In

south-east England, longshore drift is generally south-ward on the East Anglian coast, where beaches have prograded on the northern side of harbour breakwaters at Yarmouth, Lowestoft, and Southwold, while on the Channel coast eastward longshore drift has led to accumulations on the western side of breakwaters at Black Rock marina at Brighton, and at Newhaven, Rye, and Folkestone. At Rye the breakwaters built at the mouth of the Rother River have intercepted eastward-drifting shingle to form a prograded shingle beach on the western side. To the east are Camber

Sands, a pebble-free sector, the shingle reviving towards Lydd and Dungeness. Thus the Rother breakwaters have created a break in the shingle beach that was migrating alongshore. At Studland in Dorset the beach has prograded south of a training wall built in 1924 to prevent the northward drift of nearshore sand from shoaling the entrance to Poole Harbour (Bird and May 1976).

In the United States beaches have prograded alongside breakwaters built in 1930 at Santa Barbara and at Redondo in California and South Lake Worth in Florida, with concomitant erosion downdrift. Similar features are seen at Sochi on the Black Sea coast, Lagos in Nigeria, Durban in South Africa and Madras in India. Among many other examples are the interception of southward-drifting sand by the breakwaters at Praia da Barra near the Aveiro Lagoon and at Figuera da Foz, both in Portugal, and eastward drifting sand alongside the jetty built at the mouth of the Vridi Canal on the Ivory Coast when it was cut in 1950. In Australia, northward-drifting sand has accumulated along the southern sides of breakwaters at Port Adelaide and at Tweed Heads in northern New South Wales.

In some places progradation has occurred on both sides of protruding breakwaters, indicating either an alternation of longshore drift or a predominance of shoreward drift on to beaches on either side of the outflow. There has been sand accretion on both sides of the breakwaters built at Lakes Entrance in south-east Australia to stabilize the artificial outlet from the Gippsland Lakes cut in 1889 (Figure 5.24). Similar accretion has taken place on both sides of paired breakwaters at Onslow in north-western Australia, Rogue River and Siuslaw River on the Oregon coast, Newport in California, Ijmuiden on the Dutch coast and the Swina Inlet in Poland.

Beach accretion can occur in the lee of a harbour breakwater, as at Warrnambool in Victoria, Australia, where shoreward drift of fine sand occurred from the sea floor into the harbour area that became more sheltered after the breakwater was constructed, leading to beach progradation (Bird 1993a). After an offshore breakwater was built in 1934 to protect the pier at Santa Monica, California, a sandy foreland developed during the ensuing 15 years, and dredging became necessary to maintain access to the pier. Similar local progradation has occurred behind offshore breakwaters built to protect Italian beaches, notably between Rimini and Venice on the Adriatic coastline. Progradation of a beach can also be induced in the lee of a wrecked ship close to the shore, as at Sukhumi on the Caucasian Black Sea coast (Zenkovich 1967).

Accretion has taken place alongside some tidal inlets (lagoon entrances) where no breakwaters have been built, the transverse ebb and flow of currents having had the same effect as an intercepting breakwater. There is usually some deflection of an unprotected outflow channel by longshore drift, and if this continues and the gap becomes sealed off by deposition, the accreted foreland may become a lobe, which then moves on along the shore.

Figure 5.24 The artificial entrance to the Gippsland Lakes, cut through the outer barrier in 1889 and bordered by stone breakwaters. Sand accretion on either side of the breakwaters has formed coastal forelands with successive parallel foredunes, and wave patterns show the location of the looped offshore bar. Crown Copyright Reserved.

On the north-east coast of Port Phillip Bay, Australia seasonal patterns of longshore drift occur. At Sandringham, near Melbourne, breakwaters built to form a boat harbour have intercepted sand drifting south in winter in a situation whence it cannot return northward in summer. The beach has prograded within the harbour because the breakwater unexpectedly formed a sand trap (Figure 5.25).

BEACH RIDGES

Beach ridges have formed where sand or shingle have been banked up by wave action on a prograding coastline. Sandy beach ridges may originate as berms built by constructive wave action, whereas shingle beach ridges have been piled up by storm waves. The persistence of a beach ridge depends on overall progradation of the coastline and its separation from earlier ridges by swales is often the outcome of a phase of erosion. In Tasmania, Davies (1957) described the evolution of parallel beach ridges on successive wave-built berms, the outcome of alternations of cut and fill on a prograding coast. In New South Wales, McKenzie (1958) found that foredunes were initiated on successive strandlines of seed-bearing vegetation on a prograding beach. Beach ridge and foredune initiation have been much discussed, and exemplify multicausality (Hesp 1984, 1988; Bird and Jones 1988; Taylor and Stone 1996; Sanderson et al. 1998). Sandy beach ridges can be built up along the shore by storm swash. Psuty (1965) described parallel beach ridge formation by this process on the prograding shores of the deltaic coast of Tabasco, Mexico, where fluvially-supplied summer sand accretion has been built into successive ridges by winter storm swash. Sandy beach ridges are found on the east coast of peninsular Malaysia, where they are known as permatang (Teh Tiong Sa 1992).

The height and spacing of parallel beach ridges is determined by the rate of progradation (i.e. the rate of supply of sand and shingle, depending on available sources and patterns of sediment flow), the incidence of cut and fill and the upper swash limit of the waves that built them, which is modified by changes in the relative levels of land and sea (Taylor and Stone 1996). There are numerous roughly parallel shingle beach ridges on the cuspate foreland at Dungeness, Kent, each marking a former coastline, and stages in the evolution of this structure have been deduced from the ridge pattern (Lewis 1932). Similar shingle beach ridges have been used to trace stages in the progradation and southward growth of Orfordness, Suffolk, and reference has been made to beach ridges at Hoed in Denmark (Figure 5.9).

The height of the crest of a beach ridge is also related to the beach profile preceding the storm. Swash limits are higher on steep beaches than on those of gentle gradient. Moreover, the amount of beach material available to be storm-piled affects beach ridge height. If there is too little, storm swash overtops the beach, and if there is too much, storm wave energy is expended on the beach profile rather than on ridge building.

Variations in the crest levels of successive beach ridges are likely to reflect variations in the upper limit of storm swash, but they could also be related to changing sea levels. Lewis and Balchin (1940) found that the crests of some of the inner and older shingle ridges on Dungeness were two to three metres lower than those formed more recently along the eastern shore, and deduced that sea level had been rising. On the other hand some Scottish shingle beach ridge systems have ridge crests declining seaward, as in Spey Bay and on the west coast of Jura, and similar features are seen where eskers have been submerged, reworked by wave action and built into successive ridges on an emerging coast (Figure 5.26). Examples are seen near Hove in southern Sweden and around the Gulf of Bothnia, as on the island of Hailuoto in Finland (Figure 5.27). These are coasts that have been emerging as the result of postglacial isostatic recovery, and the older beach ridges formed when the sea stood at higher levels. If a coastline is emerging the crest heights of successive beach ridges are likely to decline seaward, but there are also variations associated with differing upper swash limits in each constructional phase. On the coast of Florida, Tanner (1995) considered that prograded beach ridge plains were the outcome of recurrent sea level fluctuations of between 5 and 20 centimetres.

A seaward decline in the crest heights of beach ridges is not necessarily an indication of emergence because it could have been produced by a sequence of diminishing upper swash limits (for example if the nearshore area were shallowing as the result of accretion, and incident wave heights were consequently being reduced). In a similar way a seaward rise in the crest levels of shingle ridges could result from increasing storm surge levels rather than a rising sea level.

On sandy beaches the formation of a succession of parallel beach ridges is often accompanied by the evolution of successive foredunes (p. 181). Backshore vegetation spreading forward on to the beach indicates that progradation has been taking place (Figure 5.28), especially where the vegetation canopy declines

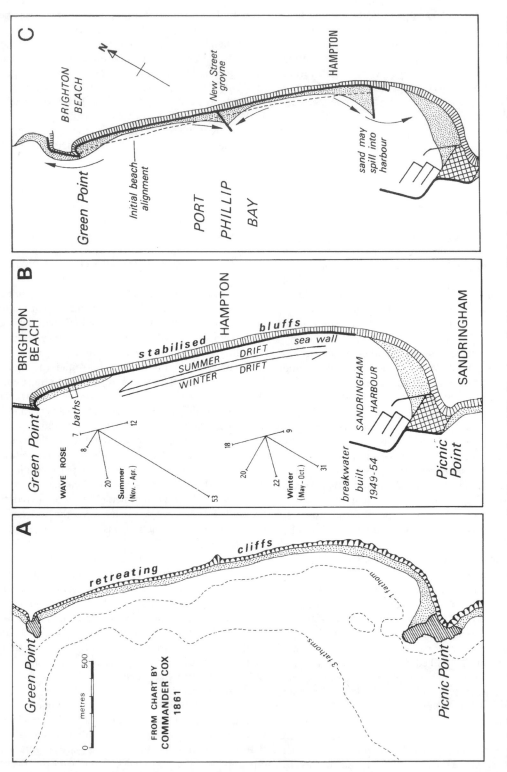

Figure 5.25 The natural configuration of the coast at Sandringham, Port Phillip Bay, Australia (A) was modified after the construction of a breakwater (1949–54) to shelter Sandringham Harbour. As at nearby Black Rock (Figure 5.5) the sandy beach is modified by northward drifting in summer and southward drifting in winter. The breakwater created a trap in which southward drifting sand has accumulated (B). Depletion of Hampton Beach led to beach nourishment in 1987 north of the New Street groyne, and in 1997 to the south. The nourished beaches have been losing sand northward in summer and southward in winter (C).

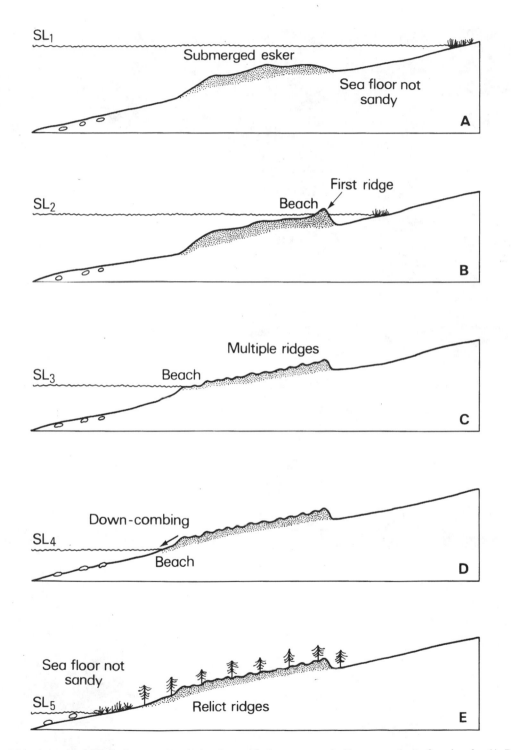

Figure 5.26 As sea level falls in the Gulf of Bothnia submerged eskers are reworked by wave action to form beaches (A, B), and as emergence proceeds multiple beach ridges have formed on a seaward slope (C). As the sea falls below the limit of the former esker, beach ridge formation comes to an end (D, E).

Figure 5.27 Beach ridges on the Finnish island of Hailuoto, formed in the way shown in Figure 5.26.

Figure 5.28 Backshore vegetation spreading behind a prograding beach in Frederick Henry Bay, Tasmania.

smoothly seaward from trees through shrubs to grassy communities, a zonation that indicates a plant succession is accompanying the deposition of new terrain.

Beach ridges can also be formed by the successive addition of spits that have grown parallel to the coast. Historical maps of South Haven peninsula on the Dorset coast indicate stages in the formation of three broad sandy beach ridges, surmounted by dunes, and the peninsula at Falsterbo, in south-west Sweden, also prograded in this way. Beach ridges of a different kind are formed on cold coasts in high latitudes, where the sea freezes and wave and nearshore processes are halted, at least in winter. When storm waves break up a winter ice fringe and drive it on to the shore, ice-pushed beach ridges are formed. These have been described from Arctic coast beaches in Alaska, northern Canada and Siberia, and from Antarctica. In summer meltwater from glaciers cuts channels across these ridges on the backshore.

CHENIERS

Cheniers are ridges of sand, shelly sand and gravel that have been deposited on coastal plains of fine-grained sediment (silt, clay or peat), particularly in deltas and marshlands. They were originally described on the deltaic plains of Louisiana where they were called cheniers because oak trees (*chênes*) grow on them. They are usually separated by broad flat swampy areas, and were emplaced by occasional storm surges as the low-lying terrain prograded.

There are numerous cheniers on the deltas of northeast China, each deposited when shelly material and fine sand were driven landward by storm swash on to a prograding marshy plain (Liu Cangzi and Walker 1989). Similar features are seen on the coastal plains of northern Australia, notably to the east of Darwin where sandy ridges rest upon, and are separated by, flat swampy deltaic lowlands. They are bordered seaward by a mangrove fringe with only minor beaches of sand with coral and shell gravel (as at Point Stuart), and it appears that sand has been washed up through the mangroves from the floor of Van Diemen's Gulf during storm surges (tropical cyclones) on a generally low wave energy coast. They are emphasized by the growth of pandanus trees, the intervening areas being bare of vegetation, apart from some sparse grassland (Figure 5.29).

Ridges of shelly sand driven landward on to salt marshes in East Anglia, notably on the Dengie peninsula in Essex, have also been termed cheniers,

Figure 5.29 Cheniers of shelly sand on a mangrove-fringed (m) coastal plain south of Van Diemen Gulf, northern Australia, dissected by tidal creeks.

and are the outcome of swash action in successive storm surges (Greensmith and Tucker 1966). A shelly chenier runs east–west across the salt marsh at Morston in north Norfolk, in the lee of the Blakeney Point spit, emplaced during a storm surge that was sufficiently strong to sweep sandy material from the floor of Blakeney Harbour up on to this marsh. Similar cheniers have been deposited in mangrove fringes on tropical coasts (p. 218).

BEACH LOBES

Many drift-dominated beaches include lobate protrusions which migrate alongshore. These beach lobes may be formed by the local convergence of longshore drift as the result of waves approaching obliquely, first from one direction then the other, or they may be the result of local and temporary progradation beside a river mouth or lagoon outlet, moving on when the gap becomes sealed by deposition (Figure 5.30). Some lobes form during occasional storms when waves arriving obliquely are strong enough to drive beach material round a headland to be deposited as a lobe on the lee side, others when shoals of sediment swept out of a river mouth by floodwaters are washed up on to the shore downdrift. However initiated, beach lobes move intermittently alongshore until they are intercepted beside a headland or breakwater or added to the end of a spit.

A succession of sandy beach lobes has formed and migrated along the coast between Somers and Sandy Point in Westernport Bay, Australia, as the result of longshore drift by waves from the south-west (Figure 5.16) (Bird 1985b). The passage of each lobe is marked by progradation, followed by erosion as the lobe moves on. Some lobes grow by the addition of beach ridges, and become travelling forelands, with beach ridges truncated on their western flanks and added succes-

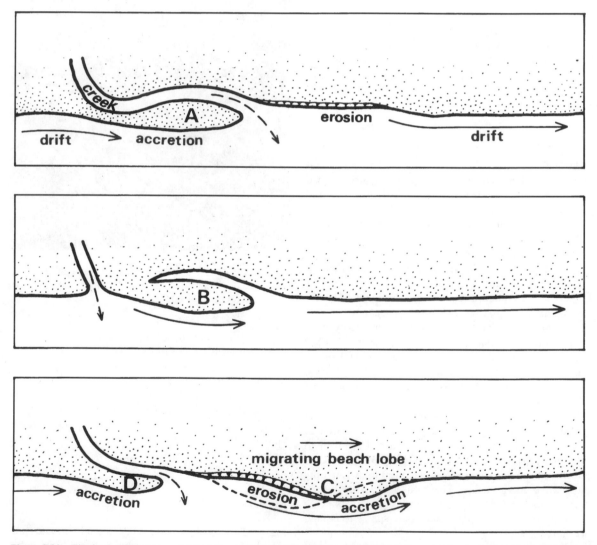

Figure 5.30   The formation of a beach lobe (A) as the result of accretion updrift of a creek outlet that is deflected by breaching to form a new outlet with the lobe downdrift (B). Beach erosion and accretion accompany the subsequent migration of the lobe (C) alongshore.

sively to their eastern flanks. Beach lobes are not found on swash-dominated high wave energy coasts.

BEACH EROSION

Beaches are eroded when they lose more sediment alongshore, offshore or to the hinterland than they receive from the various sources shown in Figure 5.2. Processes that lead to beach erosion include destructive wave action in stormy periods and the depletion of beach sediments by weathering and winnowing, as well as a diminution in inputs from rivers, cliff and shore erosion, spilling dunes and drifting from the sea floor. As the volume of beach material diminishes the beach face is lowered and cut back. Beaches that line cliffed or rocky coasts become narrower, thinner and discontinuous as losses proceed, and may eventually disappear, while on depositional coasts the previously deposited landforms are cut back. The rate of retreat of the high tide line, which is often also the seaward boundary of terrestrial vegetation communities, can be measured by comparing dated sequences of maps and charts, or air and ground photographs. Mean annual recession rates have been generally small (less than a metre per year), but there are records of beaches retreating by up to 40 metres per year, for example on the Nile delta (p. 260).

Indications of beach erosion include cliffed back-shore dunes (whereas prograding beaches are backed by beach ridges and incipient foredunes), truncated vegetation zonations (whereas prograding beaches are backed by tree canopies descending to beach level, or by shrub and grass zones on recently formed sandy terrain), or exposures of beach rock (p. 114).

Beach erosion has been reported from many coasts. Seaside resorts such as Deauville in France, Miami in Florida and Surfers Paradise in Australia are among the many that have suffered when storms have removed the beaches that attracted their visitors. In such places cut-and-fill cycles are incomplete, only part of the sand removed by storm waves being returned to the beach in calmer weather. There has been a retreat of sandy coastlines in many parts of the world during the past several decades, and in many places for a much longer period. Recession is in progress even on the shores of coasts that were previously prograded by deposition, for example along the Ninety Mile Beach in south-eastern Australia, where progradation during the past century has been confined to sectors on either side of the Lakes Entrance harbour breakwaters (Figure

5.24). Around the world there are relatively few sectors of naturally prograding sandy beach, whereas receding sandy coastlines are extensive. Only a few seaside resorts have had their beaches substantially widened by natural accretion: examples are Seaside in Oregon and Malindi in Kenya (Figure 5.4).

Between 1976 and 1984 the Commission on the Coastal Environment (International Geographical Union) assembled world-wide evidence of coastline changes over the preceding century, and found that beach erosion had become widespread. More than 70 per cent by length of beach-fringed coastlines had retreated over this period, less than 10 per cent having advanced (prograded), the balance having either remained stable or shown alternations with no net gain or loss (Bird 1985a).

Investigating the causes of beach erosion, the Commission on the Coastal Environment found that there was no single, simple and universal explanation. Some 21 factors have been identified as having initiated or accelerated beach erosion, their relative importance varying from one coast to another. Usually more than one of these factors has contributed to the onset or intensification of erosion on a particular beach. Attempts to explain the erosion of a beach should consider each of the possible causal factors, and those found to be applicable should be ranked in importance.

The causes of beach erosion listed will now be examined in greater detail.

Submergence and increased wave attack

Coastal submergence may be due to a rise of sea level, coastal or nearshore land subsidence, or some combination of land and sea movement that results in the sea standing higher relative to the land. It is likely to lead to beach erosion, for a sea level rise usually results in the deepening of nearshore waters, allowing larger waves to reach the shore and erode the beach face, thus withdrawing sand or gravel to the sea floor. Bruun (1962) suggested that on beaches which had attained a profile of equilibrium, which he took as a condition where they were neither gaining nor losing sediment over a specified period (p. 126), a sea level rise would cause erosion of the upper beach, and transference of sand or gravel from the beach to the adjacent sea floor would in due course restore the previous transverse profile in relation to the higher sea level (Figure 5.31). The transverse profile would thus migrate upward and landward, the coastline retreating further than it would have done

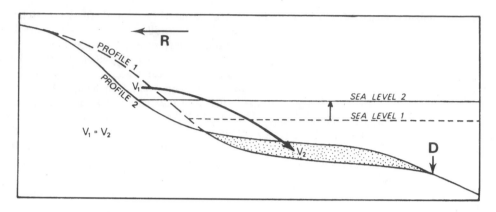

Figure 5.31   The Bruun Rule states that a phase of sea level rise on a sedimentary coast will be accompanied by recession of the beach as a volume of sediment ($V_1 = V_2$) is transferred from the backshore to the nearshore zone in such a way as to restore the transverse profile landward from D, the initial seaward boundary of nearshore sand deposits.

with an equivalent sea level rise on a rocky shore of similar profile.

This has become known as the Bruun Rule. It has been reproduced in wave tank experiments (Schwartz 1965) and is supported by observations of the backshore erosion and nearshore deposition that have taken place on beaches around the Great Lakes during and after each episode of rising water level, and on Cape Cod beaches between low neap and low spring tides (Schwartz 1967), but it is a two-dimensional model that has limitations in practice (p. 281).

Beach erosion has become widespread on coasts where the sea has been rising because land subsidence is in progress (Figure 3.10), as on the Gulf and Atlantic coasts of the United States. Land subsidence resulting from extraction of groundwater has accelerated beach erosion on the northern Adriatic coast of Italy, as on the Lidi of Venice and at Ravenna, and removal of oil and water from the Gippsland Basin may have contributed to erosion on the Ninety Mile Beach in Australia. Beaches were cut back suddenly on sectors of the Alaskan coastline that subsided during the 1964 earthquake.

Evidence from tide gauge records suggests that there has been a world-wide sea level rise of one to two millimetres per year during the past few decades, offset on some coasts by equal or greater land uplift, and varying also in relation to geophysical factors that complicate the surface topography of the oceans (p. 31). If there has indeed been a sea level rise the Bruun Rule may provide at least a partial explanation for the modern prevalence of beach erosion.

Diminution of fluvial sediment supply

On coasts where beaches have been supplied with sediment carried down to the coast by rivers, erosion is likely to follow any reduction in sediment yield to river mouths as a result of reduced runoff. Such a reduction may be due to diminished rainfall or less melting of snow and ice within the river catchment, but is more often a sequel to the building of dams to impound water upstream. These have intercepted the discharge of fluvial sediment, cutting off the supply of sand and gravel to beaches at and near the river mouth. The result is the onset of erosion on beaches that were formerly maintained or prograded by the arrival of this sediment. Erosion will develop more quickly, and become more severe, on coasts where there is also strong longshore drift of sediment away from river mouths.

The best known example of such erosion is on the shores of the Nile delta (p. 260), where sandy beaches had been prograding for many centuries as the result of the longshore spread of sandy sediment delivered to the mouths of Nile distributaries. Erosion of beaches near the mouths of the Rosetta and Damietta distributaries was first noticed early in the twentieth century, soon after barrage construction began upstream in 1902. It became much more rapid and extensive after the completion of the Aswan High Dam in 1964, which resulted in large scale sediment entrapment in the Lake Nasser reservoir, and during the next few years, beach erosion on parts of the deltaic coastline attained annual rates of up to 120 metres (Sestini 1992). Some of the sediment removed from these beaches drifted eastward along the coast towards Port Said, but much has been lost offshore (Lotfy and Frihy 1993).

Similar beach erosion has occurred on other deltaic coasts following dam construction upstream. Examples include the Rhône delta in France, the Dnieper and Dniester deltas in the Ukraine, the Citarum delta in Indonesia and the Barron delta in north-eastern Australia (p. 259). On the coast of southern California, where relatively high wave energy has largely suppressed delta development, beaches nourished by longshore distribution of sand and gravel delivered by rivers have become depleted since dam construction reduced fluvial sediment yields. Barrages built on the Tijuana River in Mexico have retained about 600 000 cubic metres of sediment annually, and this has deprived beaches north of the river mouth of a sediment supply, so that erosion has ensued, particularly on Imperial Beach near San Diego (Paskoff 1994). Beach erosion on the Caucasian Black Sea coast has been attributed to the damming of rivers, and in northern Sweden, where most rivers have been harnessed for hydroelectric power production, the necessary dams have so diminished fluvial sediment yields that deltaic coasts such as that of the Indal River are eroding despite continuing isostatic uplift of the land.

Beach erosion near river mouths has followed the dredging of sand from river channels, as in the Tenryu River in Japan (Koike 1985), and the cessation of hinterland mining that had previously augmented the fluvial sediment supply to the beach. Examples of this can be seen at Par and Pentewan in south-west England, where beaches that prograded with convex profiles when mining waste was arriving are being reshaped and reduced by erosion and sediment attrition to concave profiles. Excavation of sand from the Rhine during the Second World War resulted in river-bed hollows, and while these were filling, a diminished sand supply to the delta region downstream contributed to the onset of erosion of beaches along the delta coastline.

Another cause of reduction of fluvial sediment yield has been successful soil conservation works (slope terracing, runoff management, reafforestation) in the hinterland, as exemplified by the rivers draining to the Gulf of Taranto in southern Italy. Diminished river flow during prolonged droughts has had similar consequences. In southern California, Orme (1985) found that beach erosion occurred during dry periods, especially in 1939–1968, the beaches being restored during wet years such as 1969, 1978, 1980 and 1983, when the fluvial sediment supply revived. In Queensland, Australia, the Burdekin delivered abundant sand to its mouths during floods, to be deposited as beaches and spits (p. 254), until the 1991–95 drought diminished fluvial runoff and sand supply to the coast. Beaches north of the river mouths were then eroded as south-easterly waves continued to move sand away northward.

The importance of a river maintaining a sediment supply to a beach is also illustrated where there have been natural or artificial diversions of the river mouth (p. 260). Beach erosion has occurred on the abandoned deltas of rivers that have been diverted to other parts of the coastline, where a new delta is forming, as on the Hwang Ho delta in China after the river diverted naturally during an 1852 flood. On the coasts of Greece and Turkey beach erosion has resulted from a diminished sediment yield from rivers because of long-continued soil erosion in the river catchments, where increasing areas have been stripped of all unconsolidated material, leaving bare rock widely exposed.

On the world scale much beach erosion has been caused by diminished fluvial sediment yields, but there has also been extensive erosion on beaches remote from river mouths, and for this some other explanation must be sought.

Reduction in sediment supply from cliffs

The supply of sediment from erosion of cliffs and foreshore rock outcrops may diminish for various reasons. Cliff erosion is often preceded and accompanied by subaerial weathering, and runoff after heavy rainfall may cut ravines or gullies in the cliff face, as at Black Rock Point on the shores of Port Phillip Bay, Australia (Bird 1993a). Sediment slumps or is washed down from the cliff on to adjacent beaches in rainy periods but if annual rainfall diminishes, or runoff is controlled (e.g. by the insertion of drains along the cliff crest to intercept the water flow and prevent it flowing down the cliff face) the supply of sediment to nearby beaches will diminish, and they will be depleted.

Stabilization of coastal landslides has a similar effect, as has the cessation of coastal quarrying, halting the supply of waste material to the beach. Thus beaches at Porthallow and Porthoustock on the coast of southwest England were augmented by quarry waste between 1878 and 1985, but diminished after the quarrying ceased (p. 110). It remains to be seen whether the recent revival of quarrying here will restore them.

A decline in the strength and frequency of wave attack on cliffs may occur as a result of a climatic change leading to calmer conditions in coastal waters, or when waves are reduced by nearshore shoaling or the growth of reefs. Zenkovich (1967) argued that if sea level remains unchanged, the rate of cliff recession will eventually decline as the result of the widening of the shore platform cut in front of them, across which wave action gradually weakens, and this implies a diminishing sediment yield from cliffs to beaches.

The commonest cause of a reduced sediment yield from cliffs to beaches is the construction of sea walls along the cliff base. Beach erosion at Bournemouth in England occurred during the progressive extension since 1900 of a promenade on the seafront. This was intended to halt cliff recession and preserve coastal properties, but it also cut off the sand supply which had come from the cliffs, and with little (if any) replenishment of the sand and gravel carried away by longshore drift the Bournemouth beach gradually diminished.

On the shores of Port Phillip Bay near Melbourne, Australia, similar beach erosion is a sequel to the walling and stabilization of formerly eroding cliffs of Tertiary sandstone (Bird 1993a). Other examples of beaches depleted as the result of the building of sea walls on adjacent cliffed sectors include Ediz Hook, in the state of Washington (Schwartz and Terich 1985), and Byobugaura in Japan (Sunamura 1992).

In recent decades erosion has become more severe and extensive, on the southern shore of Hurst Castle Spit, where reduction of the supply of sand and shingle because of stabilization of Barton Cliffs to the west, accelerated beach erosion. On the Norfolk coast Clayton (1989) found that erosion of 33 kilometres of cliffs in Norfolk, averaging 25 metres in height, had under natural conditions yielded 500 000 cubic metres of sediment per year, about two thirds of which was sand and gravel supplied to beaches extending for more than 60 kilometres downdrift. With artificial stabilization of 70 per cent of these cliffs during the past century (many of the coast protection works had proved to be only partially effective in halting cliff erosion) the sediment yield has been reduced by 25 per cent to 30 per cent, and beach erosion is spreading downdrift.

In situations where cliff outcrops of sandstone or conglomerate, yielding sand or gravel to nearby beaches, are backed by contrasted rock types, such as a massive resistant formation, or soft silts and clays that do not yield beach-forming sediment, cliff recession may eventually expose these. Beaches formerly derived from the sandstone or conglomerate may then no longer be maintained.

Reduction of sand supply from inland dunes

Beaches that have been supplied with sand from dunes spilling from the land on to the shore may start to erode if the sand supply is reduced or terminated because the dunes have become stabilized. This may result from the natural spread of vegetation, or from conservation works such as the planting of grasses or shrubs, the spraying of chemicals such as bitumen, or rubber compounds, or paving, road-making and building over the dune surface. Alternatively, the sand supply may run out because the whole of an available dune has moved on to the shore.

Examples of beach erosion resulting from diminished aeolian sand supply have occurred on the south-facing Cape Coast of South Africa, where the prevailing westerly winds are driving dunes along the coast, as at Sundays River. On Phillip Island, Australia, the partial stabilization of dunes that had been spilling eastward across the Woolamai isthmus to nourish the beach in Cleeland Bight (Figure 2.4) was followed by beach erosion (Bird 1993a).

Diminution of sediment supply from the sea floor

There is much evidence, particularly on oceanic coasts, that sea floor sand or gravel drifted shoreward during and since the Late Quaternary marine transgression. As sea level rose across the continental shelf, waves reworked sediment that had previously been deposited by rivers or wind action, and eroded material from weathered rock outcrops. The coarser components were swept shoreward, and deposited as sand or gravel beaches. As the marine transgression came to an end continuing shoreward drift prograded the beaches, often forming successive backing beach ridges and parallel dunes (Figure 5.5A–C).

Shoreward movement of sand is still continuing where sediment is being washed in by waves and currents from nearshore shoals, but on many coasts this supply of sediment declined as the transverse nearshore profile became smooth and concave and beaches no longer received sediment from the sea floor. If there is no compensating input of sediment from other sources (such as rivers) progradation comes to an end, and with continued input of wave energy the transverse nearshore profile migrates landward, so that the beaches are progressively consumed. Many beaches that prograded earlier in Holocene times have passed into this condition, and are now being cut back by erosion, continuing wave action driving the transverse concave profile landward (Figure 5.5D). This may be the explanation for beach erosion on the shores of St Ives Bay and at Newquay on the north coast of Cornwall and at Braunton Burrows in Devon.

The onset of erosion has come at different times in different places because the development of the concave profile and the cessation of shoreward drift of sediment from the sea floor have not occurred simultaneously along the coastline. Tanner and Stapor (1972) described this change on beaches as a transition from an economy of abundance to an economy of scarcity of sediment supply, and the sequence has been described as a maturing of the system. They illustrated the sequence of events by reference to the erosion that had developed on the beach ridge plain at Cabo Rojo, south-east of Tampico in Mexico. It also provides an explanation for erosion on the Ninety Mile Beach in south-eastern Australia, which borders a sandy coast formerly prograded by accretion of sand supplied from the adjacent floor of Bass Strait, and is now being cut back by marine erosion (Figure 5.32). There is still plenty of sand in the nearshore area, but the transverse profile has become smooth and concave, and there is no longer shoreward drift of sand to this beach.

Beach erosion could also result from diminished production of shelly deposits washed in from the sea floor because of ecological changes, such as the destruction of shell fauna by pollution. It is possible that this has reduced shelly beaches fringing the shores of Corio Bay, near Geelong, Australia. The supply of sand and shingle from the sea floor has diminished where increased growth of seagrasses or other marine vegetation has impeded shoreward drift and trapped the sediment offshore.

Extraction of sand and shingle from the beach

Where sand or gravel has been removed from a beach for use in road and building construction the shore profile has been artificially lowered, allowing larger waves to attack the beach more strongly during storms. This has increased beach erosion at Klim, on the north

Figure 5.32 Erosion (arrowed) of the Ninety Mile Beach, on the outer barrier of the Gippsland Lakes in south-eastern Australia, with accretion (+) on either side of the Lakes Entrance breakwaters (see Figure 5.24).

coast of Jutland, Denmark, and there has been similar depletion of beaches on the south coast of England, where pebbles have been quarried from the western end of Chesil Beach at West Bay on the Dorset coast for use in industrial filters, and from nearby Seatown for ornamental purposes. Reference has been made to coastal erosion at Hallsands in south Devon following the extraction of gravel from Start Bay in the 1890s for use in the building of Plymouth Dockyard (p. 74). Beaches in Jersey and Guernsey were much reduced by the extraction of sand from beaches by the German occupying forces during the Second World War for use in building bunkers and gun emplacements, and increased backshore erosion has led to the construction of massive sea walls. Beach and backshore erosion resulting from sand extraction has occurred on Casuarina Beach, near Darwin in northern Australia. Beach quarrying may be less damaging where the sand or gravel is extracted from a sector that is prograding, notably as the result of updrift accretion alongside a breakwater, as at Timaru in New Zealand.

Beaches have been depleted by the extraction of shell sand and gravel for agricultural use as lime on farmland in Cornwall, where this is permitted below high tide level by an Act of Parliament dating from 1609. Such extraction has traditionally been on a small scale (50 to 100 tons per year) from several beaches, notably at Bude, where Summerleaze Beach has been depleted, but recent use of bulldozers to take large quantities from the beach at Poldhu, near Mullion, has severely depleted this beach and called the historical right into question. Erosion has taken place where shelly material has been taken for aggregate and lime making from beaches on islands in the Hebrides and at Kinnego Bay in County Donegal, Ireland. It is sometimes assumed that because shelly material is of marine origin it will be naturally replenished, but it is necessary to investigate the relative rates of extraction and biogenic replenishment to be sure that extracted material will indeed be naturally replaced.

Some beaches have been depleted by the extraction of mineral sands, such as rutile, tin or gold, or of diamonds, as in Namibia. In general beach quarrying has led to instability, but the effects depend on the rate of extraction and the size of the beach compartment. On some beaches the effects are almost instantaneous, on others they may take several years to become obvious. Harvesting of seaweed from beaches results in losses of adhering sand and pebbles, as when kelp is extracted from beaches on the coast of King Island in Tasmania and in California (Hotten, 1988).

Intensively used beaches at seaside resorts gradually lose sand as it is removed by visitors to the beach, adhering to their skin, clothes or towels, or trapped in their shoes. The quantities are small, but the losses are cumulative and no one brings sand to the beach. Pebbles and shells are also carried away as souvenirs by beach visitors. Regular beach cleaning operations, when bulldozers or tractors scrape or sweep seaweed and litter from the beach, lower the beach profile by compaction and also remove sediment, particularly sand. Samples taken from heaps of bulldozed seaweed on the beach at Beaumaris in Victoria, Australia, were found to contain up to 20 per cent (dry weight) of sand, and it was estimated that annual beach cleaning was removing up to 0.5 cubic metres of sand from each 100 square metres of beach surface.

Increased wave energy

Increased wave attack resulting from the deepening of nearshore water because of a rise in sea level relative to the land has already been discussed (p. 147) but nearshore water can also deepen when a shoal is removed. Such shoal migration has caused beach erosion on Benacre Ness on the Suffolk Coast, and the same effect was observed on Matakawa Island, New Zealand, when beach erosion followed the shoreward movement of a tidal channel in the approaches to Tauranga Harbour (Healy 1977). Similar features have been noted in French Guiana and Surinam (Psuty 1985). Deepening of the nearshore zone off Rhode Island, in the United States, during the 1976 hurricane permitted larger waves to reach the coastline, accelerating subsequent beach erosion (Fisher 1980).

Nearshore dredging also deepens the water, so that larger waves reach the shore. In Botany Bay, Australia, beach erosion accelerated at Brighton-le-Sands after the bay floor was dredged to provide material for the extension of a runway at Sydney International Airport. Nearshore dredging of algae and seaweed around the coasts of Brittany, in France, contributed to erosion of beaches in the Bay of the Seine (Cressard and Augris 1982). The cutting of a trench across the sea floor in St Ives Bay in Cornwall in 1994 to lay a sewage outfall was followed by erosion on nearby Porthminster Beach. At Pakiri in New Zealand, Hesp and Hilton (1996) noted weak recovery of beach and dune profiles after storms on a beach where nearshore sand extraction had occurred. On the Arctic coast of Russia increased beach and cliff erosion have been attributed to larger waves arriving as the result of nearshore deepening due to downwarping of the adjacent sea floor.

Destruction of nearshore and fringing coral reefs may initially increase the sediment supply to adjacent beaches, so that a phase of progradation ensues, but as the reefs disintegrate the nearshore water deepens and increasing wave action leads to beach erosion. Beach erosion at Colombo in Sri Lanka was partly due to greater exposure to wave attack following the decay of an old reef a short distance offshore, and there is erosion on beaches backing damaged fringing coral reefs on the Perhentian Islands off north-east Malaysia, and on the east coast of Lombok, Indonesia, where reefs have been destroyed by illegal use of explosives by fishermen.

Interception of longshore drift

Where breakwaters have been built to stabilize river mouths or lagoon entrances and improve their navigability, or create boat harbours, sand or gravel drifting alongshore have been intercepted on the updrift side, and there is beach erosion on the downdrift side. There are many examples of this. At South Lake Worth, Florida, breakwaters have intercepted longshore drift of sand from the north to prograde the beach updrift while to the south beaches deprived of their longshore sediment supply were eroded. At Tillamook Bay, Oregon, a breakwater built north of the inlet has trapped southward-drifting sand, and erosion has become severe downdrift on Bayocean Spit (Terich and Komar 1974).

At Lagos in Nigeria breakwaters built in 1912 to maintain a navigable entrance to Lagos Lagoon intercepted eastward-drifting sand on Lighthouse Beach, to the west, and resulted in erosion on Victoria Beach, downdrift of the harbour (Usoro 1985, Ibe 1988). By 1975 the coastline had advanced more than 1300 metres seaward alongside the break-water, but Victoria Beach had retreated by up to 1300 metres (despite nourishment) (Figure 5.33). Eventually sand accretion will extend out to the end of the break-water, whereupon the natural eastward drift to Victoria Beach will resume, but there will then be problems in maintaining a navigable entrance to the port of Lagos (Ibe et al. 1991).

Until about a century ago there was a shingle beach, maintained by eastward drifting, beneath the chalk cliffs between Dover and Deal but this has almost disappeared as the result of the interception of shingle drifting alongshore by the breakwaters at Dover Harbour. Protruding areas of land claimed from the sea can have a similar effect to breakwaters. At Map Ta Phut, near Rayong on the coast of Thailand, a wide protrusion of land claimed for port and industrial development has led to updrift accretion of beach sand and erosion downdrift. Interception of drifting shingle by groynes to retain a beach for the seaside resort of Brighton in England was followed by beach depletion downdrift to the east. The longshore supply of sand and shingle can also be interrupted by the growth of a fringing coral reef or some other depositional feature, as at Wewak in New Guinea. At Channel Islands Harbour in California an offshore breakwater induced accretion on the beach in its lee, but reduced the supply of sand and gravel to downdrift shores, resulting in beach erosion.

A change in the angle of incidence of waves

Beach erosion can be initiated by a change in the angle of approach of the dominant waves, either because of breakwater construction or because of growth of near-shore reefs or islands or the formation or removal of shoals. A change from a largely swash-dominated beach alignment to a drift-dominated beach alignment accelerated longshore drifting at Kunduchi in Tanzania, where the sandy coastline had prograded under the dominance of easterly swell, but the growth of reefs and shoals reduced the effectiveness of these ocean waves, and allowed locally-generated and previously sub-dominant south-easterly waves to supervene, resulting in severe erosion (Bird 1985a). A similar change in wave incidence following the construction of the Portland Harbour breakwater in Victoria, Australia, may have contributed to the onset of beach erosion at adjacent Dutton Way (Bird 1993a). Extension of the harbour breakwater at Albissola on the Ligurian coast of Italy modified the angle of wave approach in such a way that the adjacent sandy beach at Albissola Marina was re-shaped, the eastern part being eroded as the western part widened by accretion (Piccazzo et al. 1992). At Sorrento on Port Phillip Bay, Australia, Point King Beach has been re-shaped, with erosion at the eastern end and accretion at the western end, as the result of oblique waves generated by passing ships.

Intensification of obliquely-incident wave attack

Wave attack on one sector of a beach may intensify as a result of the lowering of the beach profile on an adjacent sector, allowing stronger waves to arrive obliquely, and thus accelerate beach erosion. This

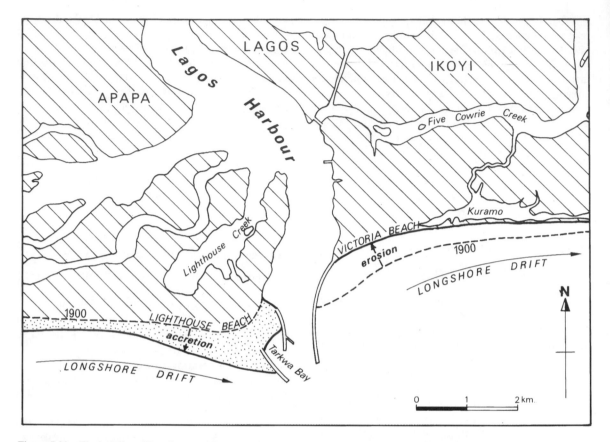

Figure 5.33 The building of breakwaters alongside the entrance to Lagos Harbour, Nigeria, interrupted the eastward longshore drift of sand in such a way as to cause extensive accretion updrift on Lighthouse Beach and erosion downdrift on Victoria Beach.

often occurs after sea wall construction, the lowering of the beach profile by reflected storm waves allowing larger oblique waves to attack the neighbouring coastline. The outcome is that beach erosion spreads alongshore, and if sea walls are extended laterally to counter this, a 'domino sequence' of cumulative beach erosion is produced, with each new sector of sea wall on a set-back alignment. This has happened at Point Lonsdale in Australia (Bird 1993a).

Increased losses of beach sediment to the backshore

Losses of sediment from the beach face to the hinterland occur when onshore winds blow sand inland (Figure 5.34), or when storm surges wash beach sand and gravel on to the backshore, or over into lagoons, swales or swamps. Alternatively, sediment brought alongshore by winds, waves and currents may be swept into lagoon entrances or river mouths. If losses

from the beach face are not compensated by the arrival of fresh supplies of beach sediment the profile is lowered and the coastline recedes.

In South Australia, beach erosion accelerated after landward movement of backshore dunes followed reduction of their vegetation cover by burning, grazing and trampling. The lowered backshore was then cut back more quickly because of the diminished volume of sand to be removed by wave attack. Overwash during storm surges has eroded beaches by driving back sandy barrier islands (p. 177) on the Atlantic coast of the United States (Fisher 1980), and shingle formations at Blakeney Point and Chesil Beach in England (Steers 1964).

Increased storminess

An increase in the frequency and severity of storms in coastal waters may result in the erosion of beaches that

Figure 5.34 Sand is moving inland from a beach as drifting dunes on the shore of Encounter Bay, South Australia. The beach is dissipative, with waves losing energy as they break across a shallow nearshore zone.

were previously stable or prograding. Beach profiles are cut back and steepened by storm waves until they attain a concave form adjusted to the augmented wave energy. A series of storms in quick succession is particularly destructive because the second and subsequent events occur on beaches already reduced to a concave eroded profile. Worsening beach erosion on shores of the North Sea and on the Atlantic coast of the United States in recent decades may be partly due to an increasing frequency of storms, but detailed long-term weather records are necessary to demonstrate that storminess has indeed increased and it is difficult to separate this factor from other causes of beach erosion. Tsunamis may have a similar effect to storms, and where very large waves steepen the nearshore profile and deepen nearshore water beach erosion will ensue.

Attrition of beach material

On relict beaches (no longer receiving a sediment supply) agitation of the beach by wave action leads to gradual attrition, and so to a reduction in volume of beach sediment. Erosion of Four Mile Beach, North Queensland, Australia, occurred after the fluvial sand supply (from the Mowbray River) was cut off by coral reef growth, and the relict beach sediment has been reduced to very fine sand by attrition. The beach has become compacted and is now firm enough to land an aircraft, drive a bus or car, or ride a bicycle. As sediment calibre is reduced, such beaches are more likely to lose the increasingly fine sediment by winnowing and removal, either landward into backshore dunes or seaward to bars and bottom deposits. The gradual lowering and flattening of the profile of Four Mile Beach has been accompanied by increased penetration by waves, and the onset of erosion along the seaward margin of backshore dunes. On the Hebridean island of Barra in Scotland the volume of a shelly beach was reduced because it was crushed and compacted by vehicles driven along it. As it became lower and flatter, the upper beach was eroded by increased wave scour (Figure 5.7).

Erosion due to beach weathering

Beach erosion can result from a reduction of beach volume by chemical weathering, including the decay and removal of ferromagnesian minerals and the dissolving of calcareous beach sand grains in rainwater, stream seepage or sea spray. Solution proceeds more rapidly on calcareous beaches in high latitudes because cold water is less saturated with carbonates, and thus more corrosive (p. 78). As beach volume diminishes the beach profile is lowered and larger waves attack the backshore. This factor has probably contributed to the beach erosion at Port Douglas, mentioned in the previous section.

Erosion because of increased scour by wave reflection

Waves breaking against a solid structure, such as a sea wall built of concrete, stone blocks, steel sheeting or timber, are reflected, and form seaward currents that carry sediment away from the foot of the wall (Figure 5.35). This has been observed on many coasts where sea walls have been built behind a beach to halt cliff recession (Figure 5.36) (Kraus and Pilkey 1988, Tait and Griggs 1990). Storm waves at high tide then overwash the beach to splash against the wall, and their reflection causes further beach erosion and the eventual undermining and destruction of the sea wall. The outcome is usually the building of a new set-back sea wall, but at Porthcawl in Wales in 1934 a new sea wall was set forward of one that had been ruined (Carter 1988). Boulder ramparts are less reflective, some of the wave energy being absorbed by percolation between the boulders, but beaches in front of them are nevertheless depleted.

On the Queensland coast in Australia the sandy beach at the resort of Surfers Paradise was reduced by a series of cyclones in the 1960s and a large boulder wall was built to safeguard the coastal hotels and apartments. The boulder wall then caused scour by reflecting the waves, which lowered the beach further, and prevented the natural restoration that had occurred after previous such erosion by cyclones, before the boulder wall was inserted.

Reflection scour following the construction of sea walls has lowered beaches in Jersey in the Channel Islands. On the west coast St Ouen's Bay had a wide sandy beach, backed by dunes, formed as the result of shoreward drift of sand during the Late Quaternary marine transgression (Figure 5.5A–C), but by the nineteenth century it was eroding as the result of a diminu-

tion in sea floor sediment supply (Figure 5.5D), and the building of a sea wall to halt coastline recession resulted in further lowering and flattening of the beach (Figure 5.5E), which is now submerged at high tide and remains wet with groundwater seepage at low tide.

Erosion accompanying migration of beach lobes

Where beach lobes (p. 121) form and migrate downdrift along the coast, there is accretion as each lobe arrives, but erosion as it moves on. At Somers, on the coast of Westernport Bay, Australia, a yacht clubhouse was built on one such lobe in the 1970s, and is threatened by the beach erosion that has developed as that lobe moved on.

Erosion due to a rise in the beach water table

It has long been known that a wet sandy beach is eroded more rapidly by wave action than a dry one. Analysis of changes over 95 years at Stanwell Park Beach, near Sydney, Australia, by Bryant (1985) identified rises in the level of the beach water table as a contributory cause of beach erosion. Such a rise in beach water table may be due to the ponding or diversion of river or lagoon outlets, to unusually heavy or prolonged rainfall or to increased discharge following land use changes in the hinterland. Attempts have been made to reduce erosion and stabilize beaches by pumping water out of them, but the effectiveness of dewatering has yet to be convincingly demonstrated (Turner and Leatherman 1997). In Estonia sandy beaches lowered to the water table have been invaded by reed-swamp vegetation.

Erosion due to removal of beach material by runoff

Beach erosion can occur as the result of runoff during a period of heavy rain, or the melting of snow or ice, particularly from a backing cliff or steep slope. Beach sediment can be swept into the sea by strong runoff issuing from a stream or drain. These effects are stronger on sandy beaches, especially if they are already wet, than on gravel where runoff disappears more quickly by percolation. The seepage mentioned in the previous section also contributes to beach erosion by washing sediment seaward as the tide ebbs. Increased runoff is often due to urbanization and the construction of roads and other sealed surfaces from which water runs off quickly, instead of percolating into the subsoil, as it did before these structures were built.

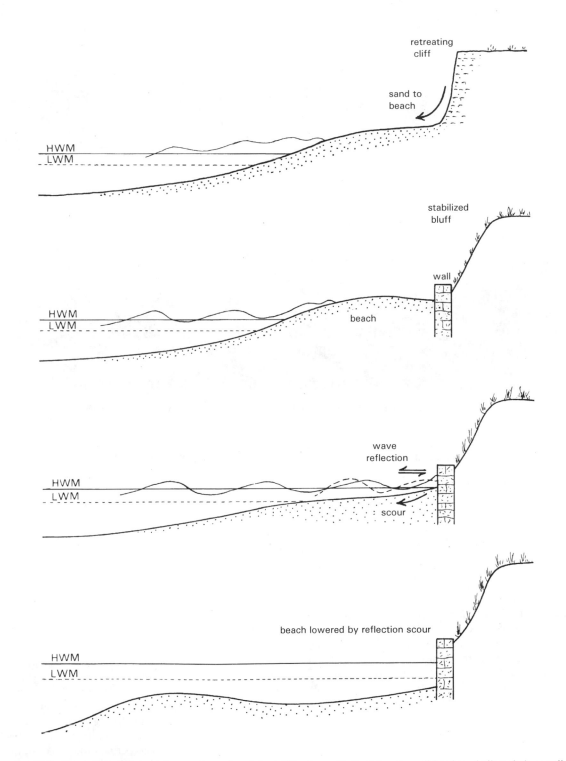

Figure 5.35 A receding cliff supplying sand to a beach is stabilized as a bluff after a sea wall has been built to halt coastline recession. Reflection of waves from the sea wall then causes beach erosion.

Figure 5.36 A sea wall and stabilized bluff at Black Rock in Port Phillip Bay, Australia, where the beach has disappeared as the result of reflection scour (as in Figure 5.35).

Erosion because of diminished tide range

Erosion by waves is more effective where their energy is concentrated at a particular level, rather than dispersed by the rise and fall of a substantial tide. It follows that a diminution of tide range will increase the effectiveness of wave action, initiating or accelerating beach erosion. Examples may be found on the shores of inlets, estuaries or lagoons that become partly or wholly cut off from the open sea by the growth of spits, or the building of structures such as weirs or barrages, so that tidal ventilation is impeded, or excluded altogether, and wave action is intensified at a particular level. This has been observed on beaches fringing coastal lagoons at the mouth of the Murray River in South Australia, which were formerly estuarine and tidal, but were separated from the sea by the construction of barrages in 1940 (p. 241).

Erosion by driftwood

Some beaches are littered with branches and trunks of undercut trees, or lumber piled on the shore. Driftwood is extensive where hinterlands are forested, as in Canada and on the Pacific coast of the United States, but it also occurs where trees have been undercut along the coast, and have fallen to the beach or floated along to nearby beaches. Driftwood piled parallel to the coastline can act protectively by impeding wave scour but where fallen trees and branches are jostled by wave action, and act as levers or battering rams, they contribute to the erosion of sand or gravel from the beach.

Erosion following removal of a sea ice fringe

On cold coasts, as in Alaska, northern Canada and Siberia, beaches are protected in the winter by the formation of a fringe of shore ice, and are subject to wave action only during the brief summer thaw. Beach erosion will increase if the climate becomes warmer, and the summer lengthens, permitting waves to reach these shores for a longer period. Many Arctic and Antarctic beaches rest on ice, or have interbedded ice (permafrost), and when this melts collapse depressions form in the beach surface and the beach sediment is loosened. With a warming climate beaches that were formerly cemented as frozen ground are mobilized by wave action, and may be removed.

THE MULTIPLE CAUSES OF BEACH EROSION

No single explanation can account for the modern prevalence of erosion of the world's beaches, or indeed for the onset or acceleration of erosion on any particular beach. It is not simply the outcome of human activities, artificial structures, a sea level rise, an increase in storminess of coastal waters or the maturing of the system as the sediment supply from the sea floor dwindled during the Holocene stillstand. Each of these factors may have contributed to beach erosion, to an extent which differs from place to place.

An example of multicausality in beach erosion is seen on the Lido di Jesolo, the beach fronting the Lagoon of Venice on the Adriatic coast. Since this island was developed as a resort in the 1950s beach erosion has been severe. The causes include coastal subsidence, augmenting a rising sea level in the Adriatic, nearshore deepening, and changes in current flow and wind and wave regimes, all of which have curtailed sandy supply from the sea floor and favoured increasingly energetic wave attack. In addition, hinterland reafforestation, reservoir construction and excavation of sand and gravel from river channels have diminished the former sediment yield from the Piave River to this coast. Artificial structures include groynes, which have reduced longshore drift, and concrete and boulder sea walls, which have increased reflection scour (p. 156).

The task of ranking the relevant factors and apportioning their contribution to beach erosion on a particular coast requires investigation of past and present patterns and rates of change on beaches and the process systems operating in coastal waters. An example of quantitative assessment of beach erosion was provided by Fisher (1980) from Rhode Island (Figure 5.37). He found that between 1939 and 1975 the beach-fringed coast had retreated at an average rate of 0.2 metres per year, in a period when sea level rise averaged 0.3 centimetres per year, and calculated that 35 per cent of linear beach recession over this period had been due to the washing of sand into tidal inlets and 26 per cent to losses of sand washed or blown over the barrier islands to form migrating dunes and washover fans. This left 15 per cent of the beach retreat accountable as the direct result of submergence, and 24 per cent lost by transference of beach sand seaward. Movement of sand over and between these barrier islands indicates that they were continuing the long-term landward migration, accompanied by the transgression of backing lagoons and marshes, that has characterized much of the Gulf and Atlantic coastline during the Holocene. A similar sequence of evolution is now being observed on the submerging coasts of the Caspian Sea.

The modern prevalence of beach erosion calls into question the idea that beaches are naturally stable, having attained some kind of equilibrium with the processes at work on them. This is difficult to sustain from geological evidence, for beaches form only a very minor proportion of the sedimentary formations found in

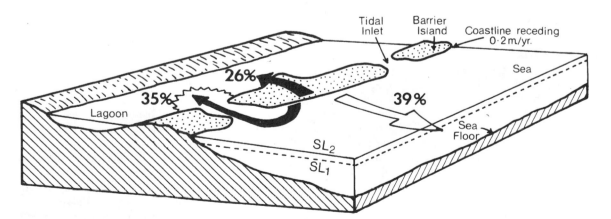

Figure 5.37 The distribution of sediment losses accompanying recession of sandy barriers in Rhode Island, United States, based on Fisher (1980).

the geological column. They have been preserved in Tertiary deposits in Texas, but they are not prominent in most depositional sequences. This is because they have been transient features, formed and reworked by waves on coasts destined either to retreat as marine planation proceeds (reducing land masses ultimately to surfaces planed down to the limit of wave erosion) or to be submerged by the rising sea and buried by younger sediments. Beaches have thus been eroded and dispersed, and it could be argued that beach erosion is natural, and that the anomaly is the widespread progradation that occurred in Holocene times, related to the unusual conditions of a stillstand following a major marine transgression (Clayton 1989). The attempt to preserve existing beaches should be considered in terms of this geological perspective.

EFFECTS OF ARTIFICIAL STRUCTURES ON BEACHES

A common response to beach erosion, especially where it threatens to undermine and destroy developed property such as roads or buildings, has been to construct, extend and elaborate sea walls. These are intended to prevent wave attack on the eroding coast, usually a receding cliff, an undermined and slumping bluff, or a truncated dune, sometimes fronted by a beach that is insufficiently high and wide to prevent waves reaching the back of the shore. Often they are initially banks of earth excavated from a parallel ditch, but when these are damaged by storm waves they are replaced by more solid stone or concrete structures. Solid walls that are designed to withstand the force of the breaking waves inevitably reflect these seaward, as backwash that scours away the beach. Boulder ramparts, also known as revetments or riprap, or artificial structures such as tetrapods, made of reinforced concrete, are less reflective than solid sea walls, but still cause backwash erosion on their seaward sides. Artificial structures designed to protect the coastline, form an increasing proportion of the world's coastline (Walker 1988). In Japan long sectors of the cyclone-prone east coast of Honshu are now protected by large sea walls and concrete tetrapods weighing up to 50 tonnes each.

Demands for the halting of coastal erosion have led to littering of the world's coastline with an array of artificial structures of various kinds. Some have been successful, but many have failed and are derelict. Some have helped to protect and maintain beaches, but many have resulted in further beach erosion. In recent decades attention has been given to artificial beach nourishment, particularly in the United States, Western Europe and Australia, as a means of countering beach erosion.

BEACH NOURISHMENT

Beaches that have been depleted by erosion can be restored by dumping sand or gravel on the shore (Schwartz and Bird 1990). Beach nourishment (replenishment, restoration, recharge, reconstruction, or fill) is artificial in the sense that the sediment has been brought to the shore by engineers. Beach renourishment is the maintenance or replenishment by deposition of suitable sediment on a beach that has been previously nourished, and the term artificial beach should only be used where there was previously no natural beach, as at Ibiza in the Balearic Islands and Praia da Rocha in Portugal (Psuty and Moreira 1990).

Beach nourishment aims to restore beaches that have been depleted by erosion, and to create a beach formation that will protect the coastline and persist in the face of wave action. Many beach nourishment projects have been at seaside resorts that had eroding beaches, and wanted them restored for recreational use, but in recent years increasing awareness of the importance of beaches in absorbing wave energy has resulted in the use of nourished beaches to protect the coastline and prevent further cliff erosion or damage to coastal property.

Prior understanding of coastal geomorphology is essential in planning a beach nourishment project. Preliminary research is necessary on the movement of sand and gravel in relation to incident wave regimes and the effects of any artificial structures on the shore sector to be treated. It is necessary to know why the natural beach was eroded and where the sediment has gone: landward, seaward or alongshore. Modelling of beach forms and processes can guide a beach nourishment project but the complexity and variability of coastal systems are such that an experimental approach, based on accumulated experience, may be more realistic than theoretical modelling (Pilkey and Clayton 1989).

Sediment used for beach nourishment can be brought from inland, alongshore or offshore sources and deposited mechanically or hydraulically on the shore (US Army Corps of Engineers 1984). Almost any kind of durable sediment of suitable grain size can be used for beach nourishment, but it should be at least as coarse as the natural beach sediment. The dumped sediment should be durable, not quickly reduced by weathering or abrasion.

Most beach nourishment projects begin by depositing sediment to form a beach terrace, which is then reshaped by waves and currents towards a natural profile, often with sand bars just offshore. It is necessary to deposit more beach material than is required to restore a beach to its natural dimensions (i.e. to form a terrace that is higher and wider than the natural beach), in order to allow for expected losses onshore, offshore or alongshore. A nourished beach may be held in place by building a retaining breakwater or a series of groynes.

It may be possible to nourish a beach by dumping sediment in places where longshore or shoreward drift will deliver it to the shore. This requires knowledge of the direction and rate of longshore and shoreward drift, taking account of variations in coastal aspect (Bird 1991). Offshore breakwaters can be used to create a pattern of refracted waves that will concentrate sand deposition and prograde the beach in their lee, as at Port Hueneme in California. A floating breakwater anchored off successive sectors of the shore can be used to induce local accretion of sand and gravel by shoreward drift of sediment to nourish a beach in stages along the coast. As the breakwater is moved away the accreted sand will begin to erode, and so a cyclic system of offshore breakwater placement is necessary to maintain such a beach.

Changes on nourished beaches are often rapid. Sand or gravel carried away by longshore drift can be intercepted and brought back, a process known as recycling or recharging. At Rye in south-east England the Kent River Board has been trucking loads of shingle taken from alongside the breakwater at the mouth of the River Rother back round to Cliff End and dumping them to restore the depleted beach. Sediment that has been withdrawn from the beach to the nearshore zone, particularly during stormy phases, can be dredged and brought back to the shore, a procedure known as backpassing.

It is generally acknowledged that nourished beaches will be eroded by the same processes that depleted preceding natural beaches, and that they will have to be replaced at intervals. Monitoring of changes on and around nourished beaches provides an understanding of the processes that erode and distribute emplaced beach material, and can guide further beach management procedures, including the insertion of groynes, the introduction of regular renourishment updrift or the repeated restoration of the profile of a beach that loses sediment offshore. Experience gained from one beach nourishment project can be applied to another, provided that the geomorphological situation is comparable.

Examples of beach nourishment projects are provided by Bird (1996a), together with a review of the principles and problems of beach nourishment. The more technical details are dealt with in the Shore Protection Manual produced by the US Army Corps of Engineers (1984) and the Delft Hydraulics Laboratory (1987) manual on beach nourishment.

6

Spits and Barriers

INTRODUCTION

Depositional features closely related to beaches, and shaped by similar processes, include spits of various kinds, bars in the intertidal and nearshore zones, and barriers built offshore, or across inlets and embayments to enclose lagoons and swamps. The shaping of these various features will now be discussed.

SPITS

Spits are beaches built up above high tide level and diverging from the coast, usually ending in one or more landward hooks or recurves (Evans 1942, Schwartz 1972). They have grown in the predominant direction of longshore drift by waves arriving obliquely to the shore, and their outlines have been shaped largely by dominant patterns of wave action. Some spits are almost straight, like the southern part of Orfordness on the east coast of England, where the mouth of the River Alde has been deflected about 18 kilometres to the south and south-west, but most end in one or more recurves, representing earlier terminations, as at Hurst Castle spit (Figure 6.1). Recurves can be formed where two sets of waves arrive from different directions, as on Hurst Castle spit, or where one set of waves is refracted in deep water around the distal end of the spit, as at Cape Henlopen on the east coast of the United States, where northward-drifting sand has been built into a recurved spit shaped by south-easterly Atlantic swell refracted into the mouth of Delaware Bay. Traces of older recurves on the landward side mark former terminations of a spit that has grown intermittently, a feature well shown by Blakeney Point, on the Norfolk coast (Figure 6.2). Salt marshes or mangrove swamps often develop on the sheltered landward side of spits, notably between the recurves.

Sediment deposited to form a spit has usually come from an alongshore source, but shoreward drift may

also have contributed. Blakeney Point is built of sand and gravel largely derived from cliffs cut in glacial drift in the Sheringham district to the west, but also includes beach material washed in from glacial drift on the adjacent sea floor. A glacial moraine deposited at the margin of the Last Glacial ice sheet, which crossed the Norfolk coastline in this area, has been sorted and rearranged in the course of the Late Quaternary marine transgression and ensuing wave and tidal current action, and built into a spit. The westward growth of Blakeney Point is indicated by successive recurves, shaped by westerly and north-westerly wave action, and the main shingle bank has been driven landward by North Sea storm surges, athwart the successive recurves. Similar features are seen on Dungeness Spit in Washington state, formed downdrift from eroding cliffs of glacial sand and gravel, diverging from the coastline and backed by recurves (Schwartz et al. 1987). Spurn Head is a 4.5 kilometre long sand and gravel spit at the mouth of the Humber estuary that has grown southward by longshore drift produced by north-easterly wave action, and been driven westward by North Sea storm surges. Much of its sediment has been carried southward from the eroding glacial drift cliffs of Holderness, but some may have been washed in from the adjacent sea floor.

Spits have formed on the coast of Burghead Bay in north-east Scotland where gravel supplied by the River Findhorn has drifted westward along the shore as the result of waves arriving from the north and north-east to form spits with recurves at their western ends. To the west Whiteness Head on the southern shores of Moray Firth is another recurved spit of well-rounded shingle, derived from glacial moraine deposits, and carried westward by longshore drift.

Hurst Castle spit in Hampshire has been supplied with sand and shingle derived from cliff and shore erosion in Christchurch Bay and carried eastward along the coast by longshore drift; again, some sediment may have been washed in from the adjacent

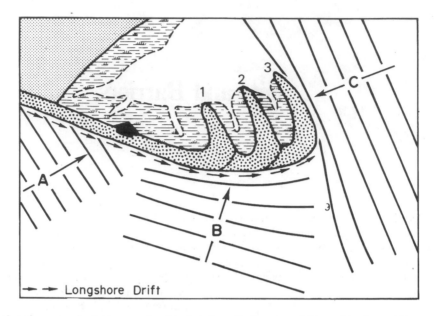

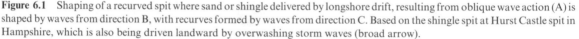

Figure 6.1 Shaping of a recurved spit where sand or shingle delivered by longshore drift, resulting from oblique wave action (A) is shaped by waves from direction B, with recurves formed by waves from direction C. Based on the shingle spit at Hurst Castle spit in Hampshire, which is also being driven landward by overwashing storm waves (broad arrow).

sea floor. The shaping of this recurved spit in relation to the direction of approach of dominant waves was demonstrated by Lewis (1931, 1938). Again, the main shingle bank has been driven landward by storm surges. King and McCullach (1971) used a computer model (SPITSIM) to simulate the processes at work here and examine the relative significance of waves approaching from various directions, the effects of wave refraction, and the influence of submarine topography, and then trace probable stages in its evolution to the present outline. When the simulation was extrapolated into the future it generated an additional elongated recurve extending north-westwards, but this is unlikely to form because of the presence of marshland in that direction. The value of this computer simulation was in identifying constraints that had not been considered, such as the morphology of the backing areas into which recurves might grow and the ways in which shoals, marshes or the backing coast can influence the shaping of a spit built by longshore drift.

Some spits have been widened by the accretion of beach material, usually forming successively-built ridges on the seaward side, and stages in their growth can be deduced from the ridge pattern. In a study of Cape Cod, Davis (1896) showed that sand eroded from cliffs of glacial drift had been built into a spit, with beach ridges marking stages in progradation at

Provincetown to the west (Figure 6.3). The beach ridge pattern showed that the spit had been re-shaped, partial truncation of the formerly prograded sector by marine erosion having released sand that formed smaller spits on the shores of Cape Cod Bay.

Many spits have beach ridge patterns indicating that they have been eroded along their outer shores, yielding sediment that has drifted alongshore to prograde their distal ends. Examples include Pointe de la Coubre and Pointe d'Arcay on the French Atlantic coast, and Farewell Spit, which has grown out from the northern end of South Island, New Zealand in this way, nourished by northward drift of fluvially-supplied sediment along the west coast of South Island, and by shoreward drift of sand from the sea floor (McLean 1978). It is now more than 30 kilometres long and up to 1.5 kilometres wide at high tide, with a wide beach backed by dunes on the seaward side and extensive intertidal sand shoals on the more sheltered landward side.

Spits have formed on the shores of deltas where sandy sediment delivered to a river mouth drifts alongshore, as on the flanks of the Ebro delta in Spain (p. 254). Sand spits on the western shore of the Mississippi delta have been derived from sectors of eroding deltaic coastline where sand has been sorted by wave action from finer material.

Evidence of the evolution of a spit can be obtained

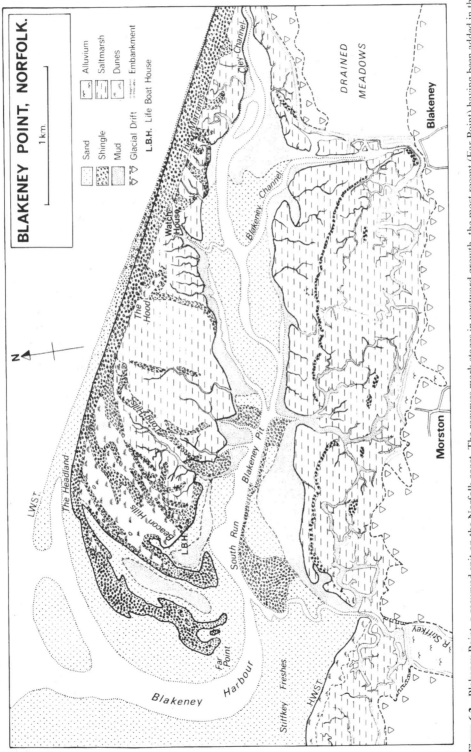

Figure 6.2 Blakeney Point, a recurved spit on the Norfolk coast. The recurves mark stages in westward growth, the most recent (Far Point) having been added in the 1970s. The spit shelters an estuarine harbour, with salt marsh on the southern side bearing sandy cheniers emplaced by waves during storm surges.

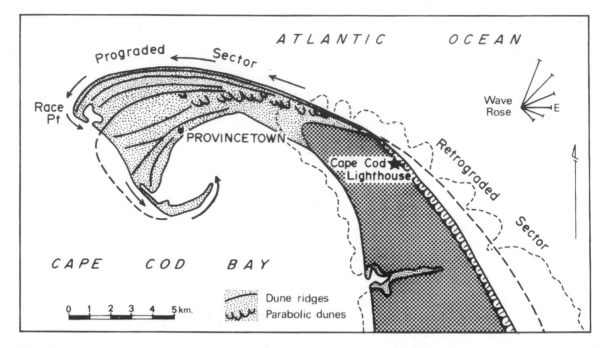

Figure 6.3   The spit at Cape Cod, Massachusetts, United States, has grown out from a peninsula of glacial drift, cliffed on its seaward side. Sand and shingle derived from the cliffs has drifted westward to form a broad spit with beach ridges marking stages in progradation. These have been truncated by wave action on the shores of Cape Cod Bay, with the formation of minor spits at Race Point and south of Provincetown.

from sequences of historical maps and air photographs. Maps made since about 1530 AD suggest that the shingle spit at Orfordness has grown about six kilometres during the past four centuries (Steers 1953), and stages in the evolution of Penouille spit, Gaspé, Quebec between 1765 and 1981 were traced from maps and air photographs by Fox et al. (1995). Historical information of this kind is rarely available outside Europe and North America, but eventually it will be possible to use successive surveys and series of dated air and satellite photographs to trace the evolution of spits. Carr (1965) used this kind of evidence to demonstrate short-term changes at the southern end of the Orfordness spit between 1945 and 1962.

The shape of a spit is influenced by the space available for its growth, and by the adjacent sea floor topography, as demonstrated by Schou (1945) in the Danish archipelago. Spits grow more rapidly across shallow nearshore areas than into deep water, and their configuration may be related to variations in fetch (exposure to wave action) determined by nearby headlands, islands or reefs. The evolution of many spits has been modified, or even halted, by the addition of

artificial structures. Dawlish Warren in Devon and Sandy Hook in New Jersey are examples of spits with complex histories that have been armoured by the building of sea walls intended to stabilize their present configuration.

Some of the best examples of spits are found on the shores of landlocked seas, lakes and coastal lagoons where sand and shingle carried along the shore have been deposited as spits where the orientation of the coastline changes, in forms related to prevailing wave conditions. Examples of this are found in the Danish archipelago, around Puget Sound, and on the New England coast, where in each case cliffing of glacial drift deposits has yielded sand and gravel for longshore drift to nourish nearby spit structures. In the Rade de Brest, a Brittany ria, wave-built sand and shingle spits built in tributary bays incorporate sediment eroded from the periglacial drift deposits (Head) that mantle bordering coastal slopes that have been basally cliffed (p. 54). Their outlines have been shaped largely by wave patterns related to the variations in fetch and wave incidence resulting from the configuration of the ria, but some have been truncated by strong tidal currents

(Guilcher et al. 1957). Truncation of spits by strong tidal currents occurs in narrow straits between the islands of Puget Sound on the north-west coast of North America, as at West Point, bordering Deception Pass in Washington.

Paired spits

On some coasts there are spits that have grown in different directions at different times. On Rattray Head in Scotland there is evidence of spit growth first to the south-east, and later to the north-west, implying a reversal of longshore drift, but it is possible that sediment came in from the sea floor and that the spits were largely swash-built. A similar alternation is seen at Sandwich in Kent, where the mouth of the River Stour was first deflected southward by the growth of the Stonar spit, then northward by a younger spit built along the shore of Sandwich Bay. This alternation may have resulted from a change in the pattern of dominant refracted waves from north-easterly to south-easterly as the result of migration of the intertidal Goodwin Sands offshore.

Paired spits often border river mouths and lagoon entrances. They have formed either as the result of convergent longshore drift or the breaching of a former coastal barrier. Spits of this kind border the entrances to tidal estuaries at the harbours of Poole, Christchurch and Pagham on the south coast of England (Robinson 1955) and Braunton Burrows beside the Taw–Torridge estuary in north Devon (Kidson 1963). These spits have been supplied with sediment from the sea floor, as well as from adjacent sectors of eroding coastline, and have been shaped by incident wave patterns. The paired spits bordering Pagham Harbour attained their present form as the result of the breaching of the shingle barrier during a storm in 1910. Paired spits are well developed alongside the tidal entrances between barrier islands, as at the tidal entrances to Corner Inlet in Australia.

Trailing and flying spits

Trailing (arrow or comet-tail) spits form in the lee of islands, as at the Plage des Grands Sables on the eastern end of the Ile de Groix off the Brittany coast and the various spits that trail leeward of high islands off the North Queensland coast (Figure 6.4) (Hopley 1971). These spits consist of beach material that has drifted from the high wave energy coast on the windward end of the island along the bordering shores to accumulate at the leeward end. Where the islands are of

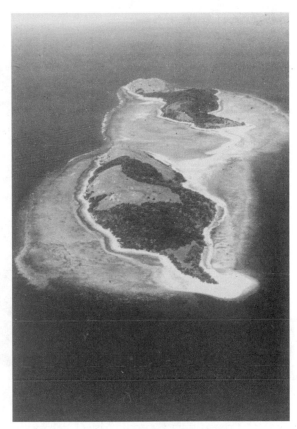

Figure 6.4 Trailing spits of sand on the leeward (north-western) shores of islands off the Queensland coast, north-east Australia.

soft material, such as glacial drift, trailing spits include sand and gravel derived from eroding cliffs on their coasts. Elongated spits trail from several islands in the Danish archipelago (Figure 6.5). On a smaller scale, trailing spits may form on the downdrift side of prominent shore rock outcrop or boulder, or in the lee of an isolated mangrove tree.

Where an island has been completely destroyed by erosion the depositional trail may persist as a flying spit, aligned at right-angles to the predominant waves. Examples are found off the Boston coast and in the Strait of Georgia, western Canada, in both cases derived from former islands of glacial drift.

TOMBOLOS

Tombolos are wave-built ridges of beach material that link islands, or attach a stack or island to the mainland.

Figure 6.5   A long trailing spit on the island of Lindholm in Denmark.

The term comes from the west coast of Italy, where these features are well developed at Orbetello. They are also found on coasts cut into glacial drift deposits, as at Nantucket in New England and in south-western Finland (Schwartz et al. 1989). Chesil Beach on the south coast of England is the western part of a double tombolo attaching the Isle of Portland to the Dorset mainland and in Australia the Yanakie isthmus ties the granitic upland of Wilsons Promontory to mainland Victoria in a similar way. Some tombolos have formed by lee-shore deposition of sediment by waves refracted round a nearshore island or stack to produce spits that grow into a linking isthmus, as at Cape Verde on the Senegal coast in West Africa. The tombolo at Tyre in the Lebanon is thought to have been formed by sand deposition on either side of a causeway built out to an offshore island by Alexander the Great in 332 BC.

Stages in the evolution of tombolos are seen in the Isles of Scilly, several of which have become linked by depositional banks of sand or shingle. On Samson a sandy isthmus links two former islands, and a shingle bank ties Gugh to St Agnes. The Ayres of Swinister in the Orkney Islands have an indented coastline with shingle tombolos and in the Shetlands the large sandy St Ninian's tombolo links St Ninian's Isle to the south-west coast of Mainland, and encloses a lagoon. Small tombolos have formed as the result of the growth of spits in the lee of offshore breakwaters, as at Rimini on the Adriatic coast of Italy.

A tombolo that is partly or wholly submerged by the sea at high tide is known as tombolino, or tie-bar. The Ilha Porcat in Santos Bay in Brazil is attached to the mainland by a tombolo, whereas adjacent Uruboqueçaba is linked by an intertidal tombolino.

CUSPATE SPITS

Cuspate spits form where beach sediment is deposited as protruding, more or less symmetrical, structures, formed where waves approaching at an angle to the shore from either direction are stronger and more frequent than those coming directly onshore. Such conditions are common on the shores of narrow straits where the transverse fetch is small. The waves generate a convergence of longshore drift, so that sediment is supplied from both directions. The sediment has usually been derived from erosion in adjacent bays, but may come from rivers updrift, or from the sea floor. Cuspate spits are also found on the shores of long and narrow lagoons, and where wave refraction has concentrated beach material in the lee of an island, breakwater or shoal (Figure 2.2C). The Anse Vata cuspate spit near Nouméa in New Caledonia consists of coral sand and gravel that has been shaped by waves refracted across a fringing coral reef. Cuspate spits have formed as the result of convergent drifting by waves refracted round offshore breakwaters in several harbours around Port Phillip Bay, Australia, notably at Middle Brighton (Figure 6.6). Breaching of a former tombolo may leave residual cuspate spits, as at Gabo Island in south-eastern Australia.

The asymmetrical spits on the north coast of the Sea of Azov (Figure 6.7) are shelly sand spits that have grown out at an angle to the coast as the result of longshore drift, usually ending in a slight recurve. They occur in a regular sequence on the Osipenko coast, where they are migrating westward. Secondary accumulation of sand on the leeward side forms a series of curved beach ridges that may be truncated by wave erosion on the eastern side (Zenkovich 1967).

CUSPATE FORELANDS

Cuspate forelands are similar to cuspate spits, but have been enlarged by the accretion of beach ridges parallel to their shores. They are known as nesses in Britain and, like cuspate spits, are found on the shores of narrow straits or in the lee of islands or shoals. Stages in their evolution can be deduced from the patterns of beach ridges. Some cuspate forelands have remained stationary and grown symmetrically, while others have been eroded on one side and built up on the other, so that the cuspate foreland has migrated along the coast as a travelling foreland.

This is true of Dungeness, a massive shingle cuspate foreland shaped by south-westerly waves from the English Channel and easterly waves arriving through the Strait of Dover. Stages in its evolution can be traced with reference to the patterns of shingle beach ridges, each of which indicates an earlier position of the coastline. There has been re-shaping of a mass of shingle that accumulated off Rye a few thousand years ago, which has been sharpened into the cuspate foreland (Figure 6.8). Beach ridges have been truncated along the southern coastline, and shingle eroded from here has drifted round the point, to be added as successive beach ridges on the prograding eastern shore. The point has thus migrated eastward.

Coastal lowlands of cuspate outline have not necessarily been formed by beach ridge accretion. On the coast of Cardigan Bay in Wales, Morfa Dyffryn and Morfa Harlech are somewhat asymmetrical cuspate lowlands. It has been suggested that they originated as spits of sand and shingle that grew out north-westward at an angle to a former steep or cliffed coast, then curved round to the north-east, enclosing an area of sandy and marshy flats. However, both incorporate glacial or glacifluvial deposits in front of an abandoned Pleistocene cliffed coastline, and the beach and dune fringe on their shores is the outcome of erosion and reworking of the margins of glacial drift forelands during and since the Late Quaternary marine transgression. Gravelly cuspate lowlands on the shores of fiords in north-west Iceland (e.g. Patreksfjördur) and on the Alaskan coast at Point Barrow originated as glacifluvial deltas.

Winterton Ness on the north-east coast of Norfolk is a sandy cuspate foreland that has been migrating southward. Evidence from successive historical maps shows that its northern coast has been cut back by erosion by waves arriving from the north-east, while the southern coast has prograded with the addition of new dune ridges. This migration was accompanied by southward longshore drift, but the movement of the Ness may also have been influenced by changing wave patterns produced by variations in offshore shoal topography. By contrast, Benacre Ness in a similar situation on the Suffolk coast, has been migrating northward as the result of accretion on its northern side of sediment supplied by the predominant southward longshore drift, and erosion on its southern side. Again, migration of offshore shoals may also have influenced this evolution (Robinson 1966).

Cuspate forelands and tombolos on the coast of Western Australia have formed in the lee of rocky islands or reef segments (Sanderson and Eliot 1996). Capes Hatteras, Lookout, Fear and Kennedy are large cuspate forelands on the Atlantic coast of the United States, each consisting of sand delivered by longshore drift and retained on a sector of convergent wave refraction behind nearshore shoals. They show erosion of the northern or eastern flank and progradation to the south. Cape San Blas in Florida is a cuspate foreland that has been nourished with sand derived from nearshore shoals deposited by the Apalachicola River. The Darss Foreland on the German Baltic coast has grown out from a promontory of glacial drift, truncated in cliffs on the western shore, with multiple sandy beach and dune ridges which indicate northward progradation during the past 3500 years (Sterr et al. 1998).

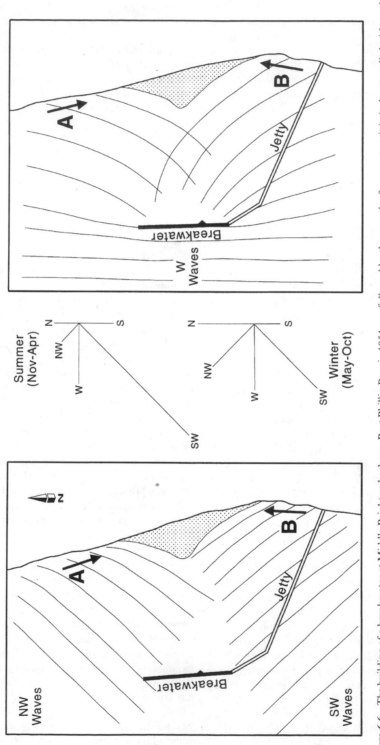

Figure 6.6 The building of a breakwater at Middle Brighton harbour, Port Phillip Bay, in 1954 was followed by the growth of a cuspate spit in its lee, supplied with sand that has drifted in from beaches to north (A) and south (B) and shaped by waves from the south-west and south-east (left) and westerly waves refracted round the offshore breakwater (right).

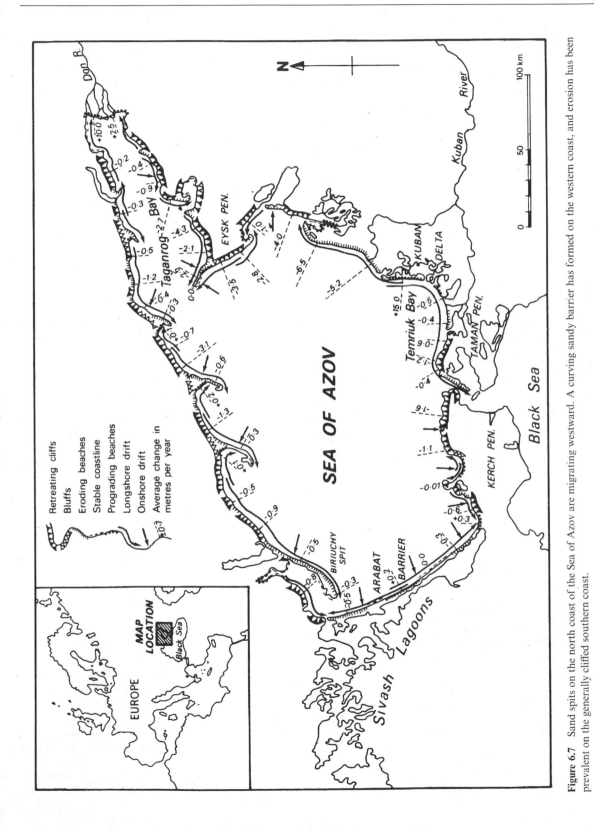

Figure 6.7   Sand spits on the north coast of the Sea of Azov are migrating westward. A curving sandy barrier has formed on the western coast, and erosion has been prevalent on the generally cliffed southern coast.

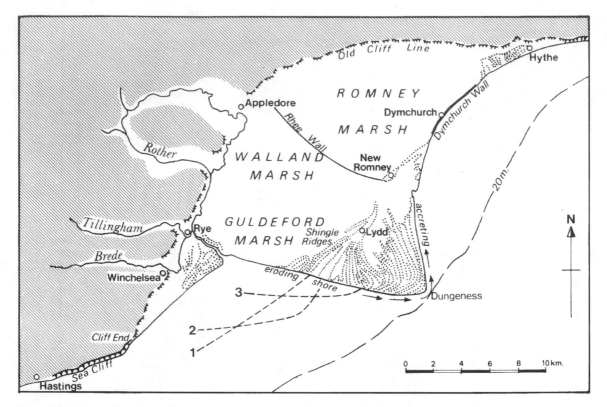

Figure 6.8 Shingle ridges (dotted) show stages in the growth of the cuspate foreland at Dungeness in Kent, formed in front of a former cliffed embayment now occupied by drained marshland. The ridge pattern indicates that the cuspate foreland grew seaward (1, 2, 3), the point migrating eastward as the southern shore was truncated by erosion and shingle drifted round to the prograding eastern shore.

COASTAL BARRIERS AND BARRIER ISLANDS

Coastal barriers and barrier islands have been formed by the deposition of beach material offshore, or across the mouths of inlets or embayments, extending above the normal level of highest tides and partly or wholly enclosing lagoons and swamps (Schwartz 1973). Barriers, thus defined, are distinct from bars, which are submerged for at least part of the tidal cycle (p. 199), and from reefs of biogenic origin, built by coral and associated organisms (Chapter 11).

Coastal barriers and barrier islands fringe about one eighth of the world's coastline. They show a variety of forms. Barrier beaches are narrow strips of low-lying depositional land consisting entirely of beach sediment, but most coastal barriers have surmounting dunes and some attain widths of several kilometres, with dunes sometimes rising more than a hundred metres above sea level. The term bay barrier describes a feature built across an embayment, and barrier island indicates a discrete elongated segment, parallel to the coastline, often recurved at both ends, and backed by a lagoon or swamp.

Coastal barriers consist of sand or gravel delivered by longshore drift or carried in from the sea floor, and deposited by wave action as beaches, behind which there may be wind-emplaced dunes. Barriers are best developed on coasts where the tide range is small, whereas barrier islands are usually formed where there is a large tide range generating strong ebb and flow currents which maintain gaps between them, preventing wave action from depositing sand or shingle to seal the intervening inlets. On the Gippsland coast in Australia spring tide range increases from about a metre at the north-eastern end of the Ninety Mile Beach, where there are no natural tidal entrances through the coastal barrier, to 2.5 metres at the south-western end, where tidal currents have maintained channels between a chain of barrier islands. This is why the Gippsland Lakes are almost completely sealed off from the sea by continuous barriers at the eastern end while Corner Inlet, a similar embayment at the south-western end, has an incomplete barrier system (Bird 1993a). A similar contrast exists between the relatively unbroken coastal barriers on the almost tideless south coast of the Baltic Sea and the chain of barrier islands (the Frisian Islands) with intervening tidal entrances on the strongly tidal south coast of the North Sea. Chains of barrier islands separated by tidal entrances are also well developed on the Gulf and Atlantic coasts of the United States and in north-western Australia, notably near the mouth of the De Grey River.

Barriers and barrier islands exemplify multicausality in that they have originated in a variety of ways, and no single explanation will account for all of these features (Schwartz 1971). Some barriers have been formed by the longshore growth of spits (in which case they may be termed longshore spits). Others result from the emergence of an offshore bar during a phase of sea level lowering, or the partial submergence of a pre-existing coastal sand ridge during a phase of sea level rise. Some are the result of shoreward sweeping of sand or gravel during and since the Late Quaternary marine transgression. Many barriers have had a composite origin (Roy et al. 1994). Longshore spits, for example, can also be driven landward.

On the West African coast Guilcher and Nicolas (1954) described an elongated barrier in the form of a longshore spit, the Langue de Barbarie, which grew as the result of southward drift of beach sand in such a way as to form and prolong a barrier deflecting estuaries, notably the mouth of the Senegal River (p. 256). Barriers that grew as longshore spits often have recurves marking stages in their growth, as on the recently formed barrier island known as The Bar on the Culbin coast, Scotland, which developed during the nineteenth century, and grew both eastward and westward, with successive terminal recurves. Similar features are seen on Tramore spit in south-east Ireland (Ruz 1987). Stages in the growth of the longshore gravel spit at Koumac, on the west coast of New Caledonia, are also indicated by recurved ridges projecting into the backing lagoon.

Other barriers and barrier islands have formed on coasts bordered by shallow seas as the result of a relatively sudden emergence, when reduction of water depth offshore caused waves to break farther out and build a bar which emerged as it was driven shoreward.

The sandy barrier island of Knotten on the south-west coast of the Danish island of Laesø may have been initiated in this way. It is possible to generate a barrier island from an emerging bar experimentally in a wave tank, but it is doubtful if it applies to many actual barriers, for the majority were initiated during the Late Quaternary marine transgression.

Sandy barrier islands on the south-east coast of the United States were probably initiated when the Late Quaternary sea rose to partially submerge pre-existing coastal dune ridges, penetrating lower land to the rear to form lagoons behind a chain of barrier islands. Sandy barriers have formed in this way and moved landward during the past few years on the coasts of the Caspian Sea, where a marine transgression is in progress (Figure 3.9).

Many coastal barriers are the outcome of shoreward sweeping of beach sediment during and since the Late Quaternary marine transgression, notably where Pleistocene deposits that had been stranded on the emerging sea floor during the preceding marine regression were collected and built up, then driven landward as the marine transgression proceeded. This shoreward drift of beach sediment may have been facilitated by minor emergences during the oscillating Late Quaternary marine transgression (p. 38), which set up broom-like shoreward sweeping.

As has been noted, Chesil Beach on the Dorset coast is a shingle barrier that stands in front of a lagoon (The Fleet) and an embayed mainland coast which has escaped marine cliffing (Steers 1953). It must have been formed and driven shoreward during the Late Quaternary marine transgression in such a way as to protect the submerging land margin from the waves of the open sea. The barrier and lagoon are underlain by a platform cut in bedrock that descends from the inner shore of the lagoon and flattens at about 15 metres below sea level. The shingle barrier is permeable, so that during storms and high tides sea water seeps through to The Fleet, washing out fans of gravel, known locally as cans, which run from the landward slopes of the barrier down into the lagoon.

Chesil Beach is still moved shoreward intermittently, when vigorous storms sweep shingle over the crest and down into the waters of The Fleet. It has overrun peaty sediments that formed in the lagoon, so that peat now outcrops on its seaward slope. Eventually it will come to rest against the mainland coast, which will thereafter be cut back as marine cliffs by storm wave activity, a stage that has already been reached at Burton Bradstock, near the western end.

The Loe Bar, near Helston in Cornwall (Figure 5.6), is a barrier 180 metres wide and up to three metres above calm weather high spring tide level. It is part of the beach that extends from Gunwalloe, north to Porthleven, and runs across the mouth of a former ria, enclosing the Loe Pool. It consists largely of fine flint and chert shingle that has been washed in by wave action from Tertiary or Pleistocene gravel deposits on the sea floor offshore, with only a small proportion of quartzite and slaty material derived from bordering cliffed coasts, and some shelly sand and gravel. The presence of flint and chert shingle on this part of the coast, and its rarity on adjacent coasts, suggests that the sea floor gravels had already been concentrated into one or more beach and barrier formation segments (along coastlines now submerged) before their final shoreward movement with the Late Quaternary marine transgression. The Loe Bar is sometimes overwashed by large waves from the south-west during storms, and by occasional Atlantic tsunamis.

Contrasts in sediment size are likely between the seaward and landward flanks of a shingle barrier where overwashing is infrequent, the wave-agitated shingle on the seaward flank being reduced in calibre by attrition, the beach face pebbles being clean and actively worn, while the less disturbed shingle on the landward slope is often stained and silty, sometimes with plant growth. Where there is shingle of similar size on the beach face, barrier crest and landward slope the implication is that overwashing has been relatively frequent in relation to the rate of attrition on the beach face.

Shoreward drift of the kind that produced Chesil Beach and the Loe Bar has also been invoked to explain the barriers of sand and shingle that have developed in front of uncliffed mainland coasts in Siberia (Zenkovich 1967), and Le Bourdiec (1958) reached the same conclusion for barriers on the Ivory Coast. Shoreward sweeping of beach sediment during the Late Quaternary marine transgression contributed to the sandy barrier islands that border the Gulf coast of the United States (Shepard 1973). In each case the barrier formations rest upon an older (Pleistocene) land surface submerged by the Late Quaternary transgression.

Some coastal barriers have remained in position during Holocene times, and may have been widened seaward by progradation with the addition of multiple beach ridges. On many barrier coasts this progradation has given place to erosion (Figure 6.9). Other barriers are (or have become) transgressive, moving landward as the result of overwashing by storm waves (as on Chesil Beach) or landward spilling of dunes.

Chesil Beach is apparently receiving little new shingle, either from onshore or longshore drift, and experiments with radioactive pebbles dumped on the sea floor failed to show any evidence of active shoreward drift to Scolt Head Island, a barrier island on the north Norfolk coast. These shingle barriers are essentially relict formations, a legacy of the Late Quaternary marine transgression. Most British barriers are of shingle (some capped by sand dunes), but in Sandwood Bay on the west coast of Scotland there is a wide barrier beach, built of sand washed in from the sea floor, enclosing a freshwater loch at the mouth of Strath Shinary, a glaciated trough.

Other barriers are still receiving sediment derived from adjacent cliffed coasts, river discharge, and shoreward drift. The barriers which extend across embayments on the New England coast, enclosing lagoons known as ponds, are built of sediment derived from adjacent cliffs cut in glacial outwash deposits. These ponded barriers are well developed on the southern shores of Martha's Vineyard and Nantucket Island off the Massachusetts coast, and on the shores of Cape Cod. In Washington state, similar features are seen in Puget Sound, where Ebey's Landing is a looped barrier (not a bar because it extends above high tide level) diverging from the coastline and curving back in to enclose a small lagoon. On the coast of Nova Scotia, where shingle barriers consist of gravel sorted from glacial drift deposits, Orford et al. (1996) related phases of shingle barrier growth, consolidation, breakdown and re-formation to sea level fluctuations. Shelly sand drifting alongshore has accumulated in the form of a curving barrier on the western coast of the Sea of Azov (Figure 6.7).

Barriers known as nehrungen on the south Baltic coast have also been partly derived from cliffed glacial deposits and carried along the shore to enclose embayments as lagoons, and partly from sediment swept in from the sea floor. By contrast, the sand and gravel barriers on the south-east coast of Iceland have been supplied largely with sediment supplied to the coast by glacifluvial streams, especially during the floods that follow the melting of hinterland ice by volcanic eruptions. Sand and gravel washed into the sea are reworked and built into barriers by the constructive action of southerly Atlantic swell. In the North Island of New Zealand sandy barriers are extensive, and mostly derived from volcanic sand carried down to the coast by rivers or eroded from cliffs of volcanic ash. In the South Island shingle barriers are commoner, derived from glacifluvial outwash deposits eroded from cliffs or swept in from the sea floor. The famous

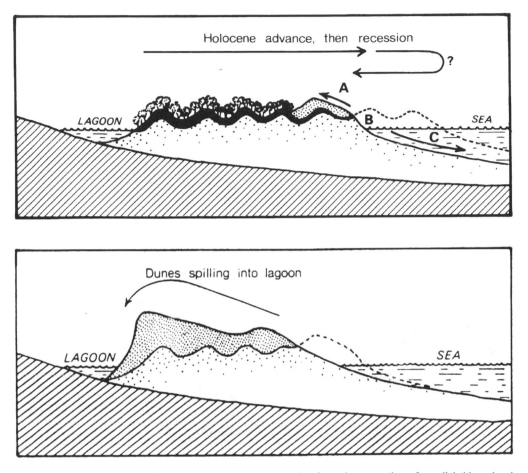

Figure 6.9 Evolution of barrier coastlines where Holocene progradation formed a succession of parallel ridges, but has been followed by recession, with losses of sand landward into spilling dunes (A), as well as alongshore (from B) and seaward (C). Such a barrier can eventually become transgressive, with sand blown (or washed) over into the backing lagoon

Boulder Bank at Nelson is exceptional in being derived from the erosion of Mackays Bluff, a cliff of weathered syenite (Healy and Kirk 1982).

On parts of the Australian coast there are multiple barriers with intervening tracts of lagoon and swamp. The inner barriers are generally of Pleistocene age, having been dissected by stream incision during the Last Glacial low sea level phase, whereas the outer barriers are usually of Holocene age, having come into existence during and since the Late Quaternary marine transgression (Figure 6.10). There are many complications, for barriers have been built, then dissected or destroyed, and subsequently rebuilt, sometimes in overlapping alignments, during Quaternary sea level oscillations.

On the Victoria and New South Wales coasts the

inner barriers are largely of quartzose sand, and have not been lithified in the manner of the calcareous sand barriers of South Australia, which are commonly preserved as relatively durable dune calcarenite. The numerous parallel ridges of dune calcarenite in south-eastern South Australia (Figure 3.4) formed successively as coastal barriers during Pleistocene times, when this part of the coast was intermittently uplifted, and subject to sea level oscillations. Between the emerged dune calcarenite barriers are corridors of swamp land that developed in former lagoons. The Coorong, behind the outer barrier on the shores of Encounter Bay, is the last in this sequence of inter-barrier lagoons, not yet filled by swamp deposits. Farther south, between Robe and Beachport, the

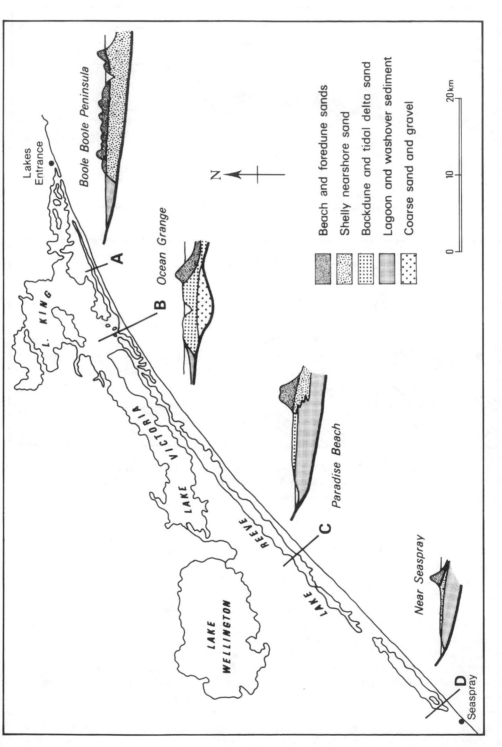

Figure 6.10   Sections through the outer barrier on the East Gippsland coast show that it formed in Holocene times by the deposition of beach and foredune sands over earlier lagoonal deposits (based on Thom 1984).

coast consists of a cliffed and dissected ridge of Pleistocene dune calcarenite, there being no Holocene outer barrier of unconsolidated sand here.

The sandy barriers in south-eastern Australia have their counterparts on similarly-oriented coasts in South Africa (Natal) and South America (between Rio and Cabo Frio). Many of these barriers have been pro-graded by the addition of successive parallel beach ridges on alignments which, like the curved outline of the present beach, result from the refracted patterns of dominant swell approaching the coast. Often the beach ridges are surmounted by parallel dunes, developed successively during progradation, as on the barrier islands in south-west Florida, where beach ridge formation has been correlated with fluctuating Holocene sea levels (Stapor et al. 1991).

The pattern of beach ridges and dunes can be used to decipher the history of barrier evolution, particularly if there are relics of former recurves indicative of stages in longshore growth. The outer barrier on the East Gippsland coast (Figure 6.10) originated as a chain of barrier islands which were extended north-eastward by longshore drift, deflecting outlets from coastal lagoons (the Gippsland Lakes, p. 233) and sealing off most of them. This barrier was subsequently widened by progradation, with successive parallel ridges added on the seaward side, until it took up the alignment of the Ninety Mile Beach along the present coast (Bird 1978a). Radiocarbon dating of shelly material obtained from drilling in these sand barriers was used by Thom (1984) to establish stages in the Holocene progradation, but the barriers have been truncated along their seaward margins by ensuing marine erosion (Figure 6.9).

By contrast, some of the barriers and barrier islands of the Gulf and Atlantic coasts of the United States are transgressive. The Outer Banks of North Carolina north of Cape Hatteras have been migrating landward as the result of spillover dunes and washover sand invading Pamlico Sound, and accompanying recession of the seaward margin. Others have prograded, with the addition of successive beach ridges, as on Seabrook and Folly Islands in South Carolina. The barrier on the shores of Encounter Bay in Australia is becoming transgressive, with numerous blowouts spilling sand over into the backing lagoon, the Coorong. Tidal inlets between barrier islands may migrate downdrift, become sealed off by deposition, or reopen during stormy epi-sodes. It is not surprising that borings in transgressive barrier formations reveal a complex stratigraphy, with overlapping sequences of beach, dune and lagoon sedi-ments, as was shown in Van Straaten's (1965) study of Holocene deposits on the Dutch coast.

On some coasts there are elongated islands that look like barrier islands but are not wave-deposited features. Often they are segments of glacial moraines, as on Long Island, New York, and Walney Island in Cumbria, which is cliffed on the seaward side. Spit growth at the limits of such islands increases their similarity to wave-shaped barrier islands. On the German North Sea island of Sylt, sediment derived from receding cliffs cut into glacial drift on the outer coast has been distributed by northward and south-ward longshore drift to dune-capped recurved spits at each end.

Several references have been made to dunes associated with beaches, spits and barriers, and the following chapter will deal with the evolution of coastal dunes.

7

Coastal Dunes

INTRODUCTION

Coastal dunes generally form where sand on the shore has dried out and been blown to the back of the beach, to accumulate above high tide level, particularly where deposition occurs against obstacles such as driftwood, or within strand litter or vegetation (Goldsmith 1985). Their growth and shaping are related to a source of sand that can be moved by wind, to wind flow characteristics, rates of aeolian transport and patterns of erosion and deposition. They are most extensive on windward coasts behind wide sandy beaches, notably on the Atlantic coasts of Europe, the Pacific coasts of the Americas, south-eastern Australia and southern Africa. Coastal dunes differ from inland (desert) dunes (Bagnold 1941) in that they are subject to a wider variety of processes, including wave action and vegetation, which influence their size, shape, evolution and persistence.

Backshore dune development is aided by frequent strong onshore winds and by the availability of a wide sandy beach as a source area: it is also influenced by wave processes, especially when storm waves trim back the seaward margins of coastal dunes. Coastal dunes are more often found behind wide gently-sloping dissipative beaches (p. 128) than behind steeper, coarser and narrower reflective beaches (Short and Hesp 1982). They are well developed where the tide range is large, as on the Atlantic coasts of Britain, where prevailing westerly winds have blown sand onshore from beaches, as in the Ainsdale dunes in south Lancashire. On the North Sea coast, dunes back wide beaches in the vicinity of Holy Island and along the shores of Moray Firth (Culbin Sands), and are extensive in Belgium, Holland, Germany, and Denmark (Bakker et al. 1990). Papers on various aspects of dune geomorphology were assembled by Nordstrom et al. (1990).

Even where the tide range is small, coastal dunes have formed where there has been a sufficient supply of accreting sand to be blown onshore from the beach.

Sandy beaches shaped by ocean swell are generally backed by dunes on the coasts of Australia, Africa, and the Americas, although some of the older dune systems (especially dune calcarenites) include sand that was carried inland by wind action from emerged coastal lowlands during low sea level phases of the Pleistocene. This has contributed to the major dune systems on the south-east coast of Australia, notably on Fraser and Moreton Islands in Queensland, behind Newcastle Bight and Wreck Bay in New South Wales, and Discovery Bay in Victoria (Hesp and Thom 1990).

On the Pacific coast of North America dunes have been derived from beaches nourished by an abundance of fluvial sand supply from the Columbia and other rivers. There are dunes on the coastal barrier at Long Island, north of the mouth of the Columbia River, and Clatsop spit, to the south. Fluvial sand swept out to the sea floor, reworked by wave action and delivered to beaches has been the source of the extensive dunes at Coos Bay in Oregon (Cooper 1958).

Some backshore dunes have not been derived from adjacent beaches, but from sources along the coast or inland. On the south-facing Cape Coast of South Africa there are backshore dunes derived from beaches upwind and driven from west to east along the coast, spilling over headlands, as at Cape Recife near Port Elizabeth. Other backshore dunes are of desert origin, as on the coast of Mauritania, where barchans deliver sand to the beach (Vermeer 1985). Coasts bordering the Sahara and other arid regions have areas where desert dunes meet and mix with dunes derived from beach sands.

Dune sands have similar characteristics to the beach sands from which they have been derived (p. 98) and generally consist of quartz, felspar, and calcareous particles (including foraminifera, bryozoa and comminuted shells and corals), sometimes with heavy minerals such as rutile and ilmenite. Where there are volcanoes or volcanic deposits on the coast or in the hinterland, dunes may be derived from deposits of volcanic ash.

Sand blown from a beach is typically fine-grained (sand grains of diameter 0.1 to 0.3 millimetres, or just below 2ø to just above 3ø, are most readily moved by wind action), well-sorted and well-rounded. Grain size analyses show that dune sands are often (but not always) finer and better sorted than beach sands, with positively-skewed grain-size distributions (p. 96). Dune sands often have highly polished grain surfaces. Their rounding may be the result of abrasion during transport by the wind.

Wind data for coastal dune studies has often been obtained from meteorological stations which may be some distance from the dune area. This has proved useful in demonstrating relationships between dune forms and dominant winds (p. 187), but for detailed process–response studies it is necessary to obtain site data using anemometers, preferably set to measure wind velocity close to the ground.

Entrainment of sand by the wind depends on near-surface air flow and surface morphology. Wind moving across a sandy beach develops a shear stress, equivalent to air density (averaging about 1.22 kilograms per cubic metre, cold or dry air being denser than warm or moist air) multiplied by wind velocity, and when this exceeds the entrainment threshold value wind energy overcomes gravitational inertia and loose sand particles are mobilized. Such mobilization is impeded where there are cohesive forces which raise the entrainment threshold. Deflation of sand from beaches is inhibited by moisture. There is no doubt that wet sand is more cohesive and less readily moved by wind action across the shore but strong onshore winds soon dry a beach surface and can drift sand even when it is raining. On some tropical beaches the formation of a surface crust by salt evaporated from sea water or spray impedes the movement of sand by wind action. Morton (1957) attributed the poor development of dunes on the coast of Ghana to the salt binding of beach sands, noting that dunes had developed at Old Ningo where the sand was unusually coarse and shelly, and not bound by salt. On some beaches a surface accumulation of pebbles or shells left where sand has been winnowed by the wind or washed away by waves, prevents further sand deflation (Pye 1980): on Magilligan Point in Northern Ireland this resulted in a reduction of sand supply to backshore dunes (Carter and Wilson 1990). In such situations wind energy may be strong and sustained enough to move much more sand than is available in source areas.

On dry, loose sand the flow of wind over sand grains causes a pressure gradient that lifts the particles, which then travel downwind by saltation (bouncing), traction (rolling) or if the wind is strong, in suspension. Sand movement by wind action is influenced by the shape of sand grains because rounder grains roll more easily, and flatter grains, such as mica flakes, are readily deflated. Dune surfaces occasionally have ripples formed by wind action, which are similar to ripples formed by nearshore currents (p. 201). An analysis of sand mobilization by wind action has been provided by Sherman and Hotta (1990).

Onshore winds sweep sand from the beach to the backshore. The sand is carried until the wind velocity diminishes, the ground surface rises, or vegetation is encountered. Olson (1958) used anemometers to compare the wind velocity profile over bare sand with that over grassy dunes, and found that the vegetation raised the upper limit of the zone of calm air near the ground. Vegetation also reduces shear stress and presents a surface roughness that promotes sand deposition. This deposition increases the backshore gradient, and the backshore is shaped into a ridge which impedes further landward transport of blown sand. The outcome is the upward growth of a vegetated sand ridge, known as a foredune. Sometimes strong winds carry fine sand over the ridge and deposit it as a thin sheet or low hummocks on the landward side.

Sand is most readily winnowed from a beach where it is fine-grained (but not necessarily well sorted) and where the grains are irregular in shape. Sand transport rates and sediment budgets on beaches and coastal dunes can be quantified by making repeated surveys of the beach face and developing dune to determine changes in volume, by trapping blown sediment in a surveyed trench or receptacle such as a cylindrical pipe (Bauer et al. 1990), or by using fluorescent tracers, as outlined by Sherman and Hotta (1990). Horikawa (1988) reviewed methods of determining beach and dune sediment budgets.

The amount of sand entrained by the wind depends on the strength of the onshore wind, and is limited by beach width. More sand can be blown from a beach to a backshore dune by onshore winds arriving obliquely, than by those that arrive at right angles to the coastline, which also generate stronger wave action. On the shore of Lake Erie, Davidson-Arnott and Law (1990) found that the width needed for maximum sand transport by a wind exceeding 50 kilometres per hour was over 40 metres.

Beach–dune interactions occur when onshore winds deliver beach sand to backshore dunes and when winds blow sand from the dunes to the beach or storm waves erode sand from backshore dunes and incorporate it in the beach (Psuty 1988). The first process tends to leave the beach sand coarser and less well sorted than the dune sand, but the second process can obscure such differentiation.

On cold coasts where there is a winter snow cover, as in Korea and eastern Canada, deflation of sandy surfaces may occur as they dry out after the spring thaw, forming dunes before annual plants can grow to stabilize the surface. Dunes have formed in this way from river channel sands on the Colville delta in the Canadian Arctic (p. 258) (Walker 1998).

Coastal dunes have formed over a variety of time scales, but most have been shaped in Holocene times from sand supplied to beaches from the sea floor (notably during the Late Quaternary marine transgression) and alongshore sources such as cliffs in soft sandstone or glacial drift deposits. On some coasts the Holocene dunes overlie, and are backed landward by dune formations that originated in the Pleistocene, some of which may have formed during phases of falling sea level, when onshore winds swept sand from emerging sea floors.

FOREDUNES

Foredunes are ridges of sand built up at the back of a beach or on the crest of a sand or shingle berm, where dune grasses have colonized, and are trapping blown sand (Figure 7.1). The colonizing vegetation acts as a baffle, diminishing wind velocity close to the ground and so creating a sheltered environment within which blown sand is deposited (Goldsmith 1989). Foredunes become higher and wider as accretion continues, depending on rates of wind-blown sand supply and coastline progradation, and may be built forward as dune terraces, flat or faintly ridged grassy and scrubby areas behind a prograded beach, as on the coast of Picardy in France.

Vegetation certainly traps and retains sand blown onshore from a beach, and the initiation of a foredune may occur along a high spring tide line where sand-trapping vegetation grows up from a seed-bearing strandline of plant litter on the beach face. There has been much discussion of whether such a strandline follows the crest of a wave-built berm of sand or shingle, or whether a foredune can be initiated from seed-bearing litter deposited on the beach face (seaward of pre-existing vegetated dunes). Foredunes run parallel to the high tide shoreline, as do berm crests

and seed-bearing litter zones, but they can also be initiated along the seaward edge of a grassy backshore terrace or an area of irregular hummocky dunes when cliffing by storm waves at high tide is accompanied by swash-borne or wind-blown sand accretion on top of the cliffed margin, forming a linear sand ridge that is colonized by sand-trapping grasses. Tussock grasses, herbs or shrubs that form circular patches tend to build mounds or hummocks of sand, whereas the close networks of stalks in more uniform grassland is more likely to build ridges or terraces in the backshore zone (Hesp 1988, Bird and Jones 1988). On the coasts of the United States foredune initiation and growth have been aided by the building of sand-trapping fences parallel to the coastline, as on Fire Island in New York State (Psuty 1990).

There are geographical variations in the plant species which act as pioneer colonists and foredune builders. In the British Isles and Europe a common pioneer dune plant is marram grass (*Ammophila arenaria*), a species that colonizes bare sand, and has been introduced to stabilize dunes in many other parts of the world. Lyme grass (*Elymus arenarius*) and sea wheat grass (*Agropyron junceum*) are also dune pioneers in Europe, the latter extending around the Mediterranean and Black Seas. Marram grass is the most vigorous of the dune-building plants, thriving on cool, moist coasts where it is able to trap and grow up rapidly through accreting blown sand. It can add up to two metres of sand to an accreting foredune in the course of a single year, but is overwhelmed if sand is supplied too rapidly. Where sand accretion is slow marram grass grows poorly, and other grasses may take its place (Ranwell 1972).

In the United States another form of marram grass, *Ammophila brevigulata*, is a common pioneer dune species, with the tussocky grass *Panicum amarum*, with *Spartina patens* common on low moist dunes. On sandy coasts in tropical regions the vegetation is generally sparser, and dominated by creeping vines such as *Ipomoea* and *Canavalia*, rather than tussocky grasses. These can build a foredune, but on prograding shores this vegetation may spread seaward and build a hummocky or ridged backshore terrace. Where there is sufficient rainfall the grassy and herbaceous dune vegetation is colonized by shrubs and trees, and eventually tropical forest (Bird and Hopley 1969).

In Australia and New Zealand the native dune pioneers are sand spinifex (*Spinifex hirsutus*), sea rocket (*Cakile maritima*) and coast fescue (*Festuca littoralis*). *Spinifex* has stolons that spread rapidly across the sand surface, putting down roots at intervals, and trapping

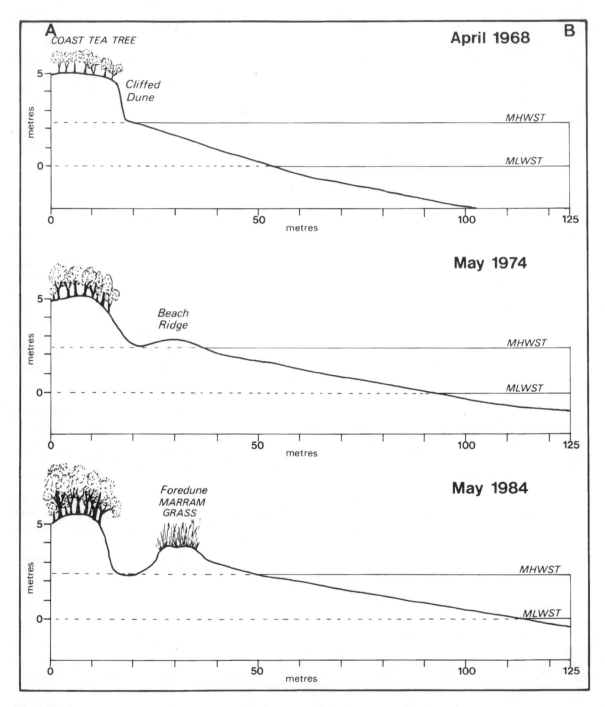

Figure 7.1   Successive surveys showing the formation of a beach ridge in front of a previously cliffed dune, and the development of a foredune when the beach ridge was colonized by marram grass.

wind-blown sand to form a widening backshore dune terrace (again hummocky, or with low ridges), whereas introduced marram grass forms tussocks that tend to build higher and narrower foredune ridges. However, foredunes were built by the native Australian grasses and herbs, prior to the introduction of European marram grass in the nineteenth century. It has been suggested that the introduction of European marram grass to Australia (often displacing native *Spinifex* and *Festuca* grasses) has altered coastal dune morphology, but foredunes were certainly built on the Boole Boole Peninsula, behind the Ninety Mile Beach, in the early Holocene, long before marram grass was introduced (Bird 1978a).

On arid coasts (as in Baja California and at the head of the Great Australian Bight) vegetation is too sparse to trap sand blown from beach to backshore, and dunes drift inland. Even on humid coasts there are sectors where coastal dunes are drifting inland because the available vegetation has proved inadequate for the trapping of sand, either because of a superabundant sand supply blown from a rapidly accreting beach (as on Sandy Cape at the northern end of Fraser Island in Queensland) or because the onshore winds are exceptionally strong, as on the north-west coast of Tasmania (Davies 1980).

BACKSHORE CLIFFING OF DUNES

Cliffing (scarping) of the seaward margins of coastal dunes and the absence of new foredunes (or the formation of new foredunes lasting at most a few years) is very widespread around the world's coastline, a consequence of the modern prevalence of beach erosion (Figure 7.2). Undercut by wave action, dune cliffs recede with sand collapsing on to the beach or into the sea, to be carried away offshore or alongshore. Sometimes clumps of dune vegetation bind the surface sufficiently to form blocks of dune sand that break away from the cliff edge and subside as irregular terracettes. Dune cliffs may stand vertical if the sand is moist and coherent, but as it dries it falls to a basal apron which (if it is not swept away by waves or

Figure 7.2 Cliffed dunes on the coast of Newfield Bay, near Peterborough in south-eastern Australia.

wind) grows as the cliff recedes, until it becomes a slope at the angle of rest of dry dune sand (about 32°).

On some coasts the formation of a dune cliff, cut back by storm waves, is followed by renewed accretion as sand is blown from the beach and banked against the cliffed dune in subsequent calmer weather. As vegetation spreads on to this accreting sand, the earlier profile is restored. Coastal dune margins may show alternations of dune cliffing and restoration during beach cut and fill cycles (p. 122). A foredune that runs parallel to the high tide line on the north-east coast of Norfolk in the vicinity of Sea Palling has revived after phases of severe erosion and breaching during storm surges, notably in 1953.

Coastal dunes generally have cliffed seaward margins in southern Britain, except where there has been local beach progradation, as at the northern part of Studland beach in Dorset (p. 105). In Northern Ireland several dune ridges were added to Magilligan Point between 1953 and 1983 during a phase when adjacent dune-fringed coasts were being cut back, but progradation has not been sustained here, and these dunes are now cliffed (Carter and Wilson 1990).

In northern Britain, where emergence is in progress as the result of continuing isostatic uplift, there are young foredunes behind beaches that are still receiving sand swept in from sea floor glacial drift deposits in the vicinity of Holy Island in Northumbria and at Tentsmuir in Scotland, where the dune fringe has advanced about 90 metres seaward since concrete tank traps were built in 1940.

During storms, the cutting of a cliff along the seaward side of a foredune may be accompanied by accretion of wind-blown and wave-washed sand on the crest and lee of the ridge, raising its altitude.

PARALLEL DUNES

On some coasts there are multiple dune ridges, usually running parallel to the coastline, that have formed successively as foredunes behind a prograding sandy beach. They differ from prograded backshore terraces in that there are intervening elongated swales or troughs. These result from sequences of cut and fill on an intermittently prograded shore. Incipient dunes form on the beach and grow into a new foredune seaward of earlier foredunes, which thus become parallel dune ridges separated by elongated swales. The seaward margin of a foredune is trimmed back by waves during a storm, forming a crumbling cliff of sand. Subsequently, during calmer weather, waves restore the beach and a new foredune is initiated along a high tide berm or strandline litter zone in front of, and parallel to, the trimmed margin of the earlier foredune, separated from it by a zone that remains unvegetated, and becomes a low-lying swale (Figure 7.1). Often the seaward slopes of parallel dunes are steep as the consequence of the cliffing that preceded the formation of the next foredune, followed by slope degradation. In the absence of such cliffing, progradation forms a widened backshore grassy terrace (Bird and Jones 1988).

The formation of swales separating parallel foredunes may be aided by the occurrence of wind eddies in the lee of each developing ridge, the swales being zones excavated by the wind so that the root systems of dune vegetation are laid bare (Figure 7.3).

Parallel foredunes are well developed on parts of the North Sea coast of Britain, as at Winterton Ness in East Anglia, on the Oregon and Washington coasts in the United States, and in south-eastern Australia, notably on the coastal barriers of East Gippsland.

Successive foredune formation parallel to a prograding coastline is well illustrated at Lakes Entrance in south-eastern Australia, where sandy forelands have developed as the result of accretion of sand on either side of jetties built beside the entrance in 1889 (Figure 5.24). The pattern of beach and dune accretion in the ensuing century has been determined by ocean swell, refracted on encountering a sand bar offshore, so that the coastline of each foreland is lobate in form. Three foredunes have formed successively, parallel to this prograding coastline, when beach accretion following storm scour provided foundations for wind-blown sand to be intercepted by colonizing vegetation and build roughly symmetrical foredunes, each about 3 metres high (Bird 1978a).

The height and spacing of parallel foredunes is a function of the rate of sand supply to the shore, the history of cut and fill and the effectiveness of vegetation in binding sand and building the dunes. Where sand supply has been rapid on a prograding shore subject to frequent storms a large number of low, closely spaced parallel dunes are formed, but on a similar shore less exposed to storms the effects of backshore erosion, which is responsible for the separation of the dunes into parallel ridges, are less frequent, and there are fewer, but larger, parallel dunes. Where the sand supply has been meagre, parallel ridges are less likely to form on a stormy shore, but a few low parallel dunes may develop on more sheltered sections of the coast.

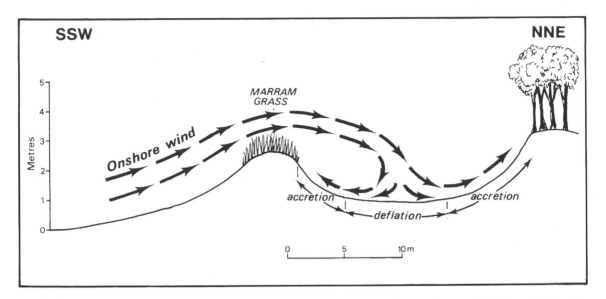

Figure 7.3 Onshore wind flowing over a grassy foredune develops eddies which may contribute to the shaping of the backing swale. Pattern determined by observing the flow of smoke from a fire on the beach.

Attempts have been made to deduce changes in sea level from the heights of successive parallel dunes and swales. An overall seaward descent of dune crests and swales could indicate that the coast was emerging during their formation, but this pattern could also have been produced by a diminishing sand supply or increasing storm frequency, without any change in the relative levels of land and sea, and so cannot be accepted as conclusive evidence for coastal emergence (p. 33) (Davies 1957). In the Netherlands, parallel dune ridges have formed during progradation on a coast where relative sea level has been rising.

Most parallel dunes have been formed by sequences of cut and fill on a prograding sandy shore, but there are exceptions. On the Falsterbo peninsula in Sweden there are parallel dune ridges that formed by wind-blown sand accretion on successively-built longshore spits, and on South Haven peninsula in Dorset three parallel dunes have developed since the seventeenth century on successive wave-built sand ridges with broad intervening low zones that have become occupied by lagoons and marshy swales (Steers 1964). On the coast of Lake Michigan, Olson (1958) traced the evolution of successive parallel dune ridges to alternating phases of backshore dune development when lake level was falling, and coastline recession when lake level was rising, overall progradation having accompanied these historical fluctuations in water level.

Continuing upward growth of a foredune gradually cuts off the supply of sand to its predecessor, which becomes relatively stable. The dune grasses are then invaded and replaced by scrub or heath vegetation, depending on the type of sand. On the coast of the Bristol Channel, notably at Berrow in Somerset, recently-formed calcareous grassy foredunes are backed by older dune topography with a rose and bramble thicket and buckthorn scrub (*Hippophae rhamnoides*), but where the dune sands are quartzose pedogenesis (soil formation) leads to the development of podzols. The upper layers are leached of shelly material by percolating rainwater, which also removes the yellow stain of iron oxides from the sand grains, leaving the surface sand grey or white in colour, while the lower layers are enriched by the deposition of iron oxides, together with downwashed organic matter, to form a slightly-cemented red-brown sand horizon, known as humate or coffee rock (p. 194). The older dune ridges on the landward side show more advanced stages in vegetation succession, often with heath or heathy woodland on sand that has been deeply leached, as on the South Haven peninsula in Dorset (Bird 1995). In south-eastern Australia parallel dunes of quartzose sand show a landward sequence from grassy foredunes through tea-tree (*Leptospermum laevigatum*) scrub to heathy forest, accompanying deepening podzolic soils.

Parallel dunes may be modified by wave overwash during occasional storm surges, which damages or destroys vegetation and sweeps sand landward into fans or sheets. Ridges and mounds are shaped into more subdued forms at the same time as the seaward margins are cut back as dune cliffs. On the Gulf and Atlantic coasts of the United States dunes have been modified in this way by recurrent hurricane surges, as illustrated by Ritchie and Penland (1988) in south Louisiana.

BLOWOUTS AND PARABOLIC DUNES

On some coasts there are unstable dunes with little or no vegetation cover. These include blowouts, parabolic dunes, and transgressive dunes (Hesp and Thom 1990). Mosaics of stable, vegetated dunes and unstable bare dunes may develop either as the result of partial stabilization of drifting sand by the arrival or introduction of vegetation (notably the planting of European marram grass during the past few centuries) or as the result of disruption of formerly well-vegetated dunes, as on the coast near Cape Arnhem in northern Australia, where dunes have been partially mobilized by the impact of trampling and grazing by introduced Asian buffalo.

Coastal dunes that have been stabilized by vegetation may subsequently be eroded and re-shaped by the winds in areas where the vegetation cover has been weakened or removed, either naturally (for example by increasing aridity or strengthening wind action) or as the result of disturbance by human activities, so that sand is no longer held in position. Formerly vegetated dunes can thus become mobile sand sheets. Parallel dunes held by vegetation have in places been interrupted by blowouts, which are unvegetated or sparsely vegetated hollows excavated by onshore winds, with sand driven mainly landward to form a looped ridge (Carter et al. 1990). Local weakening or destruction of dune vegetation may be initiated by intensive human activity, where footpaths are worn by people walking to a beach, or where trackways are formed by vehicles driving to and from the shore. Burning of coastal vegetation and excessive grazing by rabbits, sheep, cattle, or goats can destroy the vegetation and initiate blowouts. The effects of rabbits are accentuated where they burrow into the dunes, disrupting the surface, as on Blakeney Point in Norfolk. Measurements by Rutin (1992) on Dutch dunes showed dune topography being modified by local accumulation of up to 0.5 cubic

metres of sand excavated from each rabbit burrow, and by subsidence as the rabbit burrows collapse. Grazing by introduced rabbits is thought to have depleted dune vegetation on the Nullarbor coast in Western Australia, mobilizing dunes that advanced at up to 20 metres per year and overran the township of Eucla in the 1940s (Jennings 1963).

Blowouts may develop naturally during a phase of aridity when the vegetation that had held them under preceding humid conditions is weakened, resulting in mobilization of wind-blown sand. Dune erosion can also be initiated by stronger and more frequent wind scour associated with increasing storminess. Blowouts often form where the outer margin of vegetated coastal dunes is cut away by the sea during a storm, leaving an unvegetated cliff of loose sand exposed to onshore wind action. On a prograding shore, rebuilding of the beach, with new berms developing into newer foredunes, prevents much erosion, but if the coastline is gradually receding, and the beach not fully replaced, such blowouts will continue to develop and grow, especially on parts of the coast that are exposed to strong winds.

A blowout that grows until its axial length is more than three times its mean width is termed a parabolic or U-dune, and has an advancing nose of loose sand (sloping at 30°–33°) and trailing arms of partly-fixed vegetated sand on either side of an axial corridor excavated by deflation (Figure 7.4). In some areas they have been halted and fixed by vegetation as parabolic dunes within an older system of parallel dunes.

Parabolic dunes have noses of bare sand that spill in the direction of the dominant wind. The driving force of this wind shapes the parabolic form, which in turn modifies the near-surface wind flow. Analysis of wind flow patterns over a parabolic dune in the Sands of Forvie, Scotland, by Robertson-Rintoul (1990) demonstrated a crestal jet, windward and leeward eddies over the advancing nose, and spiral vortices beside the trailing arms. These wind flow patterns result from parabolic dune morphology, but then shape its further evolution. Changes in mobile or partly mobile dunes thus represent an adjustment between wind flow characteristics and the surface morphology of incoherent sand.

Elongated parabolic dunes with axes trending SW–NE and W–E are also seen in Barry Links on Buddon Ness, a large cuspate foreland on the north shore of the Firth of Tay, east of Dundee in Scotland (Landsberg 1956). On Braunton Burrows in north Devon there are blowouts and parabolic dunes, the outlines of which persist despite devegetation and damage during a phase of intensive military use.

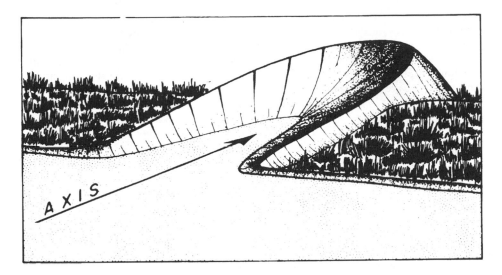

Figure 7.4   The formation of a blowout through a foredune.

Active and fixed parabolic dunes are common on the south-eastern coasts of Australia, and were described by Jennings (1957) from King Island. He showed that their movement was a response to the direction, frequency, and strength of onshore winds, the axis of each parabolic dune, defined as the line bisecting the angle between the trailing arms and directed towards the advancing nose (Figure 7.5), running parallel to the resultant of onshore winds of Beaufort Scale 3 (12–19 kilometres per hour) and over (Figure 5.14). On King Island the parabolic dunes on the west coast are moving inland eastward while those on the east coast are moving inland westward. The axial directions of numerous parabolic dunes cut through the foredunes behind the Ninety Mile Beach change as the curvature of the coastline brings in different component groups of onshore winds with different angles of onshore resultant (Bird 1978a). In north Queensland elongated parabolic (hairpin) dunes cut through previously fixed vegetated dunes have axes parallel to the prevailing south-easterly winds. They continue to grow because the hollow immediately behind the nose is kept clear of vegetation, being occupied by a lake in the wet season and drying out to expose readily deflated sand in

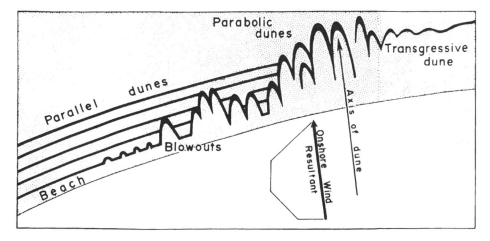

Figure 7.5   Development of blowouts in parallel foredunes and their growth into parabolic dunes and eventually transgressive dunes. The axis of a blowout or a parabolic dune runs parallel to the onshore wind resultant (Figure 5.6).

the dry season (Bird 1965). In Tasmania there are large, wide parabolic dunes on the west coast, where onshore gales vary in direction from north-west through west to south-west, in contrast with the elongated, narrow hairpin dunes on the north coast, where the onshore winds are more consistently north-westerly (Davies 1980).

TRANSGRESSIVE DUNES

In addition to blowouts and parabolic dunes there are broader transgressive mobile dunes that form where sand blown inland from a beach has been retained by vegetation. Alternatively, they may form where previously vegetated coastal dunes have been disrupted by numerous blowouts until they merge into an elongated dune, spilling inland (Figure 7.5). Parabolic dunes retain their form as long as the trailing arms are held back by vegetation, but if trampling, overgrazing or fires reduce this vegetation the parabolic form evolves into transgressive sand sheets (Hesp and Thom 1990).

Transgressive dunes are extensive on desert coasts, as in Baja California, Namibia, the Atlantic coast of Morocco, and along the shores of the Red Sea. They are also found in humid regions where the sand supply is rapid enough to prevent their stabilization by vegetation, as in the Slowinski National Park on the Polish Baltic coast (Borowca 1990). At Parangtritis on the south coast of Java, where large quantities of sand arrive periodically on the beach as a sequel to eruptions of the Merapi volcano in the hinterland, dunes are blown inland by south-easterly trade winds. On the Pacific coast of the United States large transgressive dune systems occur where there is an abundant supply of sand from beaches (particularly where the coast swings westward, so that southward drifting sand can accumulate on wide beaches) and frequently strong onshore winds to blow sand inland (Cooper 1958, 1967). These dunes are halted as precipitation ridges when they encounter forest vegetation. In south-eastern Australia transgressive dunes may have originated during Pleistocene phases of low sea level, when sand was blown landward from the emerged sea floor (Hesp and Thom 1990).

There are also long-walled transgressive dunes that form at right angles to the dominant wind and migrate downwind. They usually have an undulating crest that is often slightly sinuous, a gentle windward slope and a steep leeward slope (slipface). They advance when wind-blown sand spills down the steep leeward slope.

These masses of mobile, unvegetated sand show forms similar to dunes in arid regions, the dune crest profile at any time being a response to preceding wind conditions. A strong wind in one direction drives the dune crest one way, with a lee slope of up to 32°, but subsequent winds from other directions then modify the outline. Predominant westerly winds are driving dunes of this kind eastward along the Cape Coast of South Africa, where they spill into Sundays River (Heydorn and Flemming 1985), and similar features are seen on the coast of Uruguay near Montevideo and at Cape Howe in south-eastern Australia.

Sequences of transgressive dunes are found on the west coast of Auckland, New Zealand (Schofield 1975). In Spain, successive waves of transgressive dunes are being driven inland by the prevailing south-westerly winds on the shores of the Gulf of Cadiz (Vanney et al. 1979). The Pyla dune, in south-west France is a huge mass of sand moving from the eroding shore south of the Arcachon estuary, and spilling inland over the Landes pine forests. It was initiated during the eighteenth century, and has now attained an altitude of more than 100 metres, with a steep wall (32°) of spilling sand. A dune cliff cut by tidal scour on the estuary shore, resulting from deflection of the ebb channel by the southward growth of Cap Ferret spit, maintains the sand supply to this transgressive dune.

Extensive unvegetated sand areas may include barchans, which are sand mounds with lateral arms trailing downwind, similar to those that form inland in deserts. On the arid coasts of Baja California and south-west Peru barchans move downwind at up to 30 metres per year (Goldsmith 1985). Barchans are also found on coasts in humid regions where there are extensive areas of bare, drifting sand, as at Yanakie, near Wilson's Promontory, Australia, where the climate is cool and mean annual rainfall about 900 millimetres. Small barchans sometimes form during gales on wide beaches and unvegetated backshore sand areas, migrating downwind (onshore, seaward or alongshore). They have been described from the south coast of Honshu in Japan, the west coast of Schleswig-Holstein in Germany (Ehlers 1988), and observed at Stanley in Tasmania.

It is sometimes possible to distinguish successive waves of transgressive dunes that have migrated inland from a shore. In Newborough Warren in north Wales three parallel transgressive sand ridges have drifted inland, each backed by low-lying sandy terrain on which zones of vegetation of increasing age upwind represent stages in plant succession (Ranwell 1972).

On the Sands of Forvie in Scotland seven roughly parallel sand ridges have migrated north-eastward, one of them having buried Forvie chapel. Successive dune ridges, separated by rather broad low areas (i.e. not just narrow swales) have migrated inland behind Newcastle Bight and several other bay beaches along the New South Wales coast (Hesp and Thom 1990), and similar sequences have been described from Morro Bay, California, by Orme (1990). Broad low-lying plains can be formed within dune terrain by deflation of sand, usually to the level of the water-table, sometimes with minor dunes (residual or newly-formed) over the moist sand, or strewn with coarse residual lag deposits (Carter et al. 1990).

On some coasts formerly transgressive dunes have been stabilized by vegetation, particularly marram grass planted during the past century. During phases when larger quantities of sand were arriving, or when the vegetation was sparser, dunes were bare and mobile, drifting inland, and up and over headlands. On the north coast of Cornwall some dune systems extend below present sea level, and the wind-blown sand was derived from the emerged Atlantic sea floor during Pleistocene times, before and during the Late Quaternary marine transgression. Dunes then spilled inland, across what has become the present coastline. On the north coast of Devon and Cornwall formerly transgressive dune topography extends up to 2 kilometres inland from the high tide line at Braunton Burrows and Gwithian Sands. Subsequently, drifting dunes buried farms, villages and churches behind St Ives Bay and Perran Sands in Cornwall (Bird 1998), and at Kenfig Burrows in south Wales, where there is archaeological and documentary evidence that sand dunes were advancing inland in mediaeval times. It has been suggested that, after earlier stability, sand drifting was unusually active in these areas in the thirteenth and fourteenth centuries because of a phase of stormier climate or a minor sea level oscillation that sharply augmented the coastal sand supply (Steers 1964). Earlier dune instability is indicated at Skara Brae, in the Orkney Islands, where dunes overran an Iron Age settlement 4500 years ago. The extensive dunes at Culbin Forest on the north-east coast of Scotland drifted eastward until they were stabilized by afforestation with pine trees, which began in 1839 (Steers 1973).

The very high dunes of the coast of south-east Queensland, on Fraser Island, Stradbroke Island and Moreton Island (where at one point they exceed 275 metres above sea level) are not piled over rocky foundations. They consist entirely of wind-blown sand stabilized by vegetation in parabolic dune patterns, and borings have shown similar aeolian sand extending well below sea level. There are several sets of transgressive dunes, arranged in sequence, parallel to the ocean coast, each partly overlapping its predecessor. These have advanced away from the shore during phases of instability and dune migration related to Pleistocene phases of aridity, strong south-easterly wind action and a superabundant sand supply from the emerged sea floor during low sea level phases. They are now stabilized beneath a cover of scrub and forest (Bird 1974).

On the New South Wales coast there are several sectors where transgressive dunes of mobile, unvegetated sand are migrating inland and burying older dune topography with a scrub or forest cover. Some of these active dunes were initiated by human activities, notably the grazing of stock and the burning of vegetation, liberating unconsolidated dune sand on the coastal margin. Some blowouts and transgressive dunes originated prior to European colonization, for Captain Cook observed active dunes on the New South Wales coast and the Queensland coast north as far as Fraser Island in 1770. Aborigines may have initiated these when they repeatedly set fire to vegetation in order to hunt animals, and this weakening of the vegetation cover assisted the formation of blowouts (Bird 1974).

Dissection of previously vegetated dune topography can result in the formation of residual mounds (knobs) of vegetated sand, crowned by a clump of grass or shrubs, within areas of bare and drifting sand, as in the dunes behind Discovery Bay in south-eastern Australia. Lee or shadow dunes may form downwind of such remnant knobs, or vegetated mounds (Hesp and Thom 1990). Goldsmith (1985) mentioned a distinctive type of unvegetated dune, known as a medaño, which is steep-sided and tens to hundreds of metres high, formed where winds blow from several directions, moving sand towards the summit. Such a dune changes in shape in response to differing wind directions, but shows little if any migration. Examples have been described from Coos Bay, Oregon and south-east Lake Michigan. If colonizing vegetation prevents the arrival of sand from one direction the medaño is shaped into an elongated ridge.

The effects of human activities on mobile coastal dunes include accretion induced by sand fence construction or the planting of vegetation and erosion resulting from trampling, grazing and burning and sand mining. In the United States dune areas have been stabilized by road construction and urban development, as on the New Jersey coast, where Gares

(1990) compared processes and morphology on urbanized and undeveloped dunes, and found that wind flow patterns and the net landward movement of sand were much reduced by obstructions such as houses, roads and boardwalks, as well as by sand-trapping fences. Such urbanization may stabilize dune topography, but it can be threatened by the recession of bordering coastlines accompanying beach erosion.

CLIFF-TOP DUNES

On some cliffed coasts, onshore winds have blown sand from the beach and piled it against the cliff as a climbing dune. At Foreness Point in Kent wind-blown sand has been banked against the Thanet chalk cliffs during the past few decades. Such dunes may grow to spill inland over cliffed headlands, and where there is no longer a source of sand they become relict cliff-top dunes, as at Church Cove and Gwithian Towans in Cornwall (Bird 1998).

On parts of the Australian coast dunes have climbed up and over coastal promontories, as on Cape Bridgewater and Cape Paterson on the Victoria coast, where they form cliff-top dunes on the lee side. Jennings (1967) showed that some cliff-top dunes have formed when the sea stood at a higher level while others (notably on dune calcarenite coasts) originated as transgressive dunes moving in from the emerged sea floor during Pleistocene phases of lower sea level on to a coast that later became cliffed (Figure 7.6).

Dunes found on the top of a cliff may not have come from nearby beach sands. In arid regions some cliffs

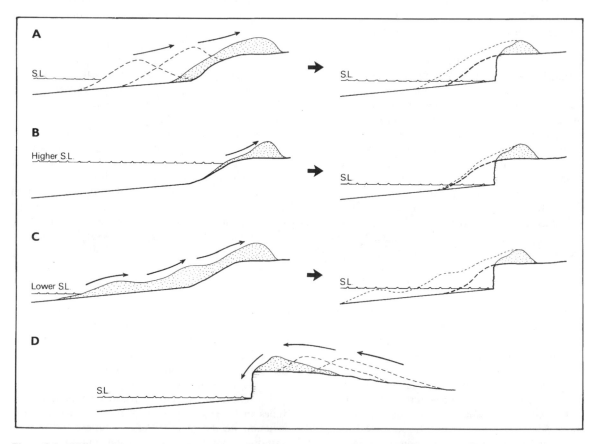

Figure 7.6   Cliff-top dunes may form A – where a transgressive dune moves up a cliff, and is later truncated by marine erosion, B – where a dune formed at a higher sea level is stranded by cliff recession after emergence, C – where a dune formed when sea level was lower is truncated by marine erosion as the sea rises, and D – where the dune has moved from inland to a position on the cliff top.

have been cut back into terrain that carries desert dunes formed by deflation of unvegetated or sparsely vegetated landscapes. There are cliff cappings of red (pindan) desert sands in north-western Australia. On the coast of Port Phillip Bay in south-eastern Australia, the grey dunes that cap the cliffs in the south-eastern suburbs of Melbourne are part of a series of elongated parallel dune ridges that formed on the coastal plateau as desert dunes during a Pleistocene arid phase (Bird 1993a).

DUNES ON SHINGLE

Similar explanations may apply where there are dunes on shingle beach ridges with no existing link with a sand supply from an adjacent beach. Sand may have arrived as a transgressive dune spilling from a nearby beach, possibly when sea level was higher or lower than it is now. In Devon, the grassy Northam dunes on the northern end of the Westward Ho! cobble spit formed when there was a contiguous sandy beach from which westerly winds supplied sand, but the sand supply from the beach diminished, the connection was broken and the dunes are now relict. Similar relict dunes stand on shingle beach ridges at Blakeney Point and Scolt Head Island in Norfolk, the link between the beach sand source having been broken by the accretion of younger shingle ridges (Steers 1960).

RATES OF DUNE MOVEMENT

Dune movement is often impeded by vegetation or obstacles, including natural rock outcrops, but where there is unimpeded drifting, measurements of dune movement have indicated that unvegetated dune fronts 10 to 20 metres high can advance downwind at rates of 1 to 10 metres per year. The 100 metre high dune at Pyla in south-western France is said to be advancing inland at an average annual rate of 1 metre, but measurements on dune fronts at Cronulla, New South Wales showed that a dune 20 metres high advanced 19 metres, and smaller dunes as much as 50 metres in a year (Chapman et al. 1982).

Rapid rates of dune advance occurred in mediaeval times in Europe, possibly because of increased storminess, but perhaps because of increased clearance and impoverishment of dune vegetation. Records of advancing dunes burying buildings, villages and farmland imply that the sand advanced rapidly over a few decades.

DUNE CALCARENITE

Where coastal dune sands are calcareous the older dunes may become lithified by internal deposition of calcium carbonate from percolating water, to form dune calcarenite, which preserves the dune topography in solid limestone. Originally described from Bermuda by Sayles (1931), the term calcarenite is strictly applied when the carbonate content of the sand exceeds 50 per cent, but dune sands can become coherent and lithified when the carbonate content is as low as 10 per cent.

Dune calcarenite topography is found on the southern and eastern coasts of the Mediterranean, in Morocco, on the Red Sea and southern Arabian coast, in western India, around the Caribbean, on the Brazilian coast, in South Africa, and extensively on the western and southern coasts of Australia (Fairbridge and Teichert 1953). It also occurs on oceanic islands such as Mauritius and Hawaii. On the Atlantic coasts of Britain Holocene dunes derived from beaches of calcareous sand contain lightly cemented horizons, but true dune calcarenite has been observed on Balta Island, a small island off the east coast of Unst in the Shetland Islands. There is a useful review by Kaye (1959).

Dune calcarenites vary in degree of lithification by precipitated carbonates, which in turn depend on leaching of carbonates from overlying calcareous sands in wet phases alternating with periods dry enough to precipitate them. Some carbonates may be brought up in rising groundwater, especially where high evaporation at the surface promotes upward capillary movement between the sand grains, and this may result in the formation of calcrete layers on the dune surface. Plant roots also draw carbonate-bearing groundwater upward, and cause precipitation and induration of enclosing calcrete pipes (rhizoconcretions) within the dune calcarenite. Subsequent erosion may expose these as 'petrified forests', as at Cape Bridgewater, Australia (Bird 1993a).

Some dune calcarenites remain relatively unconsolidated, and can be mobilized by wind action if surface vegetation (or an overlying calcrete crust) is removed. Others become consolidated as calcareous sandstone, often with laminated dune bedding (biscuit rock) or surface layers of indurated calcrete, on which karst topography may develop.

The dune calcarenites of the Nepean peninsula in Victoria, Australia, comprise a series of superimposed Pleistocene and earlier dune formations which extend more than 140 metres below, and up to 60 metres above, present sea level, a notable example of coastal dune stratigraphy (Bird 1982). They are largely the product of winnowing of sand from the emerged floor of Bass Strait during Pleistocene low sea level phases, and of shoreward drifting of sand during the Late Quaternary marine transgression. This coastline now consists of cliffs and shore platforms cut into the consolidated Pleistocene dune calcarenites, and there are unconsolidated cliff-top Holocene dunes (p. 190) (Figure 7.7).

The contrast between generally calcareous beach sands on the western and southern coasts of Australia and predominantly quartzose beach sands in the south-east (p. 104) is reflected in coastal dune morphology. There is relatively stable Pleistocene dune calcarenite topography on the western and southern coasts of Australia, but on the south-eastern coast, where the Pleistocene dune sands are quartzose, and this kind of lithification has not taken place, and the dunes remain unconsolidated, either active and mobile or retained by a vegetation cover. In South Australia there is a longshore transition in the composition of dune sands south-east from the Lower Murray, the proportion of

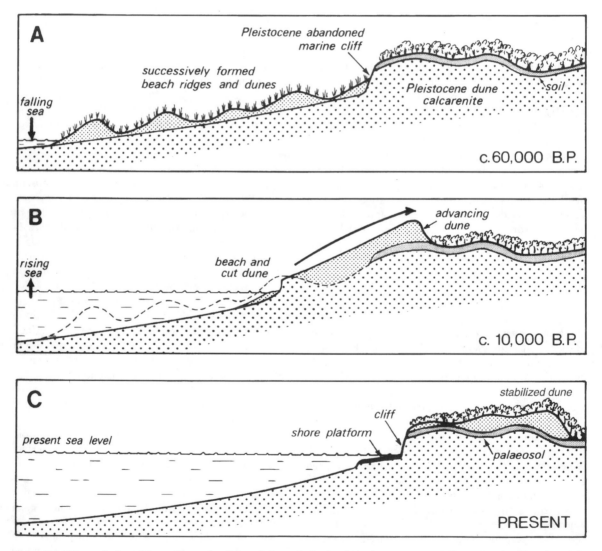

Figure 7.7   The evolution of dune calcarenite cliffs and shore platforms. A – when sea level is falling, dunes are stranded on the emerged sea floor in front of an earlier dune calcarenite coast, B – when sea level rises, these dunes are submerged and eroded, some sand being blown landward, C – when the sea has reached its present level it cuts a cliff and shore platform in the dune calcarenite, capped by Holocene dunes. The dune calcarenites were formed in earlier such cycles as sea level oscillated in Pleistocene times.

terrigenous quartzose sand diminishing as the proportion of marine calcareous sand increases south eastward (p. 137). The strongly calcareous dune ridges to the south-east are preserved in solid dune calcarenite, but as the sand becomes less calcareous lithification diminishes and the limestone ridges grade north-westward into more irregular and disrupted dune topography (Sprigg 1959). On the eastern coast of the Mediterranean there is a similar transition, the calcarenite dune topography of the Israeli coast passing southward into more irregular, mobile dunes on the coast of the Gaza Strip, where the sands have an increasing content of terrigenous (Nile-derived) quartzose sediment.

The submergence and dissection of Pleistocene dune calcarenite topography formed during low sea level phases is well-known off the coast of Western Australia, where the Late Quaternary marine transgression isolated Rottnest Island and Garden Island as it invaded Cockburn Sound and submerged the lower part of the Swan River valley (Churchill 1959). Some submerged ridges have been planed off by wave action on the sea floor, but detailed soundings have shown that segments of dune topography survive off the southern and western coasts of Australia. There are also submerged dune calcarenite ridges off the Bahamas, north-east Brazil, and Sri Lanka, and some of these have become reefs with coral and algal crusts.

MACHAIR

On the coasts of Scotland and Ireland there are areas of almost featureless low-lying calcareous sandy plain, known as machair (Ritchie and Mather 1984). Typically these lie behind a vegetated coastal dune fringe, as at Eoligarry on the northern part of the Hebridean island of Barra (Figure 5.7). Morrich More in Easter Ross is a low-lying sandy foreland on the southern shores of Dornoch Firth, with developing foredunes behind Whiteness Sands, backed by fixed and mobile parabolic dunes, then by machair.

Machair is generally of Holocene age, much of it formed between 6000 and 2000 years ago (Ritchie 1977) although there are sites where it may still be developing. It may be associated with areas of Holocene coastal emergence, for although it is extensive in Scotland and Ireland there is no equivalent behind the calcareous dunes of south-west England, although thin splays of fine wind-blown sand are sometimes seen in the lee of such dunes, as at Bantham in

south Devon. Machair can form where a coastal dune has been removed down to the level of the water table, as at Trawenagh Bay in County Donegal, Ireland (Carter et al. 1990), or where sand deflated from coastal dunes is deposited in lagoons or marshes. This is possible behind Machir Bay on Islay, and behind Traigh na Berie on the island of Lewis, where the beach is backed by dunes, which impound two small lochans, towards which the machair becomes marshy. Often machair is undergoing dissection by blowouts and gullies, as on Gualann Island on the west coast of South Uist, Balta Island in the Shetlands and behind Torrisdale Bay in north-west Scotland (Gimingham et al. 1989).

COASTAL DUNES IN THE HUMID TROPICS

Coastal dunes are best developed on coasts in the temperate and arid tropical zones. Jennings (1964, 1965) noted that in the humid tropics they are of limited and local extent, sandy coastal topography consisting of low beach ridges with little transgressive dune development. It has been suggested that the prevailing dampness of beach sands on humid tropical coasts impedes deflation and backshore dune development, particularly on microtidal sectors where the beaches are narrow. Another suggestion is that rapid colonization of coastal sand accumulations by luxuriant vegetation prevents dune development in the humid tropics, but it is doubtful if the vegetation has much effect: if aeolian sand were arriving at the backshore, dense vegetation would simply trap it and facilitate the building of a high foredune. Coconut palms and *Casuarina* trees often grow immediately behind the beach, forming a wall of vegetation, but this is partly because of the lack of strong onshore winds, and it is that lack, rather than the presence of vegetation, that explains the absence of coastal dunes (Davies 1980).

The paucity of dune development in the humid tropics is a consequence of a relatively meagre sand supply to the backshore, largely because winds strong enough to deflate beach sand and build up dunes are rare in comparison with other climatic zones. The occasional violence of winds in tropical cyclones is usually accompanied by torrential rainfall which saturates the beach surface and impedes sand transportation by the temporarily strong wind action. The formation of salt crusts on beaches in Ghana has impeded the development of coastal dunes.

Nevertheless, dunes are found on humid tropical coasts where there is a sustained sand supply and prevailing onshore winds, as at Parangtritis in southern Java (p. 102). In Fiji coastal dunes are poorly developed except in the south-west, near Singatoka, where black magnetite sands brought down by the Singatoka River have been washed up by the strong south-westerly swell to form beaches that are a source of backshore dunes. Local development of dunes occurs where there is a dry season, as on parts of the Malaysian and Sri Lankan coasts (Swan 1979), and dunes persist on humid tropical coasts where they have been inherited from earlier drier phases, as in north-east Queensland (Bird 1965, Pye 1983).

OLDER AND NEWER DUNES

Some parts of the coast show evidence of distinct phases of coastal dune accumulation, with older stabilized dunes on the landward side and newer dunes, either stable or mobile, bordering the coast. This pattern is clearly developed where the dunes are built of quartzose sand, as on King Island in Bass Strait, Australia (Jennings 1957), where there are contrasts between the topography, soils and vegetation of older and newer dunes. Older quartzose dunes (usually of Pleistocene age) have a comparatively subdued topography, and have been leached of shelly fragments and other calcareous sediment by percolating rainwater, sometimes to a depth of a metre or more, as podzols develop. They have sometimes been called grey dunes because of the colour of the leached surface sand, in contrast with the yellow newer dunes, where the surface sand still retains a colouration with sand grains coated by iron oxides.

Newer dunes (usually of Holocene age) form a coastal fringe, and are more continuous, with bolder outlines, and accretion of quartzose sand often still continuing. They are fixed by grasses or scrub, except where blowouts and parabolic dunes are developing, or where large transgressive masses of mobile sand are advancing inland. The sand is fresh and yellow or brown in colour, and has not yet been leached of its small shell content or of the iron oxides which stain the sand grains.

The junction between older and newer dunes is often well marked, particularly where the newer dunes are transgressive, advancing across the more subdued older dune topography. Coastal barriers of quartzose sand show a similar contrast, inner barriers (generally of Pleistocene age) having older dune topography while outer barriers (usually added in Holocene times, after the Late Quaternary marine transgression)

bear newer dunes. In Oregon, the inner, older gently undulating and now vegetated transgressive dunes formed before the Late Quaternary marine transgression brought the sea to its present level, when younger Holocene foredunes were added along the coast (Cooper 1958).

A contrast between older and newer dunes is also seen on calcareous dunes, where the older dunes are preserved in dune calcarenite, and newer dunes remain unconsolidated, but the contrast in vegetation is less marked, the older dunes having only a superficial layer of sand decalcified by percolating rainwater.

While older quartzose dunes are usually of Pleistocene age there are occasional examples of grey dunes of Holocene age with podzolic soils bearing heath vegetation. On the shores of Wilson's Promontory, Australia, beaches derived from weathered granite consist almost entirely of grey quartzose sand without an iron oxide staining, and in Leonard Bay Holocene dunes derived from these are already grey, and have been colonized by heath vegetation (Bird 1993a). On South Haven peninsula, Dorset, heath vegetation has spread on to quartzose Holocene dunes where leaching of sparse carbonates and iron oxide coatings has proceeded rapidly (Ranwell 1972).

The sequence of older (Pleistocene) dunes inland, or on inner barriers and younger (Holocene) dunes on the coastal fringe, or on outer barriers is widespread in Australia, but is not found in western Europe north of the Naples area in Italy. Pleistocene coastal dunes in north-west Europe were probably largely removed by glacial or periglacial processes during the Last Glacial phase, but in south-west England there are older dune sands overlying emerged Late Pleistocene shingle beaches in coastal outcrops at Godrevy Rocks, Trebetherick Point, and Fistral Beach in Cornwall and Saunton in Devon, in each case capped by periglacial Head deposits and yellow Holocene newer dune sands. The Pleistocene dune deposits are here beneath, rather than landward of, Holocene dunes (Bird 1998)

DUNE SANDROCK

Reference has been made to the formation of humate, horizons of sand cemented by downwashed iron oxides and organic matter in the course of podzol formation on older dunes. Humate is a form of sandrock that is sometimes exposed as backshore ledges of slightly more resistant material where the dunes have been dissected by wind action or trimmed back by the sea. Similar ledges may be formed of sandrock that originated where sandy swamps or peaty sand in low-lying seasonally or permanently waterlogged sites have been overrun and compressed beneath advancing dunes. This type of sandrock often contains compressed plant remains, which are not found in humate (Hails 1982).

DUNE LAKES

Hollows in dune topography that pass beneath the level of the water table are occupied by dune lakes, some of which may be intermittent, forming only when the water table rises after heavy rains, and drying out subsequently. Dune lakes are usually round or oval in shape as the result of wave and current action generated on them by wind action, the resulting configuration being related to the wind regime. Some lakes have been impounded where dunes built across the mouth of a valley have ponded back the stream, while others occupy hollows excavated by deflation during dry weather, particularly between the trailing arms of parabolic dunes, as on the Queensland coast north of Cooktown, where they provide a sand source for dunes downwind from the lake basin (Bird 1965).

There are good examples of dune lakes within the high dune topography of Fraser Island in Australia, where they stand at various levels in depressions where the presence of underlying impermeable humate (sandrock) prevents the water draining away. Several have been impounded in hollows in the older dune topography enclosed by the advance of transgressive newer dunes. Where a dune coast has been cut back by marine erosion, peat deposits that formed in dune lakes may be exposed overlying, or interbedded with, dune sands in cliff sections, as at Ocean Beach, near Strahan in Tasmania.

Infilling of dune lakes by blown or inwashed sand or by the formation of peat deposits may produce flat-floored enclaves within a stabilized dune topography. Similar flat-floored enclaves may be formed by erosion rather than deposition, where the wind has blown away dry sand to expose flat wet sand (or a humate outcrop) at the level of the water table.

8

Intertidal Landforms, Salt Marshes and Mangroves

INTRODUCTION

Much of the discussion so far has been concerned with high to moderate wave energy coasts exposed to the open sea, but within embayments, inlets, estuaries and lagoons, where wave energy is relatively low, cliffs and shore platforms become more subdued, and beaches less regular, with such features as lobes, spits and forelands. Shore forms include depositional sandflats and mudflats, shoals and vegetated areas exposed when the tide falls, and these generally become wider and more extensive as tide range increases.

Studies of the geomorphology of intertidal zones on low to moderate wave energy coasts, and of vegetated areas such as salt marshes, mangrove swamps and seagrass beds, have been dominated by work in estuaries and lagoons. However, such features also occur on open coasts where the tide range is large and the transverse shore gradient small, and it is necessary to consider them in a wider context before dealing with them in estuaries and lagoons.

Examples of coasts with extensive intertidal zones include Westernport Bay (Figure 8.1) in south-east Australia, a landlocked embayment linked to the sea by straits east and west of Phillip Island. The tide is augmented as it passes into Westernport Bay, so that tide range increases from about two metres in the entrances to more than three metres north of French Island. As the tide falls the waters subside into converging creek systems on either side of a tidal divide, and extensive intertidal mudflats are exposed, backed by a high tide shoreline that is partly cliffed, partly sandy, and extensively mangrove fringed. Westernport Bay is essentially marine, but shows estuarine features at the mouths of Bass River and a number of smaller streams. Somewhat similar features are seen in Poole Harbour, Dorset, where the intertidal mudflats front salt marshes. Southampton Water, Portsmouth Harbour, Langstone Harbour and Chichester Harbour on the south coast of England

were formed by Late Quaternary marine submergence of low-lying parts of the Hampshire and West Sussex coastal plain. They have extensive intertidal mudflats and backing salt marshes, and have persisted because the small inflowing rivers have not supplied much sediment and because this is a tectonically subsiding coast.

Intertidal sandflats, mudflats and salt marshes are extensive in Bridgwater Bay, on the coast of the Bristol Channel, the Bay of Mont Saint Michel in France, and the Bay of Fundy in eastern Canada. In the tropics, where salt marshes give place to mangrove swamps, intertidal mudflats occupy macrotidal bays, as on the north coast of Australia in Cairns Bay, where they are backed by a sandy beach ridge plain (Bird 1970).

Intertidal areas may also be rocky or bouldery, sometimes with shore platforms, as discussed in Chapter 4, sandy or gravelly on beach-fringed coasts, as discussed in Chapter 6, or sandy and muddy, particularly on sheltered coasts near river mouths and in estuaries and lagoons. Details of the morphology and sediment distribution in intertidal areas are often poorly rendered on coastal maps and charts. In Britain, for example, Ordnance Survey maps show the high and low water lines, with shaded sand areas (which may include dunes) and marsh symbols (but not muddy areas), but supplementary surveys are needed to document the pattern of intertidal sediment types.

Intertidal areas consisting largely or entirely of sandy or muddy deposits are often called sandflats or mudflats, even though they usually have a gentle transverse gradient seaward (e.g. 1:1000), often with a smooth, convex profile, and include relatively steep slopes, especially where they decline into tidal channels. They are generally contiguous with the coastline, extending from the high tide shoreline down to the low tide shoreline, but they also occur as shoals separated by deeper channels or straits. In places, sandflats and mudflats may have undulating surfaces, with bars and troughs, and shallow tidal channels.

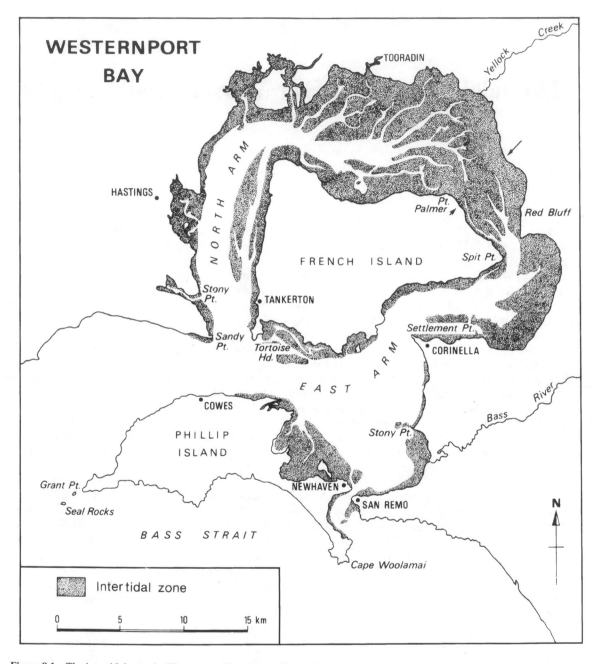

Figure 8.1   The intertidal zone in Westernport Bay, Australia.

Salt-tolerant plants that colonize intertidal sandflats and mudflats, especially in the higher parts, may develop salt marshes, mangrove swamps or seagrass beds. Before considering these it is necessary to examine the processes that shape intertidal sandflats and mudflats.

SOURCES OF INTERTIDAL SEDIMENTS

There are several possible sources of the sandy and muddy sediment (mainly silt, clay and organic matter, sometimes with a little fine sand) deposited in the intertidal zone. Some is delivered to the shore by rivers and distributed along the coast by wave and current action, some originates from the erosion of sand, silt and clay from bordering cliffs (including alluvial coastal plains and deltas), and some drifts in from the sea floor.

Fluvially-supplied sand and mud is extensive on intertidal shores bordering deltas, off river mouths, and in estuaries and lagoons. The sandy and muddy deposits in the Thames estuary, for example, include sand derived from the Reading Beds and fine-grained sediment derived from the Oxford clay and the London clay, carried downstream to the estuary and on into the North Sea. Humid tropical rivers carry vast quantities of mud derived from deeply weathered hinterlands to the coast, to be deposited in intertidal areas.

Cliffs cut into soft sand (including coastal dunes) and clay formations (including boulder clay) are sources of intertidal sand and mud around the North Sea. Muddy sediment derived from the clay cliffs of Holderness and north-east Norfolk has been dispersed by waves and currents, and deposited in The Wash and the North Sea estuaries.

Reference has been made to sand being swept in from the sea floor to beaches and sandy shores, and muddy intertidal deposits can also be supplied in this way, particularly where there are offshore outcrops of weathered mudrock or fine-grained bottom deposits such as glacial boulder clay. Muddy sediment from glacial drift on the floor of the North Sea has been delivered to intertidal areas in the Wadden Sea on the coasts of Denmark, Germany and the Netherlands, and has contributed to mudflats in eastern England.

SANDFLATS, BARS AND TROUGHS

Sandflats are exposed in front of many beaches as the tide falls, and also occur where waves and currents have carried sand onshore or alongshore and deposited it in intertidal areas fronting cliffs and rocky coasts or alluvial coastal plains and deltas. Sandflats are best developed where wave action is strong enough to have prevented the deposition of silt and clay, and they typically grade laterally into mudflats where wave energy diminishes alongshore. Alternatively, sandflats may occur seaward of mudflats, as in The Wash, where they occupy a slope between the high neap tide and mid tide lines, declining to a muddy sea floor (Evans 1965).

In general sandflats are wave-dominated features, but currents (particularly tidal currents) also contribute to their morphology. Many sandflats have bars and troughs, a bar being a ridge or bank of sand (sometimes gravel) built up by wave action parallel to the coastline to a level where it is exposed at low tide but submerged at high tide, and a trough the intervening depression. A bar is a concentration of sediment formed by breaking waves (break-point bar), material that is being carried shoreward meeting that withdrawn from the beach by backwash. Some features called bars (such as Loe Bar in Cornwall) are really coastal barriers, built above high tide level (p. 172).

Break-point bars can be shaped in wave tank experiments (Rey et al. 1995), where it is found that their size and distance from the shore are related to the dimensions of the waves, higher waves building larger bars farther offshore. In calm weather, when constructive swash is more effective, the bars move closer to the shore and become swash bars, flatter in profile, sometimes with a steeper shoreward advancing slope. Bars of this type are well developed off the Ninety Mile Beach in Australia, where in fine weather gentle waves wash them shoreward until they become welded on to the beach as berms (p. 125).

At high tide waves break across intertidal bars and troughs, generally as spilling breakers (p. 12). After the strong swash crosses the bar, the wave may re-form as it moves into the deeper water of the trough, the surf falling back from the crest as a new smooth wave front emerges. Bars are often interrupted by transverse channels formed and maintained by rip currents, which complete the nearshore water circulation by carrying water driven on to the shore by breaking waves back seaward through the surf zone (p. 14).

On sandflats where the tide range is sufficient to expose a broad foreshore at low tide there may be numerous subdued ridges (multiple bars) and troughs running parallel, or at a slight angle, to the coastline (Figure 8.2). These are known as low and ball or ridge and runnel in Britain, where they are well developed on the lower part of the beach near Formby in south Lancashire (Gresswell 1953). Their amplitude rarely attains a metre, and the ridge crests are almost flat, and often as much as 100 metres apart. They are essentially swash-built bars, formed by relatively weak wave action. Their formation as rhythmic topography is an adjustment between surface form and the oscillating turbulence produced by spilling breakers: once the bars have formed, these waves break across them, re-form as they cross the intervening swales, and break again over the next bar. The number, spacing and amplitude of multiple bars varies with the height and period of breaking waves, but bar topography, in turn, influences where and how incident waves break. The profiles of bars and troughs are modified by wave

Figure 8.2   Multiple nearshore sand bars at Rosebud, Port Phillip Bay, Australia.

action as the tide rises and falls, the bars remaining subdued as their margins show minor progradation shoreward with the rise and seaward with the ebb. Where waves have been arriving at an angle to the shore the ridges and troughs also run obliquely, as on the south-facing shore at Porth Neigwl in north Wales, where the dominant waves come in from the south-west. As the tide falls the troughs may be temporarily occupied by lagoons, which drain out by way of transverse channels: the trough sands are often rippled and the bar crests smoother. Once established, ridge and runnel patterns persist on sandflats through many tidal cycles, with minor modifications by wave action (particularly during storms and when the angle of incidence of waves changes) and currents that strengthen as neap tides grow into springs. On the Danish coast Aagard (1991) reported that multiple

bars were shaped by low frequency edge waves (p. 14) and found that the outer bars changed only during stormy periods. On the French coast south of Arcachon, Michel and Howa (1999) investigated sandy ridge and runnel on a shore where the ridge was interrupted by curving rip channels flowing out from the runnel. Sand washed in by waves was added to the ridge, while sand washed out of the runnel through rip channels by strong ebb currents was re-worked by waves and returned to the ridge.

On the Australian coast parallel sand bars of this kind are found on shores where the effects of ocean swell are weak or excluded. Up to ten parallel sand bars occupy the intertidal sandflats between Rye and Rosebud on the south-east coast of Port Phillip Bay (Figure 8.2), where at high tide incoming waves break, re-form and break again in such a way as to develop and maintain the rhythmic intertidal topography. Similar features are seen on sandy intertidal sandflats in the Darwin area and on the coast near Port Hedland in north-west Australia.

Some sandflats have bars that have grown alongshore, and are essentially intertidal spits. At Gibraltar Point on the Lincolnshire coast there are parallel sand bars which look like swash-built ridge and runnel but have instead grown from north to south as the result of deposition by currents, and possibly waves coming in from the north-east, at an angle to the coast (King 1972). Sandflats can also be re-shaped by wind action, especially in the upper intertidal zone which remains dry between the highest tides: on Morfa Harlech in north Wales the wide sandy intertidal zone grades upward into hummocky dunes across a boundary that is not distinguishable on Ordnance Survey maps.

Transverse or finger bars are usually orientated at a high angle or perpendicular to the coastline, extending out into the intertidal sandflat. They are common off dissipative sandy beaches (p. 128) on low to moderate wave energy coasts, as in Florida and parts of north Queensland. No entirely satisfactory explanation of their origin is available, but they are produced where the seaward flow of water, after waves break upon the shore at high tide, is more dispersed than in rip currents, and they may be produced by intersecting edge waves (p. 14) (Holman and Bowen 1982). Once formed, incoming waves break along them in an intersecting pattern, but this is a consequence rather than a cause of transverse bar formation.

Where wave action is strong at high tide, the intertidal sandflat may have only subdued bar topography, or be almost featureless, consisting of smooth sand packed firmly by wave action. Such areas have been used for sports such as sand yachting in the Netherlands and on the French Channel coast east of Dunkirk, and as sites for motor racing on Pendine Sands in south Wales.

Intertidal sandflats are also found on shoals, which form offshore in various ways, some as the result of sand deposition in slack water between tidal channels, others by fluvial sand deposition off river mouths, usually during floods, and others by the dissection of former sand bars. They persist where wave energy is too low to move or disperse them. They are at least partly exposed at low tide, as the ebb drains into shallowing troughs and channels, but some are subtidal. Sandy shoals emerge at low tide offshore in the Goodwin Sands off east Kent and the Brake Bank off Lowestoft in eastern England, where the sand has been derived from glacial drift deposits, re-worked by waves and currents on the sea floor.

Elongated linear sand shoals and swales have been formed parallel to tidal currents in the Strait of Dover. Intertidal sand shoals are found in the mouths of estuaries, as in the macrotidal gulfs of northern Australia. The three King Shoals where the estuary of the Ord River opens to the northern end of Cambridge Gulf are intertidal sandy and muddy ridges 15 to 20 kilometres long and about a kilometre wide, aligned parallel to the ebb and flow of tidal currents. Estuarine shoals, with intricate and variable creeks, out of which the ebb tide drains, lie behind barrier islands and spits on the north Norfolk coast and in the Wadden Sea on the coasts of Holland, Germany, and Denmark. Shoals are also found between ebb and flood channels in estuaries (p. 228).

Various kinds of ripples form on intertidal sandflats. They are typically more or less regular, at intervals of between one centimetre and a few metres, and may be formed either by wave action, or by currents, or by a combination of both processes. Wave-formed ripples are usually asymmetrical and parallel, shaped by the eddies that form as a wave passes, separating, steepening and sharpening the ripple crests. Current-formed ripples are also generally asymmetrical, with a steeper face away from the current, and associations of cross-currents generated by waves and tides over sandflats at high tide can produce decussate or rhomboidal networks of transverse current ripples. In addition, there are longitudinal ripples that form parallel to a strong unidirectional current (similar to those that form on sandy river beds), and these can be complicated by interfering wave motion and cross-current patterns.

There are also giant ripples (megaripples) on sandflats, with crests spaced at intervals of several metres and amplitudes attaining, and sometimes exceeding, one metre. They are formed by strong currents, and are usually aligned at right-angles to the tidal flow, migrating in the direction of the current, but sometimes they run diagonally or parallel to the current flow. They are found in sandy intertidal areas on macrotidal estuaries, as in the Rhine and Scheldt in the Netherlands, where Van Straaten (1953) identified 20 different kinds of megaripple. McCave and Geiser (1979) described megaripples 0.3 to 0.6 metres high and 10–15 metres long, formed by strong currents during flood tides on intertidal sandflats in The Wash. On outer sandy shoals they graded into wave-formed ridge and runnel morphology, with wave-formed ripples in the muddy troughs. Multiple bars are essentially wave-built megaripples.

Details of current ripples are discussed at length by Allen (1968), and a classification was presented by Tanner (1982).

MUDFLATS

Muddy sediment is widely dispersed by waves and currents. Strong currents keep silt, clay and organic matter in suspension, but as they slacken deposition begins, and mudflats are formed. Typically they have a slight transverse slope across the intertidal zone, increasing seaward. Mud is deposited particularly at high and low tide slack water in the upper and lower parts of the intertidal zone, the intervening mid-tide zone having more sustained current flow as the tide rises and falls, inhibiting mud deposition. This can result in a steeper lower slope, which is often sandy, as in The Wash. Wave action erodes and sorts intertidal sediment as the tide rises and falls, and mud taken into suspension is deposited in calm sheltered water as the waves weaken. While the flow and ebb of tides produce net shoreward drift and a shoreward reduction in grain size from sand to silt and clay, mudflats are sometimes backed by a beach of coarser material (sand and shells) emplaced by wave action which is often stronger at high tide, and may even cut low backshore cliffs.

Deposition of muddy sediment in the intertidal zone is impeded by frequent tidal oscillations and disturbance by wave action, but can occur as the result of flocculation, due to the electrolytic effect of mainly sodium and chloride ions in brackish estuarine or sea water (Nichols and Biggs 1985). This results in the clustering and coagulation of clay particles into flocculated silt-sized particles that settle more readily during brief episodes of diminished current flow and wave turbulence to form muddy shoals and intertidal mudflats. Mud deposition is also assisted by the effects of marine organisms, which ingest clay and organic matter and excrete it as pellets large enough to be precipitated on mudflats, and also generate sticky mucus which aggregates fine particles. Often the fine sediment is initially fluffy (fluid mud), but becomes consolidated as pore water escapes. Once deposited, mudflats are more cohesive than sandflats because of their high water content, electrolytic binding and associated organic stickiness. They are not as readily mobilized by waves and currents as sand grains, and stronger currents are required to disrupt and disintegrate them into particles that can be rolled or taken into suspension.

In estuaries, mudflats form gently-sloping banks exposed at low tide on either side of a deeper channel, steepening as they decline to low tide level. They are moulded largely by current action as the tide rises and falls, but wave action can also influence their morphology. As deposition continues, vertical accretion may be accompanied by progradation of the seaward margins of mudflats, as on the shores of the Gulf of Bo Hai, served with muddy sediment from the Hwang Ho River, and in Turnagain Arm, Alaska, where progradation of up to 12 metres per year results from the inflow of sediment from the melting Portage glacier (Klein 1985).

The surfaces of intertidal mudflats have a variable topography subject to frequent, rapid changes in response to strong wave and current action (Pethick 1996). Deposition gives place to erosion as current flow exceeds about 10 cm/sec. (when shear stress exceeds shear strength), and there are alternations between slack water deposition and mid-tide ebb or flood current scour. Strong tidal currents produce irregularities, such as mounds and banks separated by hollows and troughs. Mudflats are thus reshaped as the tide rises and falls, particularly during spring tides, and modified in stormy periods when wave action is strong. Their surfaces are generally smooth, without the ripples found on sandflats: such ripples fade out as sandflats pass laterally through muddy sandflats to sandy mudflats. However, mudflat surfaces are often disturbed by marine organisms, including burrowing worms and shells. In the tropics, mudflats sustain a rich variety of highly productive marine organisms, including cockle shells. Off the west coast of Malaysia intertidal mudflats are often hummocky with clumps of worm-secreted tubes and patches of shelly gravel sorted from the mud by wave action.

Mudflats are generally diversified by channels or creek systems, often beginning in embayments or on the flanks of tidal divides and converging in dendritic patterns, but sometimes running almost parallel to each other down to the low tide shoreline (Figure 8.3). Such creeks are better defined on mudflats, as in the southern part of San Francisco Bay, than on sandflats, as in the Dutch Wadden Sea.

Where the tide range is large, as on the west coast of South Korea, tidal currents dominate the evolution of well-defined creek systems in mudflats (Wells et al. 1990), but in microtidal areas wave action may modify them to more subdued outlines. Verger (1968) illustrated frequent changes (notably lateral migration) in the estuarine tidal creeks on mudflats off the Sée and Sélune rivers in the Bay of Mont Saint Michel in north-

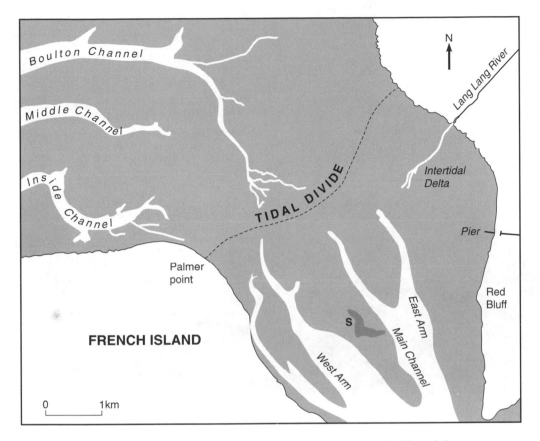

Figure 8.3   The tidal divide in Westernport Bay, Australia (location indicated by arrows in Figure 8.1).

west France. In high latitudes intertidal mudflats are scraped and grooved when waves push disintegrating sea ice and boulders shoreward, as on the shores of Gotland in Sweden (Philip 1990).

Some intertidal mudflats are actually platforms of marine erosion cut into soft clay, often with a veneer of muddy deposits. Examples of this are seen on the north-east coast of Westernport Bay in Australia, where the mudflats are backed by a metre-high receding cliff cut in dark Holocene peaty clay. On the south-west coast of peninsular Malaysia the tidal mudflats are backed by eroding microcliffs at the edge of mangroves. On the isostatically emerging eastern shore of Hudson Bay in Canada a wide tidal mudflat has been interpreted as a wave-eroded platform, with a veneer of sediment coarsening seaward (Ruz et al. 1998).

SEAGRASS BEDS

Many intertidal sandflats and mudflats carry vegetation in the form of carpets of algae, such as *Enteromorpha*, and seagrasses such as the marine wrack (*Zostera*), which partially stabilize the surface, and may trap muddy sediment to form low depositional banks or terraces (Figure 8.4). Seagrasses trap sediment because the plants are erect when the tide is high, and can diminish wave heights by up to 40 per cent and wave energy by two-thirds (Fonseca 1996). They also form a sediment-binding root network beneath the surface. Zhuang and Chappell (1991) observed the building of a seagrass terrace of silt and clay on a generally sandy intertidal substrate in Corner Inlet, Australia, as the result of wave attenuation by the plant cover. At low tide the seagrasses form a slumped carpet, but as the tide rises the stems stand erect as a sediment-filtering meadow. A seagrass bank or terrace may have vegetation spreading on its borders, or it may be sharp-edged as the result of wave or current scour. Seagrasses are not regarded as salt marshes, but terraces built by accretion in seagrass areas may in due course evolve into salt marshes.

Figure 8.4   Seagrass terrace exposed at a low spring tide at Quiet Corner, Port Phillip Bay, Australia.

Seagrasses occupied broad mudflats across the tidal divide in the north-east of Westernport Bay, Australia, an area incised by diverging steep-sided intertidal creek systems, but after they died in the 1980s the interfluvial areas were lowered by wave erosion and the creeks became wider and shallower, so that the intertidal morphology became more subdued (Bird 1993a). In Florida Bay seagrasses, mainly turtle grass (*Thalassia testudinarium*), trapped shelly deposits as well as mud, but these were released and mobilized by wave action in areas where the seagrass died (Prager and Halley 1999).

SALT MARSHES

Salt marshes are vegetated areas in the upper part of the intertidal zone on the shores of inlets, estuaries and embayments sheltered from strong wave action (Allen and Pye 1991). They can exist where wave action is sufficient to wash sediment into them, but occasional storm waves cliff their seaward margins. They can extend down to about mid-tide level, and their sub-

strate is usually muddy, but sometimes there is an admixture of sandy sediment.

Halophytic (salt-tolerant) grasses, herbs and shrubs grow between high spring tide and mid-tide level on shores exempt from strong wave action: their ecology has been described by Ranwell (1972) and Adam (1990). They are extensive on macrotidal coasts, forming wide zones on the shores of the Severn estuary and the Bristol Channel (as in Bridgwater Bay), the Bay of Mont Saint Michel in north-west France and the Bay of Fundy in Canada. They are also found between landward recurves on spits: Hurst Castle spit protects the Keyhaven salt marsh on the Hampshire coast, and salt marshes are developing in the lee of The Bar, the barrier island that began to form a century ago on the Scottish coast at Culbin Sands.

Salt marshes are less frequently inundated by the sea than sandflats and mudflats at lower levels, and currents produced as the tide rises and falls are weaker in the salt marsh. The vegetation diminishes wave action, for swards of salt marsh grass can reduce wave heights by 70 per cent and wave energy by over

90 per cent: the water velocity profile is modified in much the same way that dune grasses modify wind velocity profiles (p. 180). Salt marsh species can tolerate varying depths and durations of tidal submergence, and so spread forward to the limits thus set. As a result there is often a well-defined zonation of species parallel to the coastline, with plants such as *Salicornia* dominating the outermost zone which is most frequently submerged by the tide, and the shore rush, *Juncus maritimus*, and other salt marsh plants, occupying higher zones that are less frequently submerged. These zones could simply represent the occupation by each species of a suitable habitat that moves seaward as accretion continues, but where the vegetation is trapping muddy sediment and adding organic matter (falling leaves, decaying plants) to build up and prograde the substrate it is preparing the way for the seaward advance of the plant zones (a vegetation succession).

A distinction can be made between open marshes spreading seaward on the shores of embayments and estuaries, and vertically accreting closed marshes, formed behind sheltering spits and barriers, as between the shingle recurves on Blakeney Point and Scolt Head Island, Norfolk, which become lagoons at high tide, and then drain out through a system of converging tidal creeks (Steers 1960).

Salt marshes have similar features throughout the temperate zone, but in cold regions they can be modified by snow and ice processes. Dionne (1972) described the effects of winter snow cover and ice action on salt marshes in the St Lawrence estuary. Ice-rafted debris, including rocks, beach material, dislodged marsh fragments and driftwood, are left behind on the salt marshes after the spring melt, and when the air trapped under winter ice on mudflats finally escapes with the thaw the surface may develop small mamillated mud domes, termed monroes.

Evolution of a salt marsh

Salt marshes begin to form when vegetation (perhaps initially seagrasses) spreads from the high tide shoreline on to an accreting mudflat (Frey and Basan 1985). Often the pioneer plants (e.g. *Salicornia* spp.) form spreading clones, within which low mounds of muddy sediment are retained. As these coalesce a depositional terrace is built up by the trapping of mud washed into the vegetated area by waves and currents as the tide rises, and retained by the filtering network of stems and leaves as it falls. Terrace formation is also aided

by the development of a subsurface root network, which binds the accreting sediment. Although often characterized as tide-dominated morphology, most salt marshes are influenced by wave action as the tide rises and falls. Accretion of sediment results in the formation of a persistent depositional terrace which slopes gently from the high spring tide line to the high neap tide line, then more steeply to the mid-tide line. In the absence of salt marsh vegetation the substrate remains a mobile intertidal slope, and if the plant cover dies, or is cleared away, the depositional terrace is dissected and degraded by erosion.

Sedimentation in salt marshes is aided by the presence of adhesive algal mats on the muddy surface, and by flocculation and precipitation of clay by the salt exuded from marsh plants (Pethick 1984). When mud that adheres to leaves and stems dries off it falls to the substrate. The spread and upward growth of salt marshes is aided by an abundant supply of sediment, but salt marshes can still aggrade by the accumulation of peat derived from the decaying vegetation. Vertical accretion of up to 15 mm/yr has been measured on salt marshes in Essex and the Netherlands (Ranwell 1972), and the progressive burial of artificial markers inserted in a salt marsh on Scolt Head Island in Norfolk showed vertical accretion of up to 8 millimetres per year, with variations related to marsh elevation and inundation frequency and the retention of sediment from turbid water overflowing from tidal channel margins. Norfolk salt marshes grow vertically by accretion during ordinary high tides, but the higher parts receive sediment only in storm events (French and Spencer 1993). Examples of rapid vertical accretion in salt marshes have been documented from tidal lagoons in Argentina (Codignotto and Aguirre 1993) and on the Red River delta in Vietnam. Such accretion may be stimulated by a slow relative rise in sea level, but a rapid rise will lead to drowning and die-back of salt marsh plants.

Providing there is a supply of fine-grained sediment, and wave action is gentle, salt marshes can spread rapidly. Jacobson (1988) showed that since the eighteenth century a salt marsh had advanced on to an accreting mudflat in Maine to cover an area of eight square kilometres. The supply of mud to a salt marsh increases where fluvial sediment yields are augmented by catchment soil erosion, where the dredging of channels increases muddy sediment in suspension, or where dredged material is dumped on or near marshes, accelerating vertical accretion and progradation. An excessive rate of mud deposition may however blanket and kill salt marsh vegetation.

Salt marshes have been forming in sheltered sites around the coasts of Britain since the Late Quaternary marine transgression brought the sea up close to its present level about 6000 years ago (Pethick 1981). At St Osyth Marsh in Essex salt marsh initiation was dated by radiocarbon assay at about 4200 years BP. Early stages in the development of a salt marsh can be studied in accreting intertidal areas, especially where shelter from strong wave action is increasing because of the growth of protective spits, barriers or shoals. Evolution of salt marshes has been documented on the north Norfolk coast from initial colonization of muddy or sandy areas in the upper intertidal zone by individual halophytic plants, many of which expand vegetatively as circular clones that eventually coalesce in a broad sward (Steers 1960, Pethick 1980). Pioneer plants such as *Salicornia* spp. and *Spartina* spp. (p. 210) begin to trap muddy sediment and organic material (peat and shells) and thereby build up their substrates. Other species then colonize, and in due course the salt marsh terrace attains high spring tide level, where it is submerged only rarely by the sea. Sedimentation then proceeds very slowly, but with peat accretion and the accumulation of strand debris the surface may be raised to a level where rain and runoff leach out the superficial salt, and the marshland can be invaded by reedswamp communities dominated by such species as the common reed, *Phragmites communis* (p. 213). The vegetation sequence then continues with the invasion of the landward edge of the salt marsh by less halophytic species, reeds and rushes, and then swamp scrub, and eventually land vegetation. Examples of this succession are seen on the shores of Westernport Bay in Australia (Figure 8.5) (Bird 1993a), and a transition from salt marsh to willow and alder scrub occurs locally around Poole Harbour in Dorset (Bird and Ranwell 1964), but these later stages in vegetation succession are rarely seen because of hinterland drainage and land reclamation. Succession from salt marsh to freshwater swamp and land vegetation is accelerated by coastal emergence, as on the shores of the Gulf of Bothnia in the Scandinavian region of isostatic uplift following deglaciation. On the other hand the poor development of a freshwater swamp transition behind salt marshes on the Atlantic coast of the United States may be a response to a rising sea level, the salt marshes having failed to aggrade at a sufficient rate to attain the level where other vegetation can colonize (Kearney and Stevenson 1991).

On the west coast of Britain salt marshes are generally firmer than those on the east coast because of higher proportions of sand in the muddy sediment.

Salicornia spp. are again the pioneers, but later stages are dominated by grasses such as *Puccinellia* which form a sward on a depositional terrace of sandy mud, dissected by winding tidal creeks.

Similar stages in evolution can be traced in salt marshes on other coasts in the temperate zone. Salt marsh genera are very widespread, but there are some variations in species around the world. For example, the woody shrub *Arthrocnemum halocnemoides*, extensive in salt marshes in Australia, does not occur in Europe or North America.

Salt marsh terraces

Some salt marshes form a single terrace that extends outward from the high spring tide shoreline to the high neap tide shoreline, then descends as a slope that passes seaward into a muddy surface continuing below the mid-tide line. Others form a terrace ending seaward in a short, steeper slope, or in a microcliff that may be up to a metre high. Where the tide range is large there is sometimes an upper (mature) salt marsh of firm clay ending seaward in a microcliff, below which is a lower (pioneer) salt marsh terrace on soft accreting mud (Pethick 1992). A double terrace of this kind borders the Solway Firth, where the Ordnance Survey maps show an upper salt marsh landward of the high water line and a lower salt marsh to seaward, as in the Nith estuary near Dumfries. Similar features are seen on the northern shores of Walney Island, in Cumbria, and in Loch Gruinart on Islay in Scotland. It is possible that the upper terrace has been cliffed and cut back during a stormy phase, and that the lower terrace represents a stage in rebuilding. In the Severn estuary there are at least three salt marsh terraces, representing a cycle of marsh erosion and accretion. Accretion is most rapid (12.1 millimetres per year) on the lower terrace, submerged by every high tide, slower (6.4 millimetres per year) on the middle terrace, and slowest (2.3 millimetres per year) on the higher terrace, which is inundated only by high spring tides (French 1996).

Sections through salt marsh terraces (exposed in the banks of tidal creeks or in cliffs at the seaward edge) generally show stratified deposits, with layers of fine sand or organic material within the mud. These variations are related to wave conditions, storm waves washing fine sand into the salt marsh and mud accretion continuing as the tides rise and fall in calm weather. In the Severn estuary Allen (1996) found that the grain size of salt marsh sediments diminished from fine sand to silt and clay landward from the edge of the

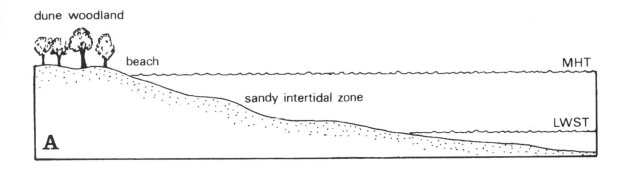

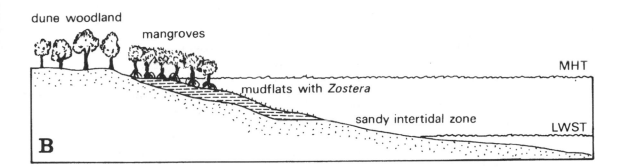

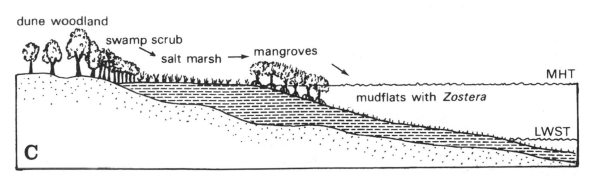

Figure 8.5   Evolution of a mangrove-fringed salt marsh terrace on the shores of Westernport Bay, Australia. A – sandy shore at the end of the Late Quaternary marine transgression (about 6000 years ago), B – muddy deposition on the shore (with *Zostera* seagrass) is followed by the formation of a mangrove fringe, spreading seaward, C – as the mangrove fringe advances a depositional terrace is formed and occupied by salt marsh, which is invaded by swamp scrub vegetation on the landward margin. MHT – Mean high tide, LWST – Low water at spring tide.

marsh as the result of sorting of sediment washed in from the seaward side, and that there was a similar diminution vertically through the aggraded salt marsh terrace because of progradation. However, there are often storm-carried sediments on the upper salt marsh, including sand and organic litter (Stumpf 1983).

Upward and outward growth of salt marshes can be accelerated by an increase in the rate of sedimentation of the kind which occurred in Cornish estuaries in the eighteenth and nineteenth centuries when river sediment yields were augmented by mining waste (p. 258). On the south coast of England rates of accretion have been very slow in areas where excavations made in salt marshes (e.g. for salt manufacture) have persisted for many decades, as at Budleigh Salterton in Devon. In the Medway estuary in Kent large quantities of clay have been cut for brick-making and cement production, leaving numerous pits and access canals which have not yet been infilled, although clay extraction ceased at least 30 years ago (French 1997).

Rates of accretion of sediment in salt marshes can be measured by laying down identifiable layers of coloured sand, coal dust, or similar material on the marsh surface, and returning to put down borings and measure the thickness of sediment added subsequently. Measurements made on salt marshes in Britain have shown that vertical accretion has been relatively slow at the upper and lower limits of a marsh, and more rapid in the intervening zone, where salt marsh vegetation forms a relatively dense sediment-trapping cover and is regularly invaded by sediment-laden tidal water (Steers 1960). Møller (1963) measured changes in marshland topography on the Danish coast, based on maps prepared in 1941 and 1959–62 on a scale of 1:10000 with marshland surfaces contoured at five centimetre intervals. Patterns of vertical and lateral erosion and accretion were thus located. On the north Norfolk coast Pethick (1981) found the upper limit of accretion on salt marshes to be about 2.4 metres above Ordnance datum (OD). Accretion is also slower away from tidal creeks, especially where they turn parallel to the coastline. In salt marshes bordering the Severn estuary French (1996) used evidence from heavy metal profiles and lead (^{210}Pb) dating to define distinct sedimentary units (between planes dating from 1840–50, 1936 ± 7, 1971 ± 4 and 1958 ± 4), and show that vertical accretion of three to four millimetres per year has been proceeding at about the same rate as sea level rise in the area. Very rapid silt accretion has been taking place on marshland in the Yangtze delta, where Yang (1999) measured accretion rates of up to 43 centimetres per year, and found that accretion in the middle marsh

was 1.5 times that in the lower marsh and twice that in the upper marsh.

Seaward margins of salt marshes

The outer (seaward) edge of many eroding salt marsh terraces is a muddy microcliff up to a metre high. Examples of this are seen on the Burry Inlet marshes in south Wales, where the microcliff drops sharply to Llanridian Sands, and on the Dengie peninsula in Essex, where the microcliff has been retreating at up to 10 metres per year. Salt marshes on the southern shores of the macrotidal Bay of Mont Saint Michel are undergoing dissection by ebb runoff and rapid erosion along their seaward margins (Figure 8.6). In some sites cliffing is accompanied by continuing vertical accretion of muddy sediment in the backing salt marsh vegetation, building up the terrace even though seaward advance has come to an end.

Cliffing may result from the lateral migration of a tidal channel, undercutting the edge of a salt marsh, but as it is very widespread (there are now only a few sites where salt marshes are spreading seaward), some more general explanation is required. It may be that, as on the sides of developing tidal creeks, seaward margins become oversteepened and cliffed, particularly during occasional storm wave episodes (Figures 7.4, 7.5). Cliffing of this kind can be repaired if there is an abundant supply of sediment to restore the profile, permitting vegetation to spread again, but if there is a sediment deficit a microcliff will persist. Guilcher (1981) recognized cyclic patterns of marginal erosion and accretion in the salt marshes bordering estuaries in Brittany.

Alternatively, the cliffing of seaward margins of salt marsh terraces could be a response to a rising sea level, deepening the adjacent water and allowing larger waves to attack the shore. This would also explain the widening and shallowing of the tidal creeks that is occurring in the salt marshes of southern England. Salt marshes in the Lagoon of Venice, where mean sea level rose 22 centimetres between 1908 and 1980, also show marginal die-back and receding cliffed edges, as well as dissection by widening tidal creeks. Wave energy and tide range have increased with the deepening of the lagoon, but the salt marshes are aggrading at an average of 1.54 centimetres per year with sediment washed in from the eroding marsh margins and the lagoon floor on high tides (Day et al. 1998).

Allen (1989) found that salt marsh microcliffs were bolder, often vertical, on sandy mud, as in the Solway

Figure 8.6   Dissected salt marsh on the shores of the Bay of Mont Saint Michel, north-western France.

Firth, and more subdued on soft mud in the Severn estuary. Where the top of the microcliff was bound by plant roots, recession was by way of calving, toppling and rotational slides.

Salt marsh creeks

As salt marsh terraces build up in the form of a sedimentary wedge the ebb and flow of the tide may maintain a system of tidal creeks, the dimensions of which are related to the volume of water flowing up and down them as the tide rises and falls (Pethick 1992). Typically dendritic and intricately meandering, they are channels within which the tide rises until the water floods the marsh surface, then they become drainage channels into which some of the ebbing water flows from the salt marsh. They are thus like minor estuaries, particularly where they receive fresh water from hinterland runoff or seepage from bordering beaches and dunes. However, it should be noted that some of the alternating submergence and drainage of a salt marsh results from inflow and outflow across the seaward fringe

rather than through creek systems. In the early stages tidal creeks are relatively wide and shallow in cross-section (like the broad channels on intertidal mudflats), but as salt marsh terraces rise and expand they become narrower and deeper, and their banks higher and steeper, with frequent local slumping (Figures 8.7, 8.8). Blocks of compact mud, often with clumps of salt marsh vegetation, collapse into the creek, especially where the banks are burrowed by crabs. Some tidal creeks are fringed by natural levees formed by deposition of sediment as the rising tide overflows, especially where such plants as *Halimione* spp. have colonized the bordering banks. This pattern is more often found on the lower (and younger) seaward fringes of salt marshes, the inter-creek areas becoming flatter as sedimentation proceeds.

Dendritic tidal creek systems give place to straighter, parallel tidal creeks on some salt marshes, as on the shores of Loch Gruinart in Islay and at Morrich More, south of Dornoch Firth in eastern Scotland. Straight parallel creeks across salt marshes are more often found where the tide range is large, the transverse gradient small, or where the rate of seaward spread of

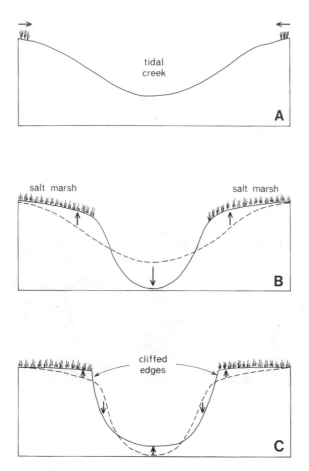

Figure 8.7   The evolution of a salt marsh creek. As the salt marsh forms it builds up a bordering terrace, the edges of which become oversteepened and cliffed.

tidal creeks tend to be rounded furrows where accretion is in progress, to show asymmetry on meanders where erosion balances accretion, and to be rectilinear where erosion is dominant (Figure 8.9).

Studies of creek systems on the tide-dominated salt marshes of Scolt Head Island in Norfolk showed that the exchange of water and sediment with the bordering marshes varied with current velocity as the creek water rose to overbank levels. Vertical growth of the salt marsh led to increasingly intermittent sediment transport in the creeks, fewer tides reaching the velocity required for the entrainment of channel sediments (French and Stoddart 1992).

Salt pans

Salt pans are small shallow depressions, some of which form as residual unvegetated areas within a developing salt marsh while others (often long and narrow) are the result of the blocking of part of a tidal creek by slumping. A third category results from local die-back of salt marsh vegetation, particularly in *Spartina* marshes.

Salt pans are flooded at high tide, and remain bare of plants because evaporation makes the trapped water hypersaline. Many become well-rounded and sharp-edged as the result of scour by small waves and circulating currents generated by winds blowing across them. They are much like dune lakes (p. 195). In dry periods the water evaporates, leaving flat-floored basins. Numerous salt pans (also known as pond holes) occur on salt marshes between shingle recurves on Blakeney Point in Norfolk, and they are also extensive on salt marsh terraces, such as those bordering the Cree estuary in south-west Scotland.

The *Spartina* story

Salt marsh morphology can be modified by the arrival of a new and aggressive species, such as the hybrid *Spartina anglica*, which originated in Southampton Water in about 1870 as a cross between native *Spartina maritima* and American *Spartina alterniflora* (Carey and Oliver 1918). It spread (or was introduced) to many other British estuaries in the ensuing few decades, advancing across intertidal mudflats and rapidly building up marshland. It has been used as a method of stabilizing and land-claiming tidal flats in estuaries in various parts of the world, including the Netherlands, Denmark, New Zealand, and the Tamar estuary in Tasmania. The effects of introduced *Spartina* on

salt marsh has been rapid. In Bridgwater Bay on the shores of the Bristol Channel, where the tide range is about 10 metres, salt marsh creeks run parallel and orthogonal to the coastline, whereas in Poole Harbour, Dorset, a microtidal estuarine embayment, creek patterns in bordering salt marshes are mainly dendritic.

The morphology of tidal creeks is related to sediment type, plant cover and tide range. There are clearly defined steep-edged tidal creeks in salt marsh terraces built largely of cohesive clay, as in Poole Harbour, but they become shallower and wider where the salt marsh is sandier, as in the Welsh estuaries opening into Cardigan Bay. Salt marsh creek patterns are modified and become trellised where linear cheniers of shelly sand have been deposited by storm surges, and channels have been cut through these. In cross-section,

Figure 8.8   Perranarworthal Creek, Cornwall, bordered by a salt marsh terrace and slumping banks formed as shown in Figure 8.7.

salt marshes in China were discussed by Chun-Hsin (1985).

Early stages of *Spartina* invasion can be seen where clones are spreading on sandy intertidal areas on the shores of Lindisfarne Lagoon, in the lee of Holy Island, and on the Humber mudflats behind the spit at Spurn Head. In Poole Harbour the arrival of *Spartina anglica* in 1899 was followed by the rapid expansion of salt marshes into broader and higher terraces covered entirely by this plant (Figure 8.10). At the same time, intervening creeks and channels became narrower and deeper, indicating that there had been a transference of muddy sediment from these into the areas of spreading *Spartina*. On the north Norfolk coast experimental introduction of *Spartina anglica* modified natural salt marshes and led to the evolution of broad depositional marsh terraces in the intertidal zone. In the Dee estuary, Marker (1967) recorded the rapid spread of *Spartina* grass introduced in 1922, noting that it had become the pioneer colonist on accreting mudflats.

Introduction of *Spartina anglica* to the Tamar estuary in Tasmania, where previously there had been little salt marsh, led to the transformation of intertidal mudflats into wide marsh terraces with deep creeks. It has spread rapidly to build depositional terraces on tidal mudflats in Andersons Inlet, Victoria, where it is invaded by mangroves. There can be no doubt that *Spartina* has acted as a sediment-trapping agent, and produced intertidal landforms that would not otherwise have developed.

In Britain some of the older *Spartina* marshes show evidence of die-back, especially around pans and along their seaward margins. The ecological reasons for this are not fully understood, but die-back is often associated with nitrogen deficiency, sulphide accumulation and waterlogging. At the seaward margins where the sward dies, and sediment previously trapped is released, there is a receding microcliff. Die-back of *Spartina* along creek margins has led to the erosion of marsh edges and resulted in the widening and shallowing of tidal creeks and channels. The process may be

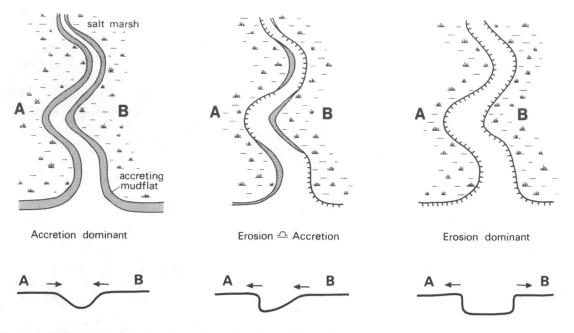

Figure 8.9   Tidal creek morphology in a salt marsh under conditions where accretion is dominant, where erosion balances accretion, and where erosion is dominant.

Figure 8.10   A depositional terrace formed by *Spartina* in Poole Harbour, Dorset.

cyclic in the sense that released mud is deposited in new or reviving *Spartina* marshes elsewhere. Within Poole Harbour there are sectors where *Spartina* is still advancing, mainly in the upper estuary, as well as sectors of die-back and erosion of the marshland, notably in Brands Bay near the marine entrance.

FRESHWATER SWAMPS

Reference has been made to freshwater swamps on the landward margins of salt marshes, where they may represent a stage in vegetation succession to land vegetation, but on microtidal or tideless coasts where salinity is low (as in the Baltic Sea) freshwater swamps fringe the shore, particularly in inlets and sheltered embayments. They are dominated by reeds (*Phragmites communis*), often with rushes (e.g. *Typha* spp.) and sedges (e.g. *Scirpus* and *Juncus* spp.) and can grow out into water about a metre deep. Freshwater swamps of this kind also reduce wave action and current flow, and promote accretion of sediment, particularly silt and clay, in such a way as to build up a depositional shore terrace. Seasonal decay of freshwater swamp vegetation produces organic matter, which is deposited with the trapped sediment, and where the sediment supply is meagre this organic matter may accumulate as a depositional terrace of fibrous peat deposits. In due course the terrace is built up to high water level, and land vegetation (scrub and forest) then moves in. The outcome is progradation of the coastline by swamp land encroachment, a process which is also important on the shores of coastal lagoons where salinity is relatively low (p. 246).

MANGROVES

Mangroves are halophytic shrubs and trees that grow in the upper part of the intertidal area on the shores of estuaries and lagoons and on coasts sheltered from strong wave action, as in inlets or embayments or in the lee of headlands, islands or reefs (Guilcher 1979). They show their greatest extent and diversity on tropical coasts, where they occupy a similar niche to temperate salt marshes. On some tropical coasts salt marshes or saline flats with little or no vegetation occupy a high tide zone landward of the mangrove fringe.

The ecology of mangroves was reviewed by Chapman (1976) and their global status by Saenger et al. (1983). A few mangrove species grow outside the tropics in the south-eastern United States, Brazil, South Africa, the Red Sea, Australia and New Zealand. Their poleward limit is in Corner Inlet, south-eastern Australia, where the white mangrove, *Avicennia marina*, is growing at latitude 38° 55′ S. The limiting factor is low temperature, for mangroves are killed by frost.

There are numerous species of mangroves, particularly in the humid tropics, where there are more than 70 in the Indo–Pacific region, but diminishing towards the latitudinal limits northward and southward. Thus in Australia, a maximum of 27 mangrove species is found beside the Daintree River estuary in north-east Queensland, but the number diminishes rapidly southward to six at Sydney, and in southern and south-western Australia there is only one, *Avicennia marina*.

Mangroves grow sparsely on rocky shores and coral reefs (their roots penetrating fractures in the rock) and on sandy substrates, but they are more luxuriant, forming dense scrub and woodland communities, on muddy substrates and shoals exposed at low tide. Where wave energy is low they spread forward to the mid-tide line, but as wave action increases along the coastline the mangrove fringe thins out and disappears. Occasionally the seaward edge of the mangroves has been cliffed by strong waves. On the other hand the longshore growth of sand bars, spits or barriers can make a coastal area more sheltered for mangrove colonization.

The width of a mangrove fringe generally increases with tide range, and on macrotidal coasts can attain several kilometres, as on the tide-dominated shores of gulfs and estuaries in northern Australia, where wide mangrove areas are backed by sparse salt marshes and saline flats flooded during exceptionally high tides and summer rains. Within the humid tropics mangroves grow to forests with trees 30 to 40 metres high, as on the west coasts of Malaysia and Thailand, in Indonesia, Madagascar and Ecuador. On drier and cooler coasts within, and just outside, the tropics mangroves generally form extensive scrub communities (Chapman 1976).

Mangrove species or associations of species are sometimes arranged in zones parallel to the coast or to the shores of estuaries or lagoons (Snedaker 1982). The zonation is accompanied by variations in substrate level related to depth and duration of tidal submergence and exposure to atmospheric conditions, notably rainfall, at low tide. It appears that each mangrove species or association of species grows best at a particular intertidal level, and that competition restricts them to their zone of optimal growth, which may have a vertical range of only a few centimetres in a horizontal zone of several metres. *Avicennia* spp. are often the pioneers, forming a seaward fringe, backed by *Rhizophora* spp. and other mangrove zones dominated by species of several genera, including *Bruguiera, Ceriops, Laguncularia, Lumnitzera* and *Xylocarpus. Avicennia* is often accompanied by *Sonneratia* and *Aegiceras* on estuary shores. As accretion proceeds, raising the substrate, the mangrove zones migrate seaward, each species zone displacing its predecessor in a vegetation succession. At the inner margin, salt marsh may replace mangroves as the substrate level rises above mean high tide level (as on the shores of Westernport Bay, Australia), but in arid regions the upper intertidal zone may become a wide, frequently desiccated and hypersaline mudflat (as in the Townsville district in north-east Australia). Above high spring tide level there is further accretion from river floods and downwash from backing slopes, and as the land level is raised hinterland vegetation moves in.

The seaward spread of mangroves is indicated by an abundance of seedlings and young shrubs on the adjacent mudflats and a smooth canopy rising landward as the trees increase in age and size (Figure 8.11). Where the seaward margin of mangroves is abrupt, with trunks and stems of mangroves visible from the sea, and any seedlings fail to survive, the mangroves are no longer advancing. A receding mangrove shoreline is indicated by exposed trunks and stems being undercut, and falling, or where the vegetation has died and there is a receding microcliff in the substrate (Figure 8.12).

Mangroves are structurally and physiologically adapted to survive in a marine tidal environment. Some mangrove species (e.g. *Avicennia marina, Sonneratia alba*) have networks of pneumatophores that rise from sub-surface root systems to project vertically out of the muddy substrate and allow the plant to respire in a waterlogged environment. Pneumatophores occupy a roughly circular zone around the stem of each plant, and can be very numerous and closely spaced, with densities of up to 300 per square metre. With the aid of these snorkel-like breathing tubes the mangroves can grow in areas that are submerged by the sea at each high tide, but the necessity

Figure 8.11   A declining canopy at the seaward edge of advancing mangroves in Westernport Bay, Australia.

Figure 8.12   A sharp mangrove edge with exposed trunks indicates that the seaward margin has retreated. Pneumatophores protrude from the bordering mudflats.

for several hours subaerial exposure between each submergence sets a seaward limit, usually close to mid-tide level. Other mangroves, such as *Rhizophora*, *Bruguiera*, *Ceriops* and *Lumnitzera* species, have subaerial prop or stilt roots that branch downward to the mud and support the stems, while others, like *Xylocarpus*, have no such structures.

Where evidence from the Holocene stratigraphy beneath mangrove areas is available, it generally indicates that they have spread to their present extent during the sea level stillstand of the past 6000 years. Before that, during the Late Quaternary marine transgression, they were confined to sheltered inlets and estuarine sites where they could migrate landward, and to sectors where they could persist on vertically accreting muddy substrates as submergence proceeded. Stratigraphic studies in northern Australia and southeast Asia have shown that mangroves were growing in accreting estuaries with a rising sea level until between 7000 and 5500 years ago, when the Late Quaternary marine transgression came to an end. They then spread

into other embayments and more exposed sites of muddy accretion along the outer coast. In the estuary of the South Alligator River mangroves spread rapidly during the Holocene stillstand, when vertical accretion proceeded with sediment washed in from sea as well as down the river, which developed intricately sinuous meanders (Woodroffe 1993). In general, mangrove swamps spread seaward on coasts where relative sea level is stable or falling, and where there is a sufficient sediment supply.

Mangroves and intertidal morphology

Sediment carried into mangroves by the rising tide is retained as the tide ebbs, gradually building up a depositional terrace between high spring and high neap tides, with a seaward slope (usually about 1:50) descending to the outer edge of the mangroves and continuing across the lower intertidal zone, which is

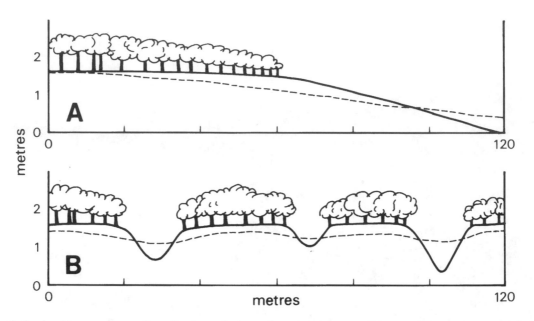

Figure 8.13   A – Mangrove terrace formation by vertical accretion as mangroves spread seaward; B – as the terrace is built up intervening creeks become deeper. If the mangroves are removed or disappear the terrace reverts to the earlier surface (pecked line). If they recolonize, the terrace and creek profiles are restored.

either unvegetated, or has patches of seagrass (Figure 8.13A).

Mangroves colonize mudflats that are slowly accreting, but once established they promote accelerated accretion within the network of stems and pneumatophores that diminishes current flow and wave action (Augustinus 1995). A mangrove fringe shelters nearshore waters from winds blowing off the land and thus reduces seaward drifting of mud from the shore, whereas shoreward drifting by waves produced by onshore winds and rising tides continues. The depositional terrace thus formed is much the same as that formed beneath salt marshes (p. 206). There is also peat accumulation, but most mangrove substrates are dominated by muddy sediment. After deposition in the *Avicennia* fringe has raised the mud surface, *Rhizophora* and other mangroves colonize, and sedimentation proceeds more slowly, but eventually the terrace is built up to high-tide level, whereupon the mangroves are displaced by the backing salt marsh or unvegetated mudflats mentioned previously. These are submerged only by infrequent high spring tides, occasional storm surges or river flooding, and sedimentation is thus very slow. As has been noted (p. 205), accretion of peat and drift litter may be required to raise the substrate to levels where it can be colonized by freshwater and land vegetation, a process that is aided by

emergence on coasts where sea level is falling relative to the land.

It is possible to measure rates of sedimentation in mangrove areas by the same methods used in salt marshes (p. 208). Marker layers can be placed on the substrate, and the depth of subsequent accretion measured by coring, but there are difficulties where burrowing crabs churn up the substrate. Monitoring of changes is more effective when measurements are made on implanted stakes. Measurements in Westernport Bay, Australia, showed that there was sustained accretion of up to 4.5 centimetres per year in the *Avicennia* fringe, in contrast with vertical fluctuations on adjacent mudflats. The pattern of mud accretion was strongly correlated with the density of pneumatophore networks, which certainly trap and retain muddy sediment. Low accretion mounds form above the general level of the muddy shore within pneumatophore networks around isolated *Avicennia* trees, and when such networks were simulated artificially by planting wooden stakes a similar accretion mound formed within the staked area (Figure 8.14). When the stakes were removed the accreted sediment was quickly dispersed. Mangroves with pneumatophores thus trap sediment as effectively as salt marsh plants (Bird 1986).

Mangroves with prop roots, such as *Rhizophora*

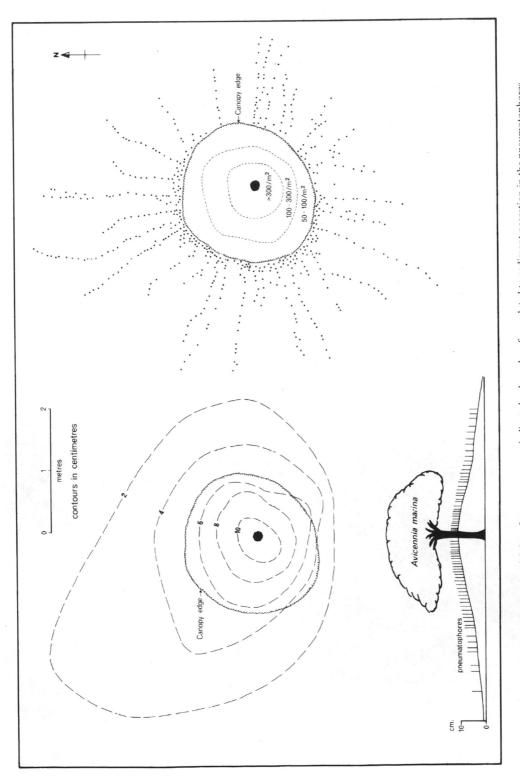

Figure 8.14   Contouring of mudflats in the vicinity of a mangrove tree indicated a domed surface related to sediment accretion in the pneumatophores.

spp., may be less effective in trapping mud, but once it has been retained it tends to become a more compact, firmer clay, coherent enough to sustain cliffs where the mangrove terraces bearing *Rhizophora* are eroded by current scour along the banks of tidal creeks, or by wave action on seaward margins. The mangrove community also generates substantial quantities of organic matter from decaying leaves, stems and roots, and from the various organisms that inhabit mangrove areas, and in due course these can form peat deposits, raising the substrate level. In Florida, *Rhizophora* is found growing on vertically accreting peat deposits.

The depositional terrace formed beneath mangroves is submerged at high spring tides, and occasional storm surges may wash sand and shelly deposits up into the mangroves to form cheniers (Figure 8.15). As the tide ebbs, runoff becomes confined to definite channels, similar to salt marsh creeks (Figure 8.13B). These may form a dendritic pattern, as in many Australian mangroves, but where cheniers are present the creek network may become reticulate, as on the Niger delta

(Allen 1965). Changes also occur within the mangroves as tidal creeks meander or migrate laterally so that mangrove trees are undermined on one bank and mangrove seedlings colonize sediment deposited on the other. As on salt marshes the spread of mangroves into an estuary slows down as tidal waters become confined to a central channel, in which ebb and flow currents are relatively strong. The channel borders may then oscillate, advancing during phases of local accretion, and receding during episodes of tidal scour, in relation to volumes of tidal ebb and flow.

The widening of a mangrove terrace also depends on a continuing sediment supply, as on deltaic shores close to the mouths of rivers or within embayments where intertidal muddy areas are extensive. Changes within river catchments, such as upstream deforestation, increase soil erosion and augment the sediment yield to the coast, thereby accelerating sediment accretion in intertidal areas and promoting the spread of mangroves. This has occurred in the Segara Anakan lagoon in southern Java (p. 244), where a greatly increased

Figure 8.15   A chenier of sand and gravel deposited in mangroves near Karembé, New Caledonia, during the storm surge that accompanied Cyclone Gyan in 1981. Mangroves have been killed in the zone where the chenier was deposited, but have survived, or are reviving, on either side.

sediment yield resulted in siltation and the rapid advance of the mangrove fringe (Bird and Ongkosongo 1980).

Cyclones (hurricanes, typhoons) may flatten a mangrove forest, but there is usually rapid regeneration among the windthrown trees and little long-term change in intertidal morphology. When the mangrove fringe is cut down and removed the substrate is usually soon lowered by erosion. This has occurred where mangroves have been cleared at Oyster Point, near Cardwell in north Queensland, Australia. In Westernport Bay, Australia, *Avicennia marina* was extensively cleared in the 1840s to obtain wood that was burnt to make barilla for soap manufacture. The depositional terraces that had formed beneath them were dissected, tidal creeks becoming wider and shallower, and degraded to a steeper transverse slope and dissected, so that the roots of former mangroves were laid bare. When *Avicennia* recolonized the dissected areas mud accretion resumed within pneumatophore networks, rebuilding the depositional terrace (Figure 8.13B) (Bird 1986).

These changes support the view that mangroves are land builders, but this has not been universally accepted (Vaughan 1910, Davis 1940, Carlton 1974). Some have suggested that mangroves merely occupy intertidal areas that become ecologically suitable as they are raised by accretion, independently of any effects of vegetation, so that the depositional terraces are landforms that would have developed even if mangroves had not been present (Watson 1928, Scholl 1968), while others have envisaged an interaction between colonizing mangroves and intertidal deposition (Thom 1967). It is possible that *Avicennia* and other mangroves with pneumatophores promote accretion and coastline progradation as they spread forward on to the intertidal zone, whereas *Rhizophora* and other mangroves without pneumatophores simply occupy suitable intertidal habitats.

Seaward margins of mangroves

Mangrove terraces are being eroded on shores now receiving little or no sediment. Low receding cliffs have been cut into their seaward margins, particularly on deltaic coasts where the sediment supply has been reduced because of dam construction or the natural or artificial diversion of a river outlet. The seaward edge of many mangrove terraces is undercut by a muddy microcliff up to a metre high, similar to that described from salt marshes (p. 208). Microcliff recession may be accompanied by continuing vertical accretion of muddy sediment in the mangroves, building up the terrace even though seaward advance has come to an end. In some places the cliffing results from lateral movement of a tidal channel, undercutting the outer edge of the mangroves, but generally it is due to larger waves reaching the mangroves as the result of deepening of the lower intertidal zone, either because of progressive entrapment of nearshore sediment drifting into the upper vegetated area, or because of continuing submergence of the coast (Guilcher 1979).

It may be that, as on the sides of developing tidal creeks, seaward margins become oversteepened and cliffed, particularly during occasional storm wave episodes. Cliffing of this kind is repaired if there is an abundant supply of sediment to restore the profile, permitting mangroves to spread again, but if there is a sediment deficit a mangrove cliff will persist and recede.

9

Estuaries and Lagoons

INTRODUCTION

During the Late Quaternary marine transgression the sea invaded valley mouths and coastal lowlands to produce inlets and embayments of various kinds. These have been given different terms, some of which overlap.

RIAS

Inlets formed by partial submergence of unglaciated river valleys are termed rias. In 1886 Von Richthofen defined a ria as a drowned valley cut transverse to the geological strike, between ridge promontories, but the Rias of Galicia, in north-west Spain, where the term originated, do not meet this strict definition (Cotton 1956). The term has come to be used as a synonym for a drowned valley mouth (usually with a branching dendritic or tree-like outline) remaining open to the sea, as in Carrick Roads (Figure 9.1) and the Tamar estuary in south west England, Chesapeake Bay in the United States, and Port Jackson (Sydney Harbour) in Australia. The straight valley-mouth gulfs on the south-west coast of Ireland, such as Bantry Bay, are also rias. The term overlaps with the broad concept of an estuary, for many rias have inflowing rivers providing fresh water that meets and mixes with sea water moved in and out by the tide.

Kidson (1971) noted that the rias of south-west England are underlain by alluvial deposits in buried channels that were cut when the rivers flowed down to lower sea levels in glacial phases of the Pleistocene. Several phases of valley incision occurred, with removal of at least some of the sediments deposited during interglacial phases. Downcutting was completed in the Last Glacial low sea level phase, and infilling has occurred during and since the Late Quaternary marine transgression (p. 31).

Rias show varying degrees of sedimentary infilling. On the north coast of Cornwall sand has been washed in from the Atlantic Ocean to fill rias, as at Padstow, where at low tide the Camel River is narrow, flowing between broad exposed sandbanks that are submerged when the tide rises. On the south coast of Cornwall rias generally remain as relatively deep inlets, such as Carrick Roads, some with a threshold sand bar at the entrance, as at Salcombe. Similar rias in Brittany, such as the Aber Benoît, have salt marshes in their upper reaches, and the Rade de Brest is noteworthy for its various sand and gravel spits (Guilcher et al. 1957). The Galician rias are generally deep marine inlets, with sandy, gravelly and rocky embayed shores, some inwashed spits, and muddy sediment and salt marshes at their heads (Castaing and Guilcher 1995).

FIORDS

Fiords differ from rias in that they are inlets at the mouths of valleys that were formerly glaciated, as on the steep mountainous coasts of British Columbia and Alaska, Greenland, Norway and Siberia. The sea lochs of western Scotland are essentially fiords. In the southern hemisphere there are fiords on the south coast of Chile, on the west coast of South Island, New Zealand (where Milford Sound is a fine example), and on the margins of Antarctica and the Falkland Islands.

The chief characteristics of fiords are inherited from prior glaciation, for they are glacial troughs scoured out by ice action well below the depths that rivers cut valleys during Pleistocene low sea level glacial phases, and submerged by the rising sea as the ice melted (Syvitski and Shaw 1995). Depths of more than 1300 metres have been recorded in Scoresby Sound, a fiord on the east Greenland coast, and Sogne Fiord in Norway is up to 1244 metres deep, but most fiords have depths of about 300 to 400 metres. Fiords are steep-sided, narrow and relatively straight compared

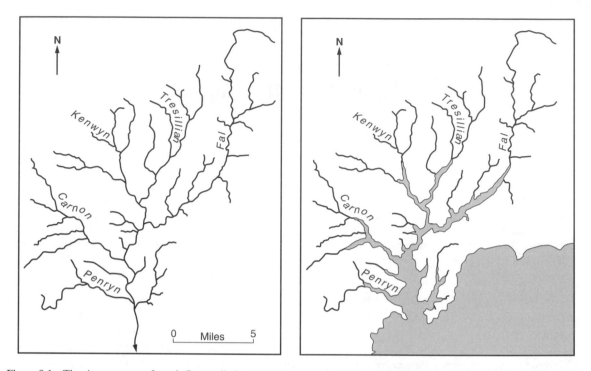

Figure 9.1   The river systems of south Cornwall about 18 000 years ago (left) and the formation of the ria at Carrick Roads by their partial submergence during the Late Quaternary marine transgression (right).

with most rias, and they have the classic U-shaped cross-profile of glaciated troughs, together with hanging tributary valleys, usually with waterfalls. The shores are generally narrow and rocky, with some talus slopes and morainic ridges. Near the seaward end there is often a shallower sill which may be a rocky feature (as at the entrance to Milford Sound in New Zealand), possibly formed because thinner ice towards the mouth of the glaciated trough scoured less deeply, leaving a sill on the seaward side of a more deeply excavated basin. Some sills may be drowned morainic banks, but most are rocky with a sedimentary capping. Sills are generally less than 100 metres deep, and were probably land isthmuses separating a lake in the glaciated trough from the sea, until they were submerged by the Late Quaternary marine transgression.

At the heads of some fiords there are ice cliffs fringing the residual glacier, but many have deltas built by glacifluvial outwash. Although their upper reaches have some of the characteristics of estuaries, with rivers discharging fresh water, most fiords remain as deep marine inlets. Some have been filled with sediment from rivers and melting glaciers. On the west coast of

South Island, New Zealand, the glaciated valleys south of Awarua Point are deep fiords whereas to the north similar valleys have been filled with glacifluvial sediment from the Southern Alps. The contrast may be due to the northern valleys having steeper catchments, or there may have been transverse tilting by tectonic movements, the fiords persisting in the downwarped southern sector.

FIARDS

Inlets formed by Late Quaternary marine submergence of formerly glaciated valleys and depressions in low-lying rocky terrain are known as fiards rather than rias. There are examples on the Swedish coast near Stockholm (e.g. Broviken), and similar inlets in coastal plains of glacial deposition are termed förden on the German Baltic coast. The Firth of Forth and other elongated firths in eastern Scotland are a closely related category of drowned valleys in formerly glaciated lowlands that are partly rocky and partly morainic. Like rias, they are also estuarine.

Beyond the limits of past glaciation many drowned valley mouths have marginal features related to periglaciation, as in south-west England, where the rias are bordered by slopes covered with solifluction (Head) deposits exposed in fringing cliffs (p. 54). The drowned mouths of deeply-incised river valleys beyond the limits of past glaciation may resemble fiords: Bathurst Channel in Tasmania is a steep-sided inlet that was formerly thought to be a fiord, but as there is no evidence that it was glaciated down to or below present sea level, and as it has none of the features of glacial sculpture that distinguish true fiords it should really be classified as a ria.

CALANQUES

The mouths of steep-sided valleys that were deeply incised during low sea level stages into the limestone plateau east of Marseilles in southern France have been submerged by the Late Quaternary marine transgression to form cliff-edged inlets known as calanques (Chardonnet 1948). Port Miou, near Cassis, has almost right-angle bends related to erosion along joint planes while Port-pins and Port d'en Vau are straighter, with beaches of inwashed sand at their heads. Calanques are also found at Wied-iz-Zurrieq and Il-Baja on the limestone coasts of the Maltese islands (Paskoff 1985), and on the Dalmatian coast. They are essentially steep-sided marine inlets on karstic coasts with few of the features of estuaries.

SHARMS AND SEBKHAS

On the Red Sea coast, long, narrow marine inlets termed sharms (sherms) have formed where wadis or valleys cut by streams in wetter Pleistocene low sea level episodes were invaded by the sea during the Late Quaternary marine transgression. Their shores are lined and their entrances constricted, and in some cases blocked, by fringing coral reefs, and they often end abruptly upstream, not necessarily in a wadi or stream inflow. A well-known example is Abhur Creek, just north of Jeddah in Saudi Arabia, and similar features are seen in Mombasa harbour in Kenya and on the east coast of Vanuatu (Castaing and Guilcher 1995).

Broader, often branched embayments are called sebkhas on the arid coasts of the Red Sea and the Arabian Gulf. They are salt-encrusted areas, often partly enclosed by sandy spits and barriers, and

invaded by the sea only during storm surges, as at Abu Dhabi. Similar embayments are seen, branching into inter-dune swales bordering King Sound in north-west Australia (Figure 9.2). Sebkhas usually show zones of different evaporite deposits (chlorides, sulphates and carbonates) in a landward sequence, formed by precipitation from sea water and groundwater (Evans and Bush 1969).

ESTUARIES

Estuaries have been defined in various ways. According to Pritchard (1967) 'an estuary is a semi-enclosed body of water which has a free connection with the open sea and within which sea water is measurably diluted with fresh water derived from land drainage', but this definition does not take account of tides or morphology, and is wide enough to include the Baltic Sea or Port Phillip Bay in Australia. Alternatively, an estuary may be defined as the seaward part of a drowned valley system, subject to tidal fluctuations and the meeting and mixing of fresh river water with salt water from the sea, and receiving sediment from its catchment and from marine sources. Channels shaped by unidirectional river flow widen downstream to estuaries subject to alternating inflow (flood currents) and outflow (ebb currents), with rapid variations in current velocity through the tidal cycle. The morphology of an estuary represents an adjustment between the capacity of its channels and creeks and the volume of water moved in and out by tidal oscillations (the tidal prism). General reviews of estuaries have been provided by Emery and Stevenson (1957), Lauff (1967), Nichols and Biggs (1985) and Perillo (1995).

Most estuaries are found on coastal plains. There is considerable overlap between the concept of an estuary and the various kinds of inlet described above. Some rias and fiords are valleys that have been almost completely drowned by marine submergence, receiving so little drainage from the land that they are essentially arms of the sea, but most are fed by rivers, the mouths of which can be described as estuarine. Where valleys have been incised into coastal plateaux, as in south-west England and Brittany, or into hilly coastal country, as in China and the North Island of New Zealand, marine submergence has produced rias which are estuaries with steep bordering slopes. Tectonic movements (folding and faulting) have influenced the shaping of estuarine inlets such as San Francisco Bay in California and Westernport Bay in Australia, and land subsidence in south-east England has resulted

Figure 9.2   A sebkha on the north-west coast of Australia.

in wide, short estuaries such as those of the Essex coast and the broad embayments known as Chichester Harbour in West Sussex, Langstone and Portsmouth Harbours on the Hampshire coast and Poole Harbour in Dorset. The Wash is a broad embayment, estuarine around the mouths of the inflowing Welland, Nene and Great Ouse Rivers. Distributary river channels on deltas are also estuarine (Morgan 1967).

As infilling proceeds, estuaries that were originally deep and branching become shallower and simpler in configuration. Funnel-shaped estuaries, widening seaward (Figure 9.3), are best developed on coasts with a large tide range. In Britain the Thames, the Humber, and the Severn show this form, as do the Rhine and the other rivers of northern Germany, the Seine, the Loire and the Gironde in France and the Tagus in Portugal. On the east coast of the United States Delaware Bay and Chesapeake Bay are much-branched coastal plain estuaries, formed where the sea has risen to invade tributary valleys. In Australia rivers draining into the macrotidal northern gulfs have large estuaries, the

Fitzroy flowing into King Sound and the Ord and the Victoria into Joseph Bonaparte Gulf.

The morphology of an estuary can be related to hydrodynamic processes, such as river flow, tidal currents associated with the incoming tide wave, wave action, chemical processes such as the flocculation of fine-grained sediment, and biological processes such as the growth of salt marsh, mangroves or seagrass vegetation and the generation of shelly deposits. These processes are complex and variable over time, and within an estuary. They have been discussed in the previous chapter in relation to the intertidal zone generally, and will now be considered in terms of the shaping of estuaries and lagoons.

Salt water penetration from the sea characterizes estuaries. The upstream limit of an estuary can be defined as the point where salinity (p. 24) falls below 0.1 ppt or where the dissolved ions (notably carbonates) become radically different from those (sodium, chlorides) found in sea water (Pritchard 1967). The seaward limit of an estuary can be taken as the point

ZONE OF RIVER DEPOSITION AND REED-SWAMP	ZONE OF TIDAL SHORES WITH SALT MARSH, SANDFLAT AND MUDFLAT	SEA SHORES

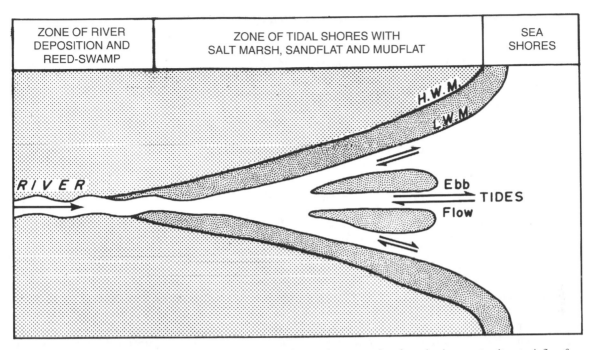

Figure 9.3   Estuary zones related to tidal conditions and salinity. Fresh water outflow from the river meets salt water inflow from the sea as it rises and falls with the tide. The intertidal zone is indicated by fine shading.

where sea water (salinity about 35 ppt) is undiluted by fresh water from rivers, but this is highly variable and may be well offshore in large estuaries or during river flooding. Alternatively, the upstream limit of an estuary can be taken as the point where tidal oscillations fade out, or where marine and estuarine sediments give place to fluvial sediments. The downstream limit may be definable in terms of the bordering coastline, or where fluvial bottom sediments come to an end. An arbitrary boundary is often used: the seaward boundary of the Severn estuary in the Bristol Channel, for example, is sometimes taken as the line between Lavernock Point on the Welsh coast and Sand Point, near Weston-super-Mare, Somerset.

Sea water is denser than river water, and as it enters an estuary it forms an underlying salt wedge which moves upstream, while the fresh river water spreads downstream over it, the junction between upper fresh and lower salt water being sharply defined. Fine-grained sediment carried out to sea by fresh water can be brought back up the estuary by a bottom current formed by the incoming salt wedge, and coarser sediment may accumulate upstream from the salt wedge limit. Such a pattern of sediment coarsening upstream has been found in Chesapeake Bay, and along the

lower Mississippi. Where the tide range is large and river input relatively small, estuarine water is less sharply stratified and there is an inflow of sediment from the sea, diminishing upstream in grain size as the current slackens, as in the Thames estuary. In wide estuaries with strong tidal currents and wave action, and only weak river discharge, vertical stratification disappears and the estuarine water becomes homogeneous. In funnel-shaped estuaries such as the Firth of Forth the inflowing tidal current tends to move towards the left and the outflowing river current to the right in response to the Coriolis Force generated by the earth's rotation (the pattern is reversed in the southern hemisphere).

Estuaries can also be classified by tide range, which strongly influences estuarine processes. A microtidal estuary is dominated by wind-generated waves and discharging river currents, and has a relatively narrow intertidal zone, a mouth often encumbered by spits and shoals or a sandy threshold, and a salt wedge that moves in and out with the tides. There are often bordering beaches and spits, and estuary-head river deltas. A mesotidal estuary has stronger tidal currents, a wider intertidal zone with mudflats and salt marshes or mangrove swamps, and multiple meandering channels and creeks. The tidal meanders may be a response to upstream movement and deposition of bottom sediments which impede the outflow of ebb currents in the channels. A macrotidal (and megatidal) estuary is dominated by tides and tidal currents, and is usually funnel-shaped, with a broad mouth and linear banks. The funnel shape is of exponentially diminishing width upstream, adjusted to accommodate the tidal prism, particularly where the length of the tide wave equals, or is a simple multiple of, estuary length, as in the Ord estuary, Western Australia, where the ratio is 1 : 4 (Wright et al. 1973). Intertidal mudflats and marshland are extensive at low tide, and there are elongated bars between sub-parallel deeper channels. Wave action is of minor importance on the high tide shoreline, but may contribute to the shaping of intertidal morphology (p. 199).

Dalrymple et al. (1992) divided estuaries into three zones: an outer, marine dominated zone with net sediment flow landward, a central zone of relatively low wave energy and convergence of marine and fluvial sediment, and an inner river-dominated zone where net sediment flow is seaward. They distinguished between wave-dominated and tide-dominated estuaries. Wave-dominated estuaries have prominent depositional sandy thresholds at the mouth, with bordering beaches and cliffs, a central zone of muddy deposition and a river delta upstream. Tide-dominated estuaries have rivers flowing into tidal channels that wind through extensive mudflats and marshlands to shoaly mouths which may have tidal deltas (the term delta is misleading, for these are produced by ebb and flood tidal currents and not by fluvial deposition) (Wells 1995).

Tides flow into estuaries in a manner determined by the configuration, the velocity of tidal currents (C) diminishing with depth ($C = \sqrt{gD}$, where g is the gravitational constant). As the tide wave (length $L = T\sqrt{gD}$, where T is the tidal period) enters the mouth of an estuary maximum velocity theoretically occurs at high tide (wave crest) and low tide (wave trough), with slack water as the current reverses at mid-tide. This changes towards the head of the estuary, where maximum inflow and outflow occur theoretically at mid-tide and slack water at high and low tides. In practice the pattern is complicated by partial reflection of the tide wave, which delays the progress of the wave crest (high tide) upstream.

In flood-dominated estuaries the narrowing and shallowing of the channel and the outflow of fresh water from the river cause the tide wave to steepen, increasing the tide range. Inflowing currents strengthen during the briefer tidal rise and outflowing currents weaken during the longer ebb, producing tidal asymmetry and a net upstream movement of sediment. This inflow of sediment produces rapid accretion on sandflats, mudflats and marshlands in the upper intertidal zone, especially during the long slack water period at high tide (p. 202). In Britain the Severn and Dee have flood-dominated estuaries. By contrast, there are ebb-dominated estuaries with stronger outflow during the briefer ebb tide and a long slack water period at low tide, resulting in net seaward movement of sediment. The rapidly falling ebb tide causes erosion in the mid-tide zone, steepening the intertidal profile and forming relatively narrow low tide channels. The Ems and Weser Rivers, draining into the Wadden Sea, are examples.

The tidal prism, or total volume of water that enters and leaves an estuary in a single tidal cycle, determines the discharge and therefore the velocities of flow across any cross section (O'Brien and Dean 1972). In microtidal and mesotidal estuaries narrowing and shallowing cause frictional drag on the incoming tide, and the tidal prism diminishes upstream. Mutual adjustment between morphology and the tidal prism results in a dynamic equilibrium that can be disturbed by storm or flood events, or by changes in sea level or the dimensions of the estuary mouth resulting from spit growth or shoal migration (Pethick 1996).

Where tide range increases upstream on macrotidal coasts the front of the tide wave may steepen into a wave (known as a tidal bore), propelled upstream, as in the Severn estuary (p. 20). Estuaries with mouths constricted by promontories or spits have impeded tidal ventilation and a diminished tidal amplitude, but increased ebb and flow current velocity at their mouths, as in the Mersey in Lancashire. Tidal currents there are strongest at high and low water (slack water at mid-tide), and sediments are carried upstream as the tide rises and deposited there as current velocity diminishes in the intertidal zone.

The meeting and mixing of fresh river and brackish sea water determine salinity regimes within an estuary. In calm weather inflowing tides bring in a wedge of brackish marine water, which moves sediment upstream along the estuary floor. As the tide turns fresh water outflow becomes dominant, initially in surface waters, where it has little effect on bottom sediments, but can discharge fine-grained sediment in suspension, forming plumes that extend off river mouths. Areas of turbid relatively fresh water extend far out to sea off the Amazon and many other rivers, especially in the tropics. The salinity distribution within an estuary can be modified by turbulence and by the effects of strong wind action and fluvial discharge, especially during phases of river flooding.

Tidal currents dominate estuarine morphology, but wind action moves water and influences estuary circulation, and also produces waves up, down and across an estuary. Wave action may become significant in shaping shore, nearshore and shoal features in large, wide estuaries. Many estuaries on high wave energy coasts, as in Australia and South Africa, are partly blocked by spits and barrier islands built up by strong wave action from the sea. Such estuaries are in fact estuarine lagoons (Roy 1984). Some are sealed off completely in the dry season, when river flow is unable to maintain an outlet, but after heavy rain river flooding builds up the water level until it spills over, reopening the entrance.

EVOLUTION OF AN ESTUARY

During the Last Glacial phase, late in Pleistocene times (p. 38), rivers drained valleys that descended to a lower sea level. The land surface at that time can often be traced as a subaerially oxidized horizon that was submerged by the sea, then buried beneath Holocene estuarine sediments. This pre-transgression land surface may have been dissected by the more deeply scoured estuary channels. It surfaces at the landward limits of the Late Quaternary marine transgression, which in the upper estuary may be a valley-side slope showing evidence of basal cliffing by wave action and beaches formed before the deposition of Holocene fluvial sediments by the inflowing river as an alluvial flood-plain. Sometimes there is a transition from intricate river meanders downstream to long, sweeping meanders on the aggraded (essentially deltaic) valley floor in the zone that was formerly submerged by the sea, but this contrast may indicate the extent of modification of river discharge by tidal oscillations or occasional wind-driven upstream flow (Bird 1978a). Further downstream the estuary may have been narrowed by shore progradation, marked by beach ridges, salt marsh or mangrove terraces, formed after marine submergence came to an end. In some places the coast has been cut back beyond the limits of the Late Quaternary marine transgression by later marine erosion, or by meandering rivers and tidal channels.

Between the pre-transgression land surface and the existing estuarine channels and bordering intertidal morphology is a body of sediment that has been deposited in the estuary in Holocene times. Some of this sediment has been brought down by the river, some has come from bordering slopes, and some has been washed in from the sea: the proportions vary from estuary to estuary (Guilcher 1967). Woodroffe (1996) gave examples of the stratigraphy of these infills in some European and Australian estuaries. The Holocene stratigraphy commonly includes interdigitating wedges of fluvial and marine sediment of varying grain size (mainly sand, silt and clay), with shell beds, peat horizons formed in salt marshes, mangroves or seagrass beds that have been submerged and buried by younger sediment. The sedimentary sequence usually shows evidence of phases of erosion resulting from the lateral migration, enlargement or reduction of tidal channels, resulting in discontinuities and lateral transitions in the thickness, texture and composition of sedimentary horizons, and embedded deposits that originated as shoals, cheniers, levees or vegetated zones. There is usually a transition from fluvially-supplied sediment in the upper reaches to inwashed marine sediment around the entrance from the sea, and from bordering beaches, dunes, cliffs and marshland down through intertidal sandflats and mudflats to central channels that carry river discharge at low tide. This pattern was demonstrated in the Yaquina estuary in Oregon by Kulm and Byrne (1967), who found seasonal variations, stronger wave energy in winter sweeping marine sediment farther upstream than in summer. As infilling proceeds, the zone of fluvial (deltaic) deposition extends downstream, a threshold of inwashed marine sandy sediment (and sand blown from coastal dunes bordering the lower estuary) grows upstream, and encroaching marshland, prograding beaches and tributary deltas narrow the estuary, suppressing tidal ventilation and reducing wave energy. The rate of infilling depends partly on fluvial sediment yields and partly on the size and shape of the estuary. Eventually, the estuary fills, and further deposition may then form a protruding delta (Chapter 10).

Stages in the filling of a tropical estuary were documented by Woodroffe (1996) with reference to the Adelaide, Mary, South Alligator and East Alligator Rivers which flow across a broad confluent delta plain to Van Diemen Gulf in northern Australia. He showed that the Late Quaternary sea level rose to flood a wide shallow embayment, submerging a land surface of oxidized sands, muddy sands and gravels. Fluvial sands were then deposited, and overlain by muddy sediment in spreading mangrove swamps, through which the river channels meander intricately.

Before considering the origins of sediments deposited in estuaries it is necessary to consider the form and dynamics of the channels that convey water and sediment through them.

ESTUARINE CHANNELS

At low tide in estuaries there are often a number of channels leading from the river to the sea, between banks of sand, silt, and clay. These channels and banks are shaped largely by tidal currents, and are subject to rapid changes in configuration. The banks often bear superficial ripple patterns, produced by the action of waves and currents on the sandy or muddy estuary floor as the tide rises and falls.

As tidal currents ebb and flow shoals move and channels migrate, and many estuaries have independent, mutually evasive ebb and flood channels (Robinson 1960). Ebb channels have a residual outflow of river water and sediment and become wider and shallower seaward while flood channels have an inflow of sea water into the estuary and become shallower upstream. Where the tidal flood begins before ebb has ended, outflow may continue in ebb channels for up to an hour after low tide, when water has begun to move up the nearby flood channels. Where the tide range is large these channels are straight and parallel, but where it is smaller they curve and interdigitate.

Currents change the position and dimensions of estuarine channels and shoals, cutting away sediment in one place and building it up in another, which is why estuarine ports often require dredging to maintain the depth and alignment of a navigable approach. A knowledge of the characteristic patterns of erosion and deposition is necessary to ensure that dredged material is not dumped where it can be washed straight back into the channel required for navigation, and in recent years use has been made of radioactive tracers to determine paths of sediment flow in the Thames estuary to guide dredging and ensure access to the port of London.

Meanders that develop in estuarine channels, as in the Humber estuary in Britain or Chesapeake Bay in the United States, are related to patterns of alternating ebb and flood currents, notably the brief phase of very high discharge that occurs during the ebb, when the channels carry water volumes comparable with those in major rivers (Ahnert 1960). Intricately meandering fluvial channels give place downstream to elongated pools with alternating cuspate bank projections and deposited bars in the intervening calmer zones (Figure 9.4). Usually there is an accompanying flaring of the estuarine channel seaward, related to the rapidly increasing volume of ebb-augmented discharge towards the river mouth. Examples of estuarine meanders are found in the rivers that drain into Joseph Bonaparte Gulf, in northern Australia.

Many ports are situated beside estuaries, and there have been modifications of morphology and tidal regimes by dredging to create or maintain navigable channels and by excavation and land fill in port areas. Insertion of gates or barriers across estuary channels to prevent storm surge flooding upstream, as in the Thames estuary, has modified tidal regimes downstream, leading to higher high tide levels on bordering salt marshes and increased wave and current scour on mudflats, salt marshes and shore embankments.

ESTUARINE SEDIMENTS

Estuaries are typically areas of active sedimentation, the area drowned by the Late Quaternary marine submergence being progressively filled, and contracting in volume, depth, and surface area until the river winds to

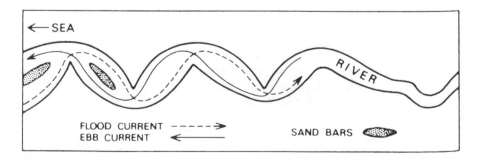

Figure 9.4 Estuarine meanders.

the sea through a depositional plain. It is possible to find stages between a deep, branching estuary, little modified by sedimentation (e.g. Port Jackson), through to estuaries that have been completely filled in (e.g. Snowy River) and deltas growing out beyond the former coastline. The rate of filling depends on the original area and volume of the drowned valley mouth and the sediment yield from rivers, downwash from bordering slopes, shore erosion and inflow from the sea. Guilcher (1967) cited the Loire estuary as an example of infilling mainly by fluvial sediment from a river carrying a large load of suspended silt and clay from the river, the Rance estuary in Brittany as one that has received substantial silt and clay from bordering slopes of Head deposits (p. 54), and the Kapatchez estuary in Guinea as one filled with sediment washed in from coastal mudflats. Enclosure by spits and barriers accelerates the natural reclamation of an estuary by diminishing wave and tidal scour and impeding outflow of sediment, but increased river discharge and/or channel narrowing can intensify current velocity and increase scour along channels. The pattern of sediments on the floor of an estuary depends on source locations, wave action as limited by fetch along or across the estuary and the rise and fall of the tide, the effects of tidal currents, and configuration, including the presence of enclosing spits and the distribution of tidal channels.

Continuing coastal submergence tends to postpone infilling by deepening the estuary and widening its mouth. On the other hand, estuaries on emerging coasts are shallowed and infilled more rapidly. On the south coast of Norway, where there is land uplift due to isostatic recovery following Holocene deglaciation, rapids develop across emerging rocky areas in the shallowing mouths of estuaries, which become elongated lakes as they are cut off from the sea.

Patterns of sedimentation also vary in relation to the relative dominance of inflowing (flood) and outflowing (ebb) tides: where flood tides are stronger (flood-dominated estuaries) sediment tends to move upriver, while in ebb-dominated estuaries sediment is scoured and carried downstream.

Sediment is moved in estuaries by wave and current action, which rolls and bounces sand grains along the floor of the estuary and lifts finer sediment into suspension. Winnowing of fine sediment by waves and tidal currents leaves zones of coarser sediment, mainly sand, gravel and shell deposits, on bordering shores, in shoals and along channel floors. Sand is readily deposited on the floors of estuaries when the current flow slackens, and sandflats form on either side of tidal channels. Mud deposition, assisted by flocculation, forms intertidal mudflats, as described in the previous chapter.

Rivers draining catchments where argillaceous formations outcrop, or where superficial weathering has yielded fine-grained material, are sources of estuarine mud. Many of the rivers of southern and eastern England have derived muddy sediment from extensive clay lowlands for deposition in their estuaries: the London clay has been a major source of Thames estuary mud, and the Mesozoic clay formations of the English Midlands have yielded muddy sediment to the estuaries of the Severn, the Humber and rivers draining into The Wash. The mud found in the rias of Devon and Cornwall has been washed in by rivers draining granite areas (where there has been extensive hydrothermal alteration of feldspars to clays, mainly kaolin) and Palaeozoic metasediment outcrops, which also weather to yield clay minerals. Much of the valley-floor alluvium and estuarine mud has come from fine-grained material washed out of the Pleistocene periglacial Head deposits (derived from weathered granites and metasediments) which mantle hillsides and are eroded from low bordering cliffs. The same deposits mantle coastal slopes, but the fine-grained sediment eroded from these is quickly dispersed by the open sea, except in low wave energy sectors, as on the muddy shores of Mounts Bay, in the lee of the Land's End peninsula in Cornwall.

The muddy sediments of the Wadden Sea are derived, at least in part, from the north German rivers laden with silt and clay derived from loess and argillaceous outcrops, but there may also have been fine-grained sediment washed in by wave action from glacial drift deposits on the sea floor. In South America the River Plate drains a pampas catchment with extensive silt and clay deposits, and delivers mainly muddy sediment to its estuary, which has only a few beaches of fine sand. In the Fly River estuary in Papua New Guinea suspended silt and clay from tropically weathered material in the catchment are flocculated and deposited on meeting brackish water (Wolanski and Gibbs 1995).

In Australia the muddy deposits that occupy the wide intertidal zone in Westernport Bay (Figure 8.1) have come partly from erosion of peaty clays and silts that had accumulated in extensive freshwater swamps to the north, and partly from the weathered mantle of clays on basalt outcrops around Phillip Island to the south. Shallow Inlet, in Victoria, is a sandy estuarine area in the lee of a dune-capped barrier spit, with a very small catchment and little fluvial sediment inflow. It contrasts with nearby Andersons Inlet, which is of similar configuration, but has been receiving muddy sediment brought down by the Tarwin River (Figure 5.15).

The oscillatory motion of waves moving across estuarine mudflats at high tide can throw silt and clay into suspension, and ebb tide currents can remove this sediment downstream or out to sea. On the other hand sediment thrown into suspension by waves on a rising tide can be carried forward by those waves, and by flood tide currents, for deposition in the upper intertidal zone or on backing salt marshes or mangrove swamps. Some of the muddy sediment thus mobilized is lost seaward during ebb tides and discharging river floods, and some is carried upstream for eventual deposition on aggrading flood plains. Muddy sediment that is retained in mudflats and marshlands may be recycled by waves and currents within the estuary as tides rise and fall, the cyclic exchange of sediments being indicated by changes in cross-profile.

In the estuaries of eastern England Pethick (1996) found that wave action from the sea into the outer parts of estuaries was erosive during stormy periods, when the upper mudflats were eroded and sediment moved to the lower intertidal zone, widening and flattening the intertidal profile, whereas in calmer weather sediment was washed back up on to the upper mudflats, restoring the earlier profile. Waves from the open sea did not reach the upper estuary, but winds generated waves, especially at high tide, which shaped the mudflats into steeper, narrower and higher profiles upstream. There was also a longer term alternation (over decades to centuries) whereby high convex mudflats shaped by deposition during flood-dominant phases produced ebb-dominant conditions, resulting in erosion of the upper mudflats to lower concave profiles until flood-dominant conditions returned, and the convex profiles were cyclically restored.

The rate of sediment yield from a river catchment is a function of lithology, weathering, and runoff, and much influenced by climate. Sedimentation can be accelerated either naturally by uplift, volcanic activity, or a climatic change in the hinterland, or artificially by man's impact, through mining and quarrying, deforestation, overgrazing, or unwise cultivation of erodible land within the catchment. Stronger runoff then augments the sediment supply to estuaries. As has been noted (p. 215) many tropical estuaries have been rapidly filled with abundant fluvial sediment from deeply weathered hinterlands, forming deltas and coastal plains. Extensive clouds of turbid water extend seaward off humid tropical estuaries, as in northern Java.

Estuary morphology and sedimentation patterns have often been modified by human activities in river catchments. Soil erosion in deeply weathered catchments in the humid tropics has supplied vast masses of mud to estuaries and coastal embayments, accelerating the progradation of mudflats and the spread of mangroves, as in Jakarta Bay, Indonesia. Infilling of estuaries with fluvial sediment has also been augmented by soil erosion in the river catchment in Chesapeake Bay on the Atlantic coast and several Oregon estuaries on the Pacific coast of the United States. On Mediterranean shores, shrinkage of estuaries during the past 2000 years results from river sediment yields being augmented by widespread soil erosion following the impoverishment of vegetation by overgrazing and unwise land use, notably in Greece and Turkey. In California, Rooney and Smith (1999) used successive charts to measure volumes of sediment accretion in Tomales Bay at intervals between 1861 and 1994, and found that changes in land use in the catchment resulted in variations in rates of soil erosion and infilling of the bay from 94 tonnes per square kilometre per year in 1861–1931 to 357 tonnes per square kilometre per year in 1931–1957. They noted that additional sediment generated by deforestation or cultivation took decades to reach the river mouth as riverbed deposits moved slowly downstream.

Mining operations have also accelerated sediment yields to rivers, as in Cornwall, where the wastes from tin, copper, and kaolin mining have been washed into the rivers and carried down to fill estuaries at Par and Pentewan (p. 102), and build the Fal delta into an arm of Carrick Roads (p. 258). On the coasts of New Caledonia estuaries are being rapidly infilled with sediment generated by open-cast mining of nickel in the steep hinterland, and in Tasmania, sandy waste from tin dredging in a hilly headwater region has been carried downstream to be deposited in the estuary of Ringarooma River (Bird, J.F. 2000).

TIDAL DELTAS AND THRESHOLDS

In open, funnel-shaped estuaries sandy sediment washed in from the sea may be deposited in the form of banks or thresholds, sometimes shaped into paired tidal deltas submerged at least at high tide, with channels diverging upstream and downstream. Tidal deltas of this kind are found between barrier lands on the Gulf coast of the United States, for example at Yass Abel, one of the entrances to Barataria Bay in Louisiana. On the Atlantic coast, where wave energy is stronger, there are flood tide deltas on the landward side of such entrances, as at Moriches Inlet on Long Island. By contrast, ebb tidal deltas have formed off inlets on the New Zealand coast, varying in form with coastline configuration, tide regime and quantity of sand available (Hicks and Hume 1996).

Where the tide range is small and wave action strong the tidal delta becomes a wider threshold, usually with a single channel winding over a broad sandy flat exposed at low tide, a smooth and gentle seaward slope and a steep inner slope, where sand is advancing into the estuary from the sea. Thresholds of this kind are found in the entrances to estuarine inlets on the New South Wales coast, notably at Narooma (Figure 9.5) (Bird 1967a).

ESTUARY SHORES

Estuary shores extend from the low spring tide line across the intertidal zone to the high spring tide line, which may be marked by a beach or the transition to land vegetation behind a salt marsh or mangrove swamp, and on to the limit of flooding, whether by rivers, very high tides, or storm surges when waves drive water on to bordering land. In passing from an open coast to the shores bordering an estuary there is a reduction in wave energy. Cliffs become less bold, beaches more irregular, and the generally broader intertidal zone is diversified by sandflats, mudflats, shoals and bars, salt marshes and mangrove swamps. As the tide rises and falls in these shallow intertidal zones wave energy is dispersed across broad shore environments.

Estuarine beaches consist of sand, gravel or shelly material washed in from the sea, eroded from bordering shores or derived from the nearby foreshore (Nordstrom 1992). These are moderate to low wave energy shores, with wave processes effective only at and about high tide and wind action on exposed sectors (Jackson and Nordstrom 1992). There is often inward longshore drift, forming beach lobes that migrate alongshore and spits that grow into the estuary.

The configuration of an estuary can be changed by wave and current action, particularly at high tide when shores exposed to strong wind and wave action are eroded and the material deposited on more sheltered parts of the shore, or on the estuary floor (Nordstrom and Roman 1996). Depositional features on estuary shores sheltered from strong wave action and current scour include the terraces built in salt marshes, mangrove swamps and seagrass beds, described in the preceding chapter. Poole Harbour, a large estuarine inlet

in southern England, has been modified in this way, with cliffed promontories and embayments occupied by salt marsh. Deltaic deposition may occur as rivers flow into estuaries, and fluvial sediment carried into an estuary may be moved to and fro before eventually being deposited on the shore near high tide level.

Changes on the shores of the Dee estuary in north-west England during the past two centuries have been documented by Pye (1996), who traced the effects of dredging and stabilization of tidal channels and of land reclamation on estuarine morphology. There has been erosion of salt marshes and intertidal sandflats and mudflats along the north-western (Welsh) shore of the estuary, and rapid accretion on the south-eastern (Wirral) shore, with lateral and vertical growth of salt marshes.

ESTUARINE MARSHES AND SWAMPS

The upper part of the intertidal zone in estuaries is usually vegetated, with salt marshes, freshwater swamps or mangroves. As has been shown in Chapter 8, these vegetation communities have colonized and stabilized the shore, diminishing the effects of waves and currents, and so promoting the accretion of sediment in such a way as to build terraces between mid tide and high spring tide that would not otherwise have developed. Typically salt marshes and mangroves are fronted by intertidal mudflats and sandflats and backed by a transition through freshwater swamps to land vegetation. Their seaward margins are gentle slopes where the vegetation is spreading seaward, or small cliffs cut back by wave action or meandering tidal channels. They are dissected by tidal creeks that continue seaward across mudflats and sandflats to deeper low tide channels. As has been explained, the existing morphology of salt marshes or mangrove swamps, the mudflats and sandflats that front them, and the systems of channels and creeks that receive tidal ebb and flow and river discharge, has been moulded largely by currents, and partly by wave action.

As infilling proceeds, bordering salt marshes and mangrove swamps encroach on the estuary, and as the intertidal sandflats and mudflats are built up into depositional terraces the central channel becomes narrower and often deeper. Thereafter river flooding brings fluvial sediments to aggrade the marshlands and form an alluvial valley floor. In terms of the division by Dalrymple et al. (1992) the inner river-dominated zone thus displaces the central low wave energy zone and the outer marine dominated zone

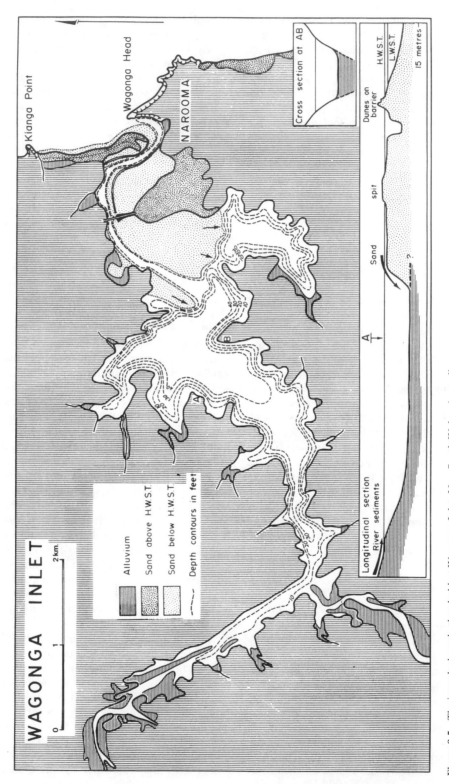

Figure 9.5   The inwashed sandy threshold at Wagonga Inlet, New South Wales, Australia.

progressively seaward. This sequence can be reversed by a relative rise of sea level if fluvial discharge and sediment yield remain unchanged.

COASTAL LAGOONS

Coastal lagoons are areas of relatively shallow water that have been partly or wholly sealed off from the sea by the deposition of spits or barriers, usually of sand or shingle, built up above high tide level by wave action (Emery and Stevenson 1957, Colombo 1977, Phleger 1981, Kjerfve 1993, Cooper 1994). This definition excludes lagoons enclosed by coral reefs, either within atolls or between fringing or barrier reefs and the mainland, because these are essentially marginal marine environments linked with the open sea at high tide. However, there are examples of coastal lagoons formed by localized tectonic subsidence, as on the north coast of New Guinea (p. 22).

Most lagoons are to some extent estuarine, being subject to tidal oscillations and the interaction of salt water from the sea with fresh water from inflowing rivers, and much that has been said about estuaries also applies to coastal lagoons. Indeed, some classifications regard coastal lagoons as bar-built or back-barrier estuaries (Pritchard 1967, Fairbridge 1980).

Coastal lagoons range in size from over 10 000 square kilometres (Lagoa dos Patos, Brazil) down to less than a hectare. There are very many small lagoons at river mouths (often termed barred, blocked or blind estuaries), some of which are of interest despite their small size. Thus Oyster Pond, on Cape Cod in the United States is a small lagoon that was studied intensively over a long period by Emery (1969), and research by Cambridge scientists has made famous the little lagoon at Swanpool, near Falmouth in south-west England (Barnes 1980).

The simplest lagoons are found where the mouth of a river has been enclosed by a wave-built barrier. Such a barrier may be breached from time to time by storm waves, or when river floods pour out over it after heavy rain, but usually it is soon rebuilt by wave action when fine weather returns. Lagoons of this type are common on oceanic coasts, where barriers have been built across drowned valley mouths by the action of strong swell.

Some lagoons are long and narrow, parallel to the coast and separated from the sea by barriers built up in front of the former coastline. The Coorong in south-eastern Australia is an example. Others show a branched configuration, elongated at an angle to the coastline, formed where river valleys have been submerged, then enclosed by a depositional barrier built up across their mouths. Drakes Estero in California is a lagoon formed where a branching inlet (ria), incised into the hilly country of the Point Reyes peninsula, has been partly enclosed by bordering sand spits (Figure 9.6). At Orbetello, on the west coast of Italy, a lagoon has been enclosed between twin barriers that form a double tombolo, attaching a former island to the mainland (Sacchi 1979), and on the Chatham Islands, east of New Zealand, barriers of dune calcarenite have formed between three high islands to frame Te Whanga Lagoon, which has a small outlet on the east coast (Bird 1993b). The largest and most complicated lagoon systems are found where broad embayments have been sealed off from the sea by successive depositional barriers. The Gippsland Lakes in south-eastern Australia (Figure 6.10) are a chain of coastal lagoons enclosed by a Holocene outer barrier (behind the Ninety Mile Beach, Figure 5.24) and separated by remnants of Pleistocene inner and prior barriers (Bird 1978a).

Coastal lagoons usually have one or more entrances from the sea, which are permanent or intermittent gaps through the enclosing barriers (Bird 1993b). The definition of a coastal lagoon implies that their entrances are narrow compared with the coastwise extent of the enclosing barrier or the lagoon. The European Wadden Sea, behind the Frisian barrier islands, is not usually considered a coastal lagoon because the combined width of the openings to the sea (including the broad gap east of Wangeroog) is more than one-third of its coastwise extent, but it can be classified as an open lagoon (Lasserre 1979). The same is true of the area behind the sandy barrier islands (Sea Islands) off the coast of Georgia, in the United States. In general, the term coastal lagoon can be applied where the width of marine entrances at high tide is less than one-fifth (20 per cent) of the total length of the enclosing barrier. A problem arises where the lagoon is elongated at an angle, rather than parallel to the coastline, so that a substantial water area is enclosed by quite short barriers or spits across a narrow seaward entrance. Features of a sheltered lagoon environment may then occur even though the marine entrance is substantially more than one-fifth of the length of enclosing barriers. Indeed there are some inlets of intricate configuration with rocky entrances that have no enclosing depositional features, and yet show lagoonal characteristics: the Knysna Lagoon in South Africa, Lake Maracaibo in Venezuela, and the large landlocked bay on

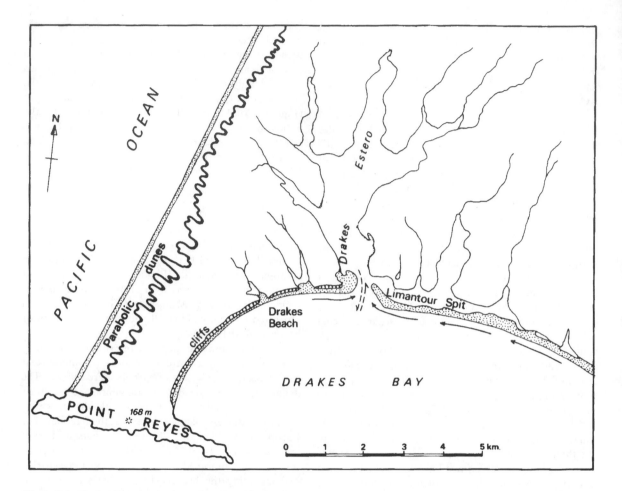

Figure 9.6 Drakes Estero, a lagoon of branched outline impounded by paired spits on the Californian coast.

Sumbawa, Indonesia are examples. On a larger scale the Baltic and Mediterranean seas also have some of the characteristics of large coastal lagoons, with relatively narrow entrances from the Atlantic Ocean.

Coastal lagoons occur on about 12 per cent of the length of the world's coastline. An inventory prepared by the United Nations Educational, Scientific and Cultural Organization (Unesco) listed about 450 coastal lagoons (minimum surface area at high tide one square kilometre) on the world's coastline. There are good examples on the Gulf and Atlantic coasts of the United States, where they and their enclosing barriers have transgressed landward in response to a continuing sea level rise, a process that can also be seen on the submerging shores of the Caspian Sea (p. 159). Coastal lagoons occur in Brazil, where they formed as the sea attained a slightly higher level, and

their enclosing barriers prograded after sea level fell (Martin and Dominguez 1993). They have formed behind storm-built barriers in south-west Iceland and on the southern shores of the Baltic Sea (Gierloff-Emden 1961). They are found around the Mediterranean, notably between Perpignan and Marseilles in France, in eastern Corsica, the northern Adriatic, Egypt, and the northwest Black Sea, including the Sea of Azov. An intermittent chain of barrier-enclosed lagoons extends from the Ivory Coast to the Cameroons in West Africa, and lagoons are also present in Natal, on the Indian and Sri Lankan coasts, on the north coast of Hokkaido, and eastern Sakhalin. They are well developed on the Arctic coasts of Alaska and north-eastern Siberia.

On the south coast of Britain there are a number of small lagoons impounded by shingle barriers, notably

Loe Pool in Cornwall, Slapton Ley in Devon, and The Fleet, behind Chesil Beach in Dorset. In the Orkney and Shetland Islands there are many small shingle barriers, known as ayres, that have been built up across the mouths of embayments to enclose, or partly enclose, lagoons, known as oyces, which rise and fall with the tide. On the Australian coast lagoons are best developed behind sandy barriers formed during Pleistocene phases of high sea level and during and since the Late Quaternary marine transgression, on sectors of the coast of Western Australia south of Perth, and intermittently along the coast of south-eastern Australia from the mouth of the Murray around to southern Queensland (Bird 1967b). New Zealand has coastal lagoons on the shores of the Bay of Plenty in the North Island, and Lake Ellesmere, the Southland lakes east of Bluff, and Okarito Lagoon on the west coast of the South Island.

Coastal lagoons are generally about 6000 years old, having formed where valley mouths or lowlands have been submerged by the sea during the later stages of the world-wide Late Quaternary marine transgression. Some also existed during Pleistocene phases of relatively high sea level, enclosed by Pleistocene barriers, parts of which survive as inner barriers, as in the Gippsland Lakes. Lagoons that were enclosed by Pleistocene barriers drained out during the Last Glacial low sea level phase, leaving subaerial basins that were flooded when the sea rose again. The Gippsland Lakes are an example of a lagoon system that drained and revived in this way. In South Australia, the Coorong is the latest in a series of long, narrow lagoons that were enclosed by successive barriers on a coast that has been uplifted during Quaternary times (p. 36). Its predecessors are marked by tracts of lagoon and swamp (now largely drained) lying between successive emerged sand barriers in the country behind Encounter Bay (Figure 3.4). The majority of coastal lagoons, however, are simply the product of barrier and spit deposition across inlets and embayments that formed in Holocene times.

Coastal lagoons have a variety of shapes and sizes, related to the configuration of the pre-existing coastline and the enclosing spits and barriers, as modified by internal erosion and deposition around their shores and on their floors. They are best developed on low-lying coasts behind shallow coastal seas. They are poorly developed on coasts dominated by high retreating cliffs, as in the Great Australian Bight, on the steep and rocky coasts of Norway, British Columbia, Chile, and southern New Zealand, on the ice-girt Antarctic and Greenland coasts and on the rapidly-emerging coasts of northern Canada and the Gulf of Bothnia. They are also rare on macrotidal coasts, such as the Bay of Fundy in Canada or the Bay of Mont Saint Michel in France, because strong tidal currents have prevented the formation of enclosing spits and barriers.

Lagoon morphology (water depth) can be mapped using conventional survey techniques (traverses with echo sounders), or from satellite multispectral band imagery (checked by on-site sample surveys). Typically depositional lagoon floors are subhorizontal, with deeper channels that may be inherited from prior submerged topography, or scoured by existing currents. Deep scour holes (colks) may be found in narrow straits, especially where there is sharp curvature in the lagoon causing current eddies.

Coastal lagoons show a wide variety of geomorphological and ecological features, but their essential characteristics are summarized in Figure 9.7. There are often three zones: a fresh water zone close to the mouths of rivers, a salt water tidal zone close to the entrance, and an intervening transitional zone of brackish (moderately saline) but relatively tideless water. The three zones may occur in a single lagoon system, as in the Gippsland Lakes. The proportions of each zone vary from one system to another: the Myall Lakes, on the New South Wales coast, consist largely of a fresh water zone; Lake Illawarra consists largely of the intermediate zone, and Wagonga Inlet largely of the salt water tidal zone. The extent of each zone depends on the relative proportions of fresh water and sea water inflow to the lagoon system and on climatic conditions, lagoons tending to be more brackish in arid regions. Lagoons completely cut off from the sea, like those on the Landes coast north and south of Arcachon, in France, are essentially fresh water lakes. In Dorset, The Fleet is a typical estuarine lagoon with salinity augmented by percolation of sea water in through the shingle barrier (Chesil Beach) at high tide and during storms, and Slapton Ley is an almost fresh coastal lake.

Lagoon entrances

Some lagoons have been completely cut off from the sea by barriers, but most have at least one marine entrance (or tidal inlet). The entrance to a coastal lagoon may be bordered by a spit or paired spits, as in the estuarine lagoons at Poole, Christchurch and Pagham on the south coast of England, or there may be several entrances separating barrier islands, as in the Dutch, German and Danish Wadden Sea.

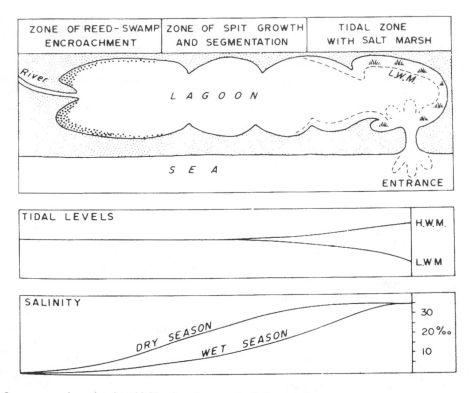

Figure 9.7   Lagoon zonation related to tidal levels and seasonal salinity variations.

Some lagoon entrances are residual gaps that persisted between spits or barrier islands where the lagoon was never completely sealed off from the sea. Others are the outcome of breaching, either by storm waves or floodwater spilling out of the lagoon. Many lagoon entrances are artificial, having been excavated and stabilized, usually between bordering breakwaters, to facilitate navigation or hasten the drainage of flood-waters to the sea. In Italy the three entrances to the Lagoon of Venice have been artificially stabilized by breakwaters up to two kilometres long, the intervening barrier islands having been armoured by large sea walls (Figure 9.8).

Lagoon entrances are larger, more numerous, and more persistent on barrier coastlines where relatively large tide ranges generate strong currents, as on the German North Sea coast, than where tidal action is weak, as on the long sandy barriers which fringe the southern shores of the Baltic Sea. Lagoon entrances are often located on parts of the coastline where wave action is relatively weak, and the action of inflowing and outflowing currents therefore more effective. On wave refraction diagrams, such sections are indicated where there is a marked divergence of wave orthogonals, as at the head of the embayment in Figure 2.2B. Persistent lagoon entrances are found alongside rocky headlands as at the entrance to the Tauranga lagoon at Mount Maunganui on the north coast of New Zealand. Others are in the lee of islands or reefs where wave action is weakened and the ebb and outflow currents are sufficient to maintain a gap. The entrance to Lake Illawarra in New South Wales is protected by Windang Island immediately offshore, while the entrances to several other lagoons on the New South Wales coast are situated at the southern end of sandy bays, close to rocky headlands, where the dominant southeasterly ocean swell is much refracted, and therefore weakened (Bird 1967b).

The configuration of a natural lagoon entrance (like that of an estuary mouth) is the outcome of a contest between the currents that flow in and out and the effects of onshore and longshore drift of sand or shingle which tend to seal them off. The current velocity depends on the tidal prism, the volume of water that enters and leaves a lagoon as the tide rises and falls (p. 223), and the mean cross-sectional area of the

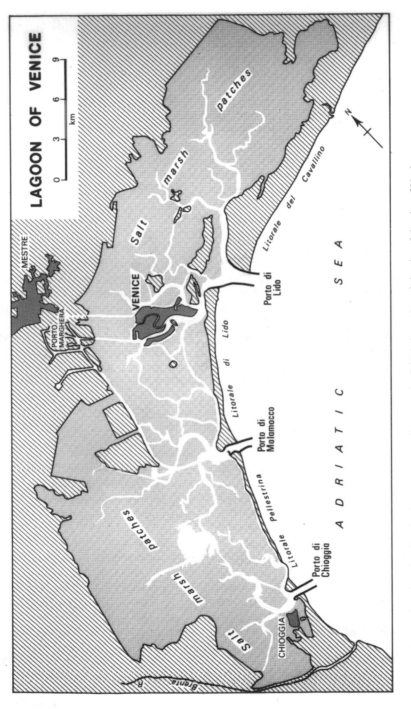

Figure 9.8 The Lagoon of Venice, showing the three entrances, the network of tidal channels and the site of the city of Venice.

entrance increases with the volume of the tidal prism (Pethick 1994).

Currents are generated through entrances in several ways. There are tidal currents produced by tides entering and leaving the lagoon, their strength increasing with tide range. There are currents due to outflow from rivers, particularly after heavy rain when floods build up the level of the lagoon so that water pours out through the entrance, and there are currents generated by wind action, onshore winds driving sea water into the lagoon and offshore winds driving lagoon water out to sea. Strong currents tend to maintain the dimensions of a lagoon entrance, the cross-sectional area varying in relation to the volume of water passing through, being widened or deepened during episodes of floodwater discharge. When the outflow is weak the entrance may be modified by wave action. Waves that arrive parallel to the coast move sand from the sea floor shoreward on to beaches and into a lagoon entrance, and waves that reach the shore at an angle cause longshore drift, supplying sand to build spits that deflect a lagoon entrance. The growing spit impacts transverse ebb and flow of tidal currents against the farther shore, which is cut back by current scour.

The position and dimensions of lagoon entrances thus change frequently in response to variations in the processes at work on them. Migrating entrances punctuate the sandy outer barrier on the Atlantic coast of the United States: the Outer Banks of North Carolina have had a long and complicated history of breaching, enlargement, reduction, migration and closure of tidal inlets to Pamlico Sound as sand moved to and fro along the coast (Dunbar 1956).

Lagoon entrances may show seasonal variations, shallowing or becoming sealed in dry seasons when outflow is weak, and reopening, widened or deepened, when the wet season brings greater volumes of water outflow from the lagoon. This sequence is well known on the South African coast, where most lagoons are sealed off by sand deposition in the relatively dry winter season (May to August) and entrances are reopened in the summer when rains in the hinterland increase fluvial discharge into the lagoons until the water spills over into the sea. In south-eastern Australia lagoon entrances are usually reduced in size or sealed off altogether in dry summer periods, then reopened or enlarged in the wetter winters. When a barred entrance has persisted for several seasons, local people may reopen it by digging an outflow channel, which widens and deepens as the head of water disperses.

Lagoon entrances are generally backed by partly or wholly submerged shoals or fans of inwashed sediment,

similar to the thresholds seen in estuaries. There may also be tidal deltas, lobate or triangular shoals deposited on the inner or outer sides of a lagoon entrance by inflowing or outflowing tidal currents. Examples of thresholds have been documented from the south coast of New South Wales, as at Burrill Lake (Jennings and Bird 1967) and at the entrance to the Murray-mouth lagoons (Bourman and Harvey 1983, Walker and Jessup 1992). Such features are less common on the seaward side, where stronger wave action disperses outwashed sediment offshore or alongshore. At Lakes Entrance, Victoria, an outlet from the Gippsland Lakes, cut in 1889, was initially deep, but a sandy threshold has formed on the landward side and is growing into Lake King, evidently nourished by sand inflow caused by the dredging of the looped sand bar off the lagoon entrance.

Migration of lagoon entrances has occurred on many barriers, and is well illustrated on the Danish North Sea coast where the barrier enclosing the lagoon known as Ringköbing Fiord (the term fiord includes lagoons in Danish terminology) has had an entrance in various positions since 1650, with a tendency to migrate southward (Figure 9.9). This variable entrance became sealed off altogether when it was replaced by an artificial canal at Hvide Sande, maintained between stone jetties; a sluice was added in 1931 to reduce sea water inflow. Sand accumulation on the northern side of the breakwaters is indicative of the southward drift which formerly diverted the lagoon entrance.

In New Jersey curving channels lead from lagoons into the rear of the sandy barrier towards entrances that have been deflected along the coastline by the longshore drift of sand, and behind the Ninety Mile Beach in Australia similar curved channels lead towards the sites of tidal entrances to the Gippsland Lakes that have since been sealed off by sand deposition.

Lagoon entrances thus show a variety of features. At one extreme is the permanent entrance, natural or man-made, which allows a perennial unhindered exchange of water, sediment, dissolved materials and organisms between the lagoon and the open sea. The lagoon then shows estuarine characteristics, such as a transverse salinity gradient declining towards the mouths of inflowing streams and an inwardly diminishing tidal ventilation, as in the Peel-Harvey Inlet in Western Australia (Hodgkin et al. 1981). Tide ranges within a lagoon diminish rapidly away from the entrance(s), the more remote sections of large lagoon systems being unaffected by marine tides. In the Gippsland Lakes the range of spring tides at Lakes Entrance is slightly less than a metre, but at Metung, 10 kilometres to the

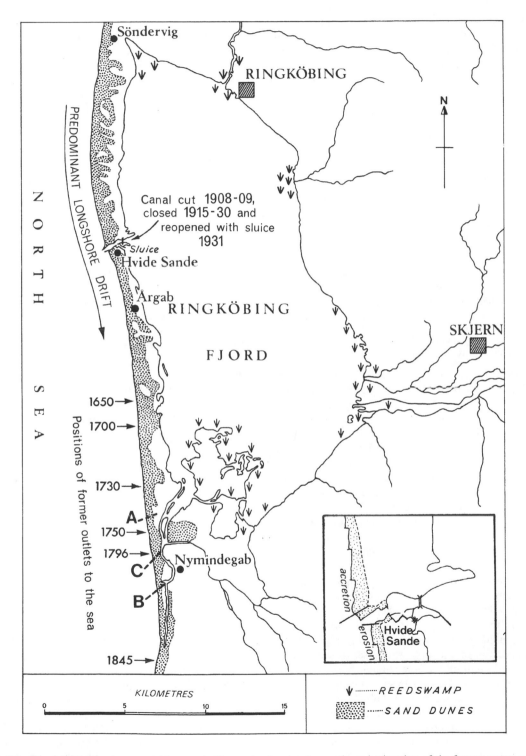

Figure 9.9   Ringköbing Fiord, a coastal lagoon in Denmark, showing the southward migration of the former entrance as a result of longshore drift of sand, and (inset) the artificial entrance cut at Hvide Sande.

west, it is less than 30 centimetres, and tides are not perceptible further away from the entrance, in Lake Victoria and Lake Wellington. There are, however, irregular changes of level due to heavy rain or river flooding, and oscillations (termed seiches) during and after periods of strong wind. Similar variations have been observed in other coastal lagoons.

At the other extreme is the completely enclosed lagoon with an impermeable barrier preventing exchanges with the sea. As has been noted, such lagoons tend to become fresh water lakes in humid environments and hypersaline in arid regions. Between these extremes are lagoon entrances that vary in form, dimensions and location, and are sometimes completely sealed off. These variations modify the extent to which tides invade a lagoon (tidal ventilation), river flood levels, marine incursion, salinity regimes and related ecological conditions within a lagoon system and can influence patterns of sedimentation and geomorphological change.

Tides and salinity in lagoons

The hydrological characteristics of a coastal lagoon are determined partly by its configuration, partly by the dimensions of entrances from the sea, and partly by the balance between atmospheric precipitation, fresh water inflow from rain and rivers and evaporation on the one hand, and salt water inflow from the sea, related to tide range and tidal ventilation of the lagoon, on the other. Winds blowing over a lagoon lower the level at the windward end and build it up to leeward, so that when the wind drops normal level is restored, often by way of seiches of diminishing amplitude.

Entrance dimensions also influence the pattern of salinity in a coastal lagoon. As in estuaries, salinity is determined by the meeting and mixing of fresh water from rain and rivers and salt water from the sea, and generally diminishes from the lagoon entrance towards mouths of the rivers. In regions with seasonal variations of precipitation the salinity regime is also seasonal, for in the dry season sea water flows in through the entrance to compensate for the diminished river flow and the loss of fresh water by evaporation but in the wet season evaporation losses are reduced and the lagoon is freshened by rain and rivers. In humid regions lagoons typically have estuarine salinity regimes, salt water inflow from the sea being diluted by rainfall and fresh water runoff into the lagoon system. However, on arid coasts lagoons may lose more water

by evaporation than they receive from rainfall and runoff, and if the inflow of sea water is insufficient to prevent the development of high concentrations of salt they become hypersaline, or even dry out altogether as saline flats. This has happened in the Laguna Madre, on the Texas coast, which has shallow hypersaline areas and saline flats with gypsum and algal mats (Fisk 1958). Similar features are seen at the southern end of the Coorong in South Australia, away from the lagoon entrance, where tidal ventilation is too weak to prevent the development of high salinity in evaporating water. During the summer months hypersaline conditions develop in enclosed lagoons such as Lake Eliza and Lake St Clair on the coast of South Australia, and in the desiccated lagoon behind Israelite Bay in Western Australia saline evaporite deposits have formed where a lagoon has dried out completely.

Salinity conditions influence modes of sedimentation in lagoons. Clay carried in suspension in fresh water is flocculated and precipitated by the electrolytic effect of sodium chloride in solution when saline water is encountered (p. 202). Inflowing fresh water is often brown or grey with fine-grained sediment in suspension, and becomes clear when salt water is encountered, the junction often being sharply defined, with coagulated sediment raining down. Salinity is also of ecological importance affecting the development and distribution of shore vegetation around lagoons and thus influencing patterns of sedimentation in encroaching swamps.

Lagoons with more restricted or temporary entrances are less influenced by tidal movements and more protected from the effects of waves from the open sea. Geomorphologically they may resemble inland lakes. In New South Wales, for example, Lake Macquarie is essentially a marine lagoon, with almost tideless shores, a barrier that excludes ocean waves, and water salinity similar to that of the open sea, except in small areas of dilution near the mouths of inflowing streams. Marine influences are also much reduced in lagoons where the entrance is in the form of a long, winding channel through the enclosing barrier, as in the Myall Lakes, also in New South Wales (Thom 1965). Usually the water is fresh or slightly brackish, but during droughts salinity increases as sea water spreads in along the connecting channel.

Some coastal lagoons have been sufficiently cut off from the sea by enclosing impermeable barriers to become fresh water lakes such as Slapton Ley in south-west England, while the Murray-mouth lagoons in South Australia (Figure 9.10) have become fresh after their entrances were artificially sealed by barrages.

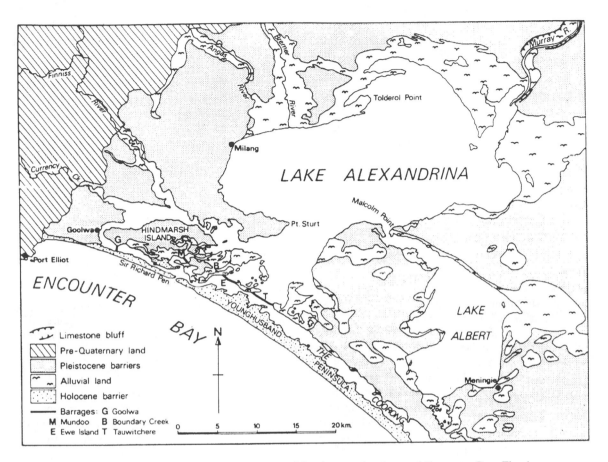

Figure 9.10 The Murray-mouth lakes, enclosed by coastal barriers on the shores of Encounter Bay. Five barrages were constructed across gaps in the inner barrier in 1940 to prevent tidal incursion and to control salinity.

In New Zealand, the Waitangitaona River changed its course during a flood in 1967 to flow to the mouth of Okarito Lagoon, a brackish lagoon which was thus freshened. By contrast, the cutting or enlarging of an entrance to a lagoon that was previously fresh or slightly brackish results in a salinity increase that can have geomorphological as well as ecological consequences. This has happened in the Gippsland Lakes (p. 246) and two West African lagoons: Lake Nokoué in Benin, after the opening of a marine entrance at the port of Cotonou, and the Ebrié lagoon on the Ivory Coast after the cutting of the Vridi Canal through the enclosing barrier in 1950 (Hinschberger 1985). Reduction of runoff from rivers by reservoir construction or soil conservation schemes in the hinterland can also increase salinity in coastal lagoons, and may have contributed to the recent rise in salinity in the Gippsland Lakes.

Evolution of coastal lagoons

The initial form of a coastal lagoon depends on the shape of the inlet or embayment enclosed and the inner shores of the barriers that enclosed it. Some lagoons were originally broad embayments (e.g. Lake Illawarra in New South Wales), others show the much branched form of submerged valley systems (e.g. Lake Macquarie in New South Wales) and in some the enclosing barriers incorporate high islands (e.g. the Tuggerah Lakes in New South Wales). The evolution of coastal lagoons has been influenced by the geological and geomorphological history of the coastal area, and the sequence of changes in the levels of land and sea which have resulted in coastal submergence, forming the inlets and embayments. The growth of coastal barriers shaped the initial morphology, and determined the position and dimensions of entrances from the sea.

Once enclosed, coastal lagoons are modified by erosion and deposition. The inner shores of the enclosing barriers are usually simple in outline, but there may be protrusions where recurved ridges, which marked stages in the prolongation of a spit which became an enclosing barrier, or promontories formed where sand has blown over as an advancing dune, or been washed over or through low-lying sections of the barrier by storm waves or exceptionally high tides. Washover fans are trimmed and re-shaped by lagoon waves and currents, and may evolve into cuspate spits or forelands on a lagoon shore. Deposition is also common in the zones behind barrier islands where tides flowing in from neighbouring entrances meet, as in the Wadden Sea and on the southern side of Scolt Head Island, where deposition forms a tidal divide with sandflats and mudflats. As accretion proceeds this may become marshy.

Sediment is carried into a coastal lagoon by rivers, by tidal currents entering from the sea, and by winds blowing sand from bordering coastal dunes. Deposits include material of organic origin, such as shells, guano and peat, and in arid regions chemically precipitated salt, calcite and dolomite. In one way or another, most lagoons are being gradually filled in and will be replaced by depositional coastal plains.

Lagoons fed by rivers receive sediments ranging from coarse sand to silt and clay, some of which may be deposited in deltas. In the sheltered waters of the Gippsland Lakes small deltas of silt and clay have been built by the Latrobe and Avon Rivers into Lake Wellington, and the Tambo and Mitchell Rivers into Lake King (Bird 1978a). The deltas of the Latrobe, Avon and Tambo Rivers are cuspate in form, while the Mitchell delta, built into the more sheltered water in the northern part of Lake King, consists of elongated silt jetties similar to those built by the much larger Mississippi River in the United States. As on the Mississippi delta, the growth of all of these deltas was assisted by the presence of reedswamp on their shores, in which sediment washed down to the river mouths during floods was trapped and retained. Where the reedswamp fringe has disappeared as the result of salinity increase deltaic sediment is no longer retained by shore vegetation, and waves are attacking the unprotected shores of the deltas. The Tambo delta and the Mitchell delta in Lake King show advanced stages in dissection, but the Latrobe and Avon deltas, in the less brackish waters of Lake Wellington, remained reed-fringed and were still growing by the addition of intercepted silt and clay until the 1970s. Now only patches of reedswamp remain, and these deltas are no longer growing. Evidently the presence of reedswamp promoted sedimentation so that deltas could be built at these river mouths, but when the reed fringe disappeared the deltas were no longer stable, and erosion began to consume them. Figure 9.11 uses outlines from an 1848–49 survey (when the delta was reed-fringed) and air photographs taken in 1940 (when the reed fringe had disappeared) and 1990 to show the shrinkage of the Mitchell River silt jetties.

Sand and gravel deposited as the river enters a lagoon may be added to lagoon beaches and spread around the shore by wave action, while the finer sediment is carried out into the lagoon and deposited on the floor, progressively reducing the depth.

Fluvial sediment yields to lagoons may be modified and accelerated by the reduction of vegetation cover, the onset of soil erosion or mining activities in the river catchment. As has been noted, deforestation and agricultural development in the catchment of the Citanduy River in southern Java greatly increased the flow of muddy sediment from that river into the Segara Anakan lagoon, which is rapidly silting (Figure 9.12). The rivers that drain into the Gippsland Lakes formerly delivered mainly silt and clay, but deforestation and soil erosion in the catchment has resulted in downstream movement of sand and gravel. So far little of this has reached the lagoons, but successive floods are moving the sand and gravel downstream, and this coarser sediment will eventually be deposited on the lagoon floors and shores. Lake Wellington, which is now bordered by a mainly swampy, eroding shore will then become a lagoon fringed with sandy beaches. In north-eastern Tasmania sand generated by tin dredging has moved down valleys into the George River, which is now building a sandy delta into the coastal lagoon at George Bay. Medea Cove, formerly an arm of the George Bay lagoon, has been largely filled with sandy mining waste washed in by Golden Fleece Creek to form broad sandflats and mudflats, colonized by rushy salt marsh and riverine scrub (Bird, J.F. 2000).

On the other hand, there could be a reduction of sediment yield to a coastal lagoon where a dam constructed on an inflowing river is intercepting sediment in the reservoir, or by successful soil conservation works in the hinterland.

Subsidence of coastal regions, as in the northern Adriatic or along the Gulf and Atlantic coasts of the United States, may deepen and maintain coastal lagoons, delaying their infilling. Cavazzoni (1983) concluded that subsidence had deepened the water in the Lagoon of Venice and allowed larger waves to erode the lagoon floor. On the north coast of New Guinea

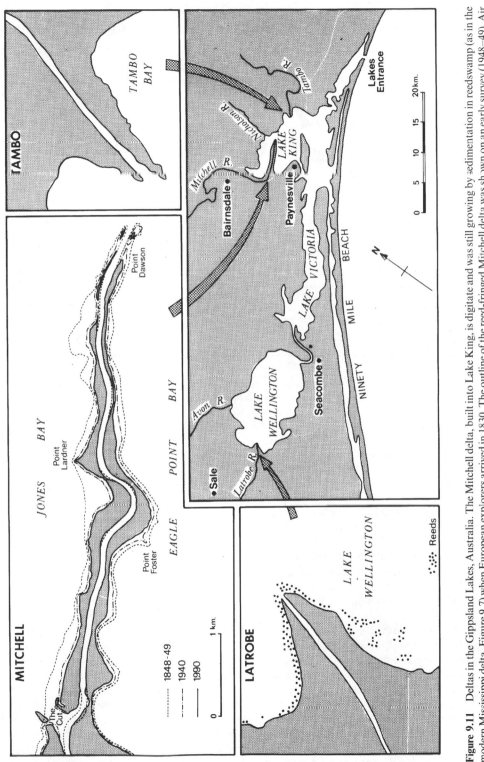

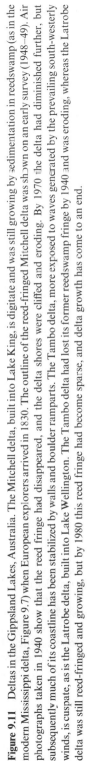

Figure 9.11   Deltas in the Gippsland Lakes, Australia. The Mitchell delta, built into Lake King, is digitate and was still growing by sedimentation in reedswamp (as in the modern Mississippi delta, Figure 9.7) when European explorers arrived in 1830. The outline of the reed-fringed Mitchell delta was shown on an early survey (1848–49). Air photographs taken in 1940 show that the reed fringe had disappeared, and the delta shores were cliffed and eroding. By 1970 the delta had diminished further, but subsequently much of its coastline has been stabilized by walls and boulder ramparts. The Tambo delta, more exposed to waves generated by the prevailing south-westerly winds, is cuspate, as is the Latrobe delta, built into Lake Wellington. The Tambo delta had lost its former reedswamp fringe by 1940 and was eroding, whereas the Latrobe delta was still reed-fringed and growing, but by 1980 this reed fringe had become sparse, and delta growth has come to an end.

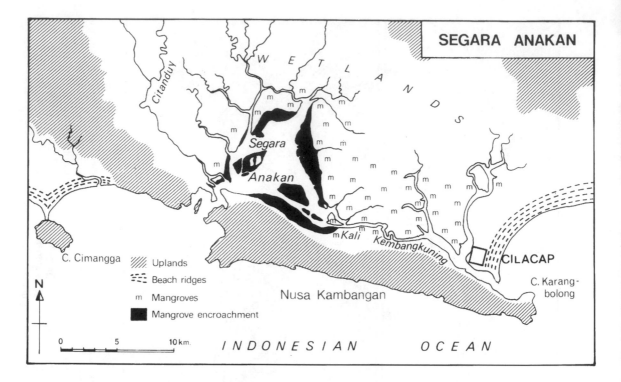

Figure 9.12 The Segara Anakan is an estuarine lagoon in southern Java which has been shallowing rapidly as the result of inwashing of sediment from the Citanduy and other rivers. In recent decades mangroves have advanced along the lagoon shores.

the 1907 earthquake in the Torricelli Ranges resulted in local subsidence on the coastal plain and the formation of a lagoon up to two metres deep in an area that previously carried villages and coconut plantations, behind a sandy coastal barrier at Sissano, near Aitape (Bird 1985b). There was extensive, though brief, marine submergence here during the 1998 tsunami. Along the northern flank of the Sepik delta in New Guinea subsidence has widened and deepened an old meandering channel to form the chain of lagoons known as the Murik Lakes behind a narrow sandy barrier.

Accumulation of inwashed sediment, organic deposits such as peat or shells, and precipitated salts, results in the shallowing and shrinkage of lagoons. The following sections show how changes in configuration are related to the effects of wind-generated waves and the currents produced by rivers, wind action and tides within the lagoon, and how ecological conditions, particularly water salinity and temperature, are important in the geomorphological evolution of coastal lagoons. According to Isla (1995) coastal lagoons in

high latitudes are dominated by physical processes, and in low latitudes by biological processes.

Rounding and segmentation

As a barrier develops to enclose a lagoon, sea waves are excluded and the effects of marine salinity and changes of tide level are reduced. Winds blowing over the lagoon generate waves and currents that are related to the direction and strength of local winds and the lengths of fetch across which these winds are effective. Long, narrow lagoons have the strongest wave action in diagonal directions, along the maximum fetch. The lagoon shore may be cliffed by wave attack, yielding sediments that are carried alongshore by waves arriving obliquely. Waves coming in at an angle move sediment to and fro along lagoon shore beaches, eroding embayments and depositing spits and cusps, which may grow to such an extent that the lagoon becomes divided into a series of small, round, or oval lagoons, linked by narrow straits, or sealed off

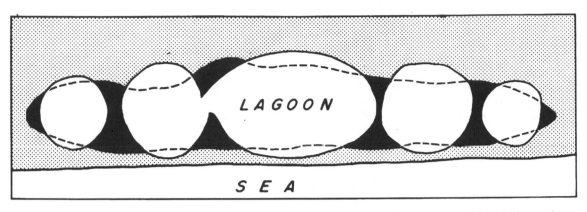

Figure 9.13 Segmentation of a long, narrow lagoon into a chain of smaller, rounded lagoons separated by paired or coalescent spits, as the result of erosion and deposition by wave action. The shape of the segmented lagoons is related to waves generated by local winds, and reflects the wind regime.

completely (Figure 9.13). This process was called segmentation by Price (1947), who described it from lagoons on the Texas coast, and the growth of spits into cusps and cuspate forelands has led to segmentation of Koozata Lagoon on St Lawrence Island, Alaska (Fisher 1955) and the Kosi lagoon in Zululand (Cooper 1994). Zenkovich (1959) described a similar process at work in the Shagany lagoon on the Ukrainian Black Sea coast, where sediment moving along the shore had been built into spits and bay barriers, rounding and smoothing the initially irregular configuration.

Segmentation is essentially an adjustment of lagoon shape to patterns more closely related to the waves and currents generated within a lagoon. Wind-driven currents play a part in smoothing the curved outlines of the shore in the later stages of segmentation and may also maintain the connecting straits between segmented lagoons. With equivalent winds from all directions, the segmented lagoons produced by wave action would be circular, but they are more often oval, with a long axis parallel to the direction of the prevailing winds.

Segmentation takes place most readily in tideless lagoons or in parts of lagoons where the tide range is small, for tidal currents interfere with the wave processes by deflecting growing spits and forelands so that they trail towards, or away from, the point of tidal entry and the coalescence of opposing spits is prevented. Changes of tidal level prevent continuous wave action at a particular level, so that a neat adjustment of coastlines to wave resultants is less likely. Instead, an initially elongated lagoon with straight parallel shores may develop a meandering outline as

the result of tidal flow, as in the lagoon at Cananéia in southern Brazil.

In the Gippsland Lakes segmentation is illustrated by the growth of a recurved spit on the eastern shore of Lake Wellington, which has been almost isolated from Lake Victoria, except for the link maintained by currents through intervening McLennan Strait. The oval outline of Lake Wellington reflects the prevalence of westerly winds. In adjacent Lake Victoria erosion of embayments and growth of intervening cuspate forelands are further signs of segmentation in progress. Also in the Gippsland Lakes, Cunninghame Arm shows a series of cuspate spits which, towards the eastern end, have grown to such an extent that they have almost cut off a chain of shallow pools, the Warm Holes, linked by connecting creeks maintained by wind-driven and tidal currents. On the South Australia coast, the lagoons between Robe and Beachport (Lake Eliza, Lake St Clair, and Lake George) have been formed by the segmentation of a long narrow lagoon, originally similar to the Coorong farther north. Many lagoons are bordered by growing spits and scoured embayments indicating incipient segmentation. The Lagoa de Araruama, near Cabo Frio in Brazil, has cuspate spits (e.g. Punta do Aceira) that have been enlarged as a result of emergence, which shallows lagoons and hastens segmentation.

Another form of segmentation occurs where tidal inflow from more than one marine entrance meets behind barrier islands. Deposition in the meeting zone can produce a tidal divide (p. 204), and eventually a land isthmus segmenting the lagoon, as in the Laguna Guerrero Negro, Mexico (Phleger 1969).

Swamp encroachment

As shown in the preceding chapter, vegetation has a strong influence on patterns of sedimentation. This is illustrated on coastal lagoon shores by the process of swamp encroachment. Near tidal entrances to lagoons salinity conditions are similar to those in estuaries, with foreshores and banks of sediment exposed at low tide, which may be colonized by salt marshes or mangroves. These can spread forward and build up depositional terraces. Salt marshes can spread from tidal lagoon shores, and mangrove encroachment has been described from Lagos lagoon in Nigeria (Webb 1958) and the Segara Anakan in southern Java (p. 244).

Away from the entrance, where the water is brackish and tidal fluctuations diminish, the salt marsh or mangrove fringe is much reduced. Lagoon shores in this intermediate zone may be unvegetated and bordered by beaches of sand or gravel. Towards the mouths of rivers where the water is relatively fresh, reedswamp dominated by species of *Phragmites*, *Scirpus*, and *Typha* may colonize the shore, and spread into water up to 1.5 metres deep. Reedswamp promotes sedimentation by trapping silt and floating debris, and by contributing organic matter (peat) so that new land is built up. Reedswamp is then invaded by swamp scrub communities, followed by swamp forest as the substrate aggrades (Figure 9.14). In Slapton Ley, Devon, reedswamp that has spread across a lagoon is being replaced by a natural vegetation succession to willow scrub and woodland.

Swamp encroachment is only possible where ecological conditions are favourable. If the reeds and rushes are not present, sedimentation occurs across the lagoon floor, and is not concentrated near the shore, so that the lagoon becomes shallower instead of shrinking. Strong waves or current scour impede reedswamp encroachment, the reed fringe that is broad on sheltered parts of a lagoon shore, thinning out where wave exposure is greater. Reedswamp can be reduced by cutting, or damaged by boat scour or water pollution, but the limiting factor in coastal lagoons is usually salinity. The reedswamp fringe is best developed in fresh water around river mouths, and thins out, disappearing towards marine entrances, where the lagoon becomes more saline.

The Gippsland Lakes were formerly relatively fresh, with shores extensively fringed by encroaching reedswamp (Figure 9.15), but in 1889 the intermittent natural outlet was replaced by a permanent artificial entrance cut to improve navigation in and out of the Gippsland Lakes, and this allows unrestricted inflow of

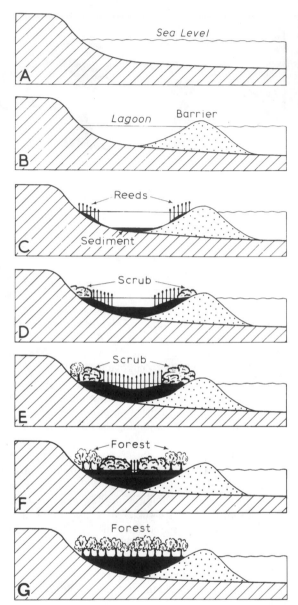

Figure 9.14 The formation of a coastal lagoon by barrier formation (A, B) is followed by reedswamp encroachment (C) and a vegetation succession to scrub (D, E) and forest (F, G) as the lagoon fills with sediment and peat deposits.

sea water in dry seasons. Over subsequent decades salinity has increased in the Gippsland Lakes, and the reedswamp fringe has died back, surviving only in relatively fresh water areas close to river mouths. Without its protection the lagoon shores, including deltaic silt jetties built by inflowing rivers, have been

Figure 9.15   Reedswamp encroachment on the shores of the Gippsland Lakes (see Figure 9.14C).

eroded (Figure 9.16), and sediment removed from them deposited offshore. The change to more brackish water has led to the Gippsland Lakes becoming larger and shallower.

This sequence is reversed where lagoons that have been sealed off from the sea, either naturally by the development of barriers or artificially by the insertion of weirs and barrages, have diminished salinity. Such freshening stimulates the spread of reedswamp around lagoon shores and initiates new patterns of sedimentation, as in the coastal lagoons at the mouth of the Murray River in South Australia (Figure 9.10), which were naturally brackish, estuarine systems with several natural entrances from the sea. In 1940 barrages were built to exclude sea water, and the Murray-mouth lagoons have freshened. Salt marshes that formerly bordered their shores have been replaced by reed-swamp, which is spreading forward into the lagoons and trapping sediment that would otherwise have been dispersed or swept out to sea. The sealing off of natural entrances to the Murray-mouth lakes thus reversed the geomorphological sequence seen in the Gippsland Lakes.

Similar changes are taking place in the Etang de Vaccarès, a lagoon on the Rhône delta in the Camargue region of southern France, where sluices now exclude sea water and freshening has resulted in the rapid spread of reedswamp. Kalametiya Lagoon in southern Sri Lanka showed a similar response following the building of a sluice at the entrance to keep out sea water, and increased inflow of fresh water discharged from irrigated ricefields in the hinterland has resulted in extensive swamp encroachment (Mahinda Silva 1986).

Lagoon configuration

Lagoons bordered by encroaching reedswamp contract in area until they are completely occupied by swamp land, whereas those without shore vegetation are reshaped by waves and currents as segmentation proceeds. Eventually, after many changes in configuration, coastal lagoons receiving sediment are filled in, and replaced by coastal plains, across which rivers and residual creeks wind, uniting to pass out to sea through the tidal entrance. Sediment patterns on coastal plains sometimes show the outlines of former lagoons that have been extinguished by deposition, as

Figure 9.16   Shoreline erosion of previously formed swamp land on the shores of the Mitchell River silt jetties, Gippsland Lakes (see Figure 9.11).

in the Anzio district in Italy and on the Sussex coastal plain.

Lake Reeve, behind the Ninety Mile Beach in Australia, is an elongated shallow lagoon which often dries out completely. Its shores have advanced by the accretion of low beach ridges of shelly sand, emplaced by the small waves generated across the shallow lagoon, which collect shelly and sandy material from the lagoon floor. The spacing of these contraction ridges is partly related to sparse salt marsh vegetation that spreads forward from the lagoon shore, especially when the water level is low.

On parts of the Gulf and Atlantic coasts of the United States lagoons have become narrower, and in places extinguished, by transgressive, enclosing barriers that have been driven landward (p. 173). Napier Lagoon on the east coast of North Island, New Zealand, emerged and drained out during the 1931 earthquake.

Coastal lagoons may be reopened, reviving a coastal embayment, where erosion breaches and removes the enclosing barrier. Guichen Bay, at Robe, on the South Australia coast and Rivoli Bay, at Beachport, farther south, were at one stage enclosed lagoons comparable with Lake Eliza, Lake St Clair and Lake George, which lie behind a barrier of calcarenite on the intervening coast. Reefs and islands of calcarenite in Guichen Bay and Rivoli Bay indicate the former extension of the enclosing barrier northward and southward (Bird 1967b). The breaching of similar calcarenite barriers on the coast of Western Australia has led to the formation of Cockburn Sound, and lunate embayments have also developed where dune calcarenite barriers have been breached on the coast of Israel near Nahsolim, and on the north coast of Puerto Rico (Kaye 1959).

In terms of the geological time scale, coastal lagoons are ephemeral features, likely to be replaced by depositional plains or opened as coastal embayments, depending on the subsequent evolution of the coastal region in which they have developed.

10

Deltas

INTRODUCTION

Deltas have been built where sediment brought down by rivers has filled the mouths of valleys drowned by the Late Quaternary marine submergence to form a depositional formation that protrudes from the general outline of the coast (Wright 1985). They have formed where the rate of sediment accumulation at the river mouth has exceeded the rate at which sediment is eroded and dispersed by waves and currents. The volume of sediment deposited in the world's deltas in Holocene times is enormous, but collectively they occupy only about 1 per cent of the world's coastline.

The term delta was introduced by the Greek scholar Herodotus in the fifth century BC to describe the large alluvial lowland at the mouth of the River Nile, which resembled the Greek letter Δ. It became a geomorphological term for depositional lowlands formed around river mouths, even if (like the Rhine delta) they do not protrude from the general outline of the coast. The extensive deltas built by the several large rivers that flow to the north coast of Java have coalesced to form a wide deltaic plain (confluent deltas).

DELTA COMPONENTS

Most deltas have subaqueous and subaerial components, above and below the low tide line (Coleman 1981). The subaqueous component includes a nearshore sea floor plain sloping gently out to a more steeply sloping delta front that declines to a flatter prodelta apron on the sea floor. These features have been formed largely by deposition of sediment by outflowing rivers, the calibre of sediment generally decreasing from sand and silt on the nearshore sea floor plain and delta front, to clay in the prodelta. These subaqueous features have all advanced seaward as the delta prograded. The nearshore sea floor plain may however include segments that are essentially

wave-cut, formed where parts of the delta have been cut back by marine erosion.

The subaerial component (above sea level) consists of a lower and an upper delta plain. On some deltas the river divides into distributaries that diverge across a delta. The upper delta plain has been built above high tide level by the deposition of alluvial sediment, and may include natural levees alongside river channels, declining into backswamp depressions. Some deltas have grown to enclose former high islands as bedrock hills, as on the Klang delta in Malaysia.

In the lower delta plain the river channel becomes tidal. Distributaries reach the coast as salients between embayments, which contain salt marshes or mangrove swamps and beaches built by wave action along the shore. In the coastal fringe there may be beach ridges, dunes and cheniers.

On some large deltas there are active zones where vertical accretion and seaward progradation are continuing, and abandoned zones in which there is no longer river deposition, where the coastal fringe may be submerging and eroding.

DELTAIC PROCESSES

Deltas have been built at the mouths of rivers delivering an abundant water and sediment yield to the coast, derived from runoff and erosion in extensive drainage basins and depending on their climate, geology and topography. The Mississippi drains 3.3 million square kilometres per year and the Amazon 5.9 million square kilometres. Deposition of fluvial sediment takes place in and around river mouths as the velocity of river flow diminishes on entering the sea, but most deltas also incorporate sediment drifting alongshore, and marine sediment moved in from the sea floor. In general deltaic sediments show gradations from coarser material (sand) to finer silt and clay downstream along channels, resulting from a diminution in flow velocity (Morgan 1970).

Deltas are found on coasts in various climatic zones. In the humid tropics they carry luxuriant vegetation including mangrove swamps, except where the vegetation has been cleared to make way for agriculture and aquaculture. Sedimentation is influenced by chemical and biological processes. On cold and arid coasts delta vegetation is sparse, and physical processes predominate. Climate within river catchments influences runoff and river regimes, which with geology and topography determine the nature and rate of sediment yield downstream to the coast.

Natural levees border river channels on alluvial valley floors and deltas, and are backed by low-lying, often swampy or flooded depressions (Russell 1967). These are the outcome of an unequal building up (aggradation) of the alluvial plain. When the river rises and overflows its banks the flow of water is most rapid along the line of the river channel and much slower on either side so that the coarser load of sand and silt carried by floodwaters is relinquished in the zone immediately adjacent to the river channel, where water velocity diminishes, and only the finer clay particles are carried into the calmer water beyond, to be deposited on the valley-floor plain. On macrotidal deltas such as the Irrawaddy, natural levees may be built in a similar way by the overflow of water flowing up-channel during high spring tides. As natural levees are built up along the sides of the river channel they develop gentle outward slopes, passing down lateral depressions, known as levee flank or backswamp depressions, as in the Mississippi valley. These depressions are floored by clay deposits which settle from floodwaters, and are often occupied by freshwater swamp vegetation, which can build up peat deposits. In dry regions repeated evaporation of water from enclosed backswamp depressions leads to a concentration of salt, increasing soil salinity so that they are occupied by salt marshes, or even unvegetated saline flats. Deflation from dry depressions may lead to the building of silt or clay dunes, as on the Senegal delta (Tricart 1956), and river channels that dry out during periods of low flow may provide a source of wind-blown sand for dunes built leeward of river channels during periods of low water flow (p. 181). Overbank splays of sediment are deposited where the river has flooded over or through the natural levees.

Distributaries may form as the result of breaching of natural levees during floods, particularly after deposition of sediment on the floor of a river channel has lifted the river, so that it spills out over its banks and finds a new outlet (Russell 1967). Alternatively, a river mouth may be split into two or more channels by the formation of shoals that grow up as islands. There may be two or three diverging distributary channels, as on the Rhône delta, or more complex bifurcations, sometimes rejoining or anastomosing, as on the Volga delta: the more complicated patterns are best developed where the offshore gradient is very low, or where the fluvial sediment yield is coarse (sand and gravel). Distributaries wax and wane: some may become major river channels while others silt up. On the Rhône delta the former main outlet is now the relatively unimportant Petit Rhône, the present river discharging through the Grand Rhône channel farther east. The Nile delta downstream from Cairo used to have many distributaries, but all but two of the outflow channels are now defunct (Figure 10.1).

River channels on deltas are influenced by tides. Tidal ebb and flow currents tend to maintain river mouths, whereas longshore drift may divert them or seal them off altogether. Where the tide range is large delta distributaries widen seaward to funnel-shaped estuarine outlets, such as those of the Rhine (Berendsen 1998).

River-mouth processes depend partly on nearshore water depth, the velocity of discharge diminishing in shallow water because of bottom friction. Fluvial sediment is deposited as the current slackens, forming one or more bars in shapes depending on the type of discharge, the influence of tidal currents and the effects of wave action. Where the nearshore water is relatively deep a turbulent jet flow forms a lunate bar across the river mouth, often of coarse sand or gravel on the inner side, grading to fine sand, silt and clay on the seaward slope. The pattern of deposition then follows the pattern described by Gilbert (1890) from a section through an emerged and dissected Late Pleistocene delta on the shores of the former Lake Bonneville, now reduced to the Great Salt Lake in Utah. He described almost horizontal radially dispersed bottom-set beds, overlain by inclined (10°–25°) coarser foreset beds deposited on a steep prograding bar front, capped by horizontal or landward dipping topset beds behind the bar crest.

Where the nearshore zone is shallower and bottom friction stronger the river mouth has a triangular middle ground bar between divergent channels, a pattern characteristic of the microtidal Mississippi mouths. Strong tidal currents can divide this into linear shoals diverging slightly off a funnel-shaped river mouth, with intervening mutually evasive channels, some ebb-dominant and some flood-dominant, as in the macro-

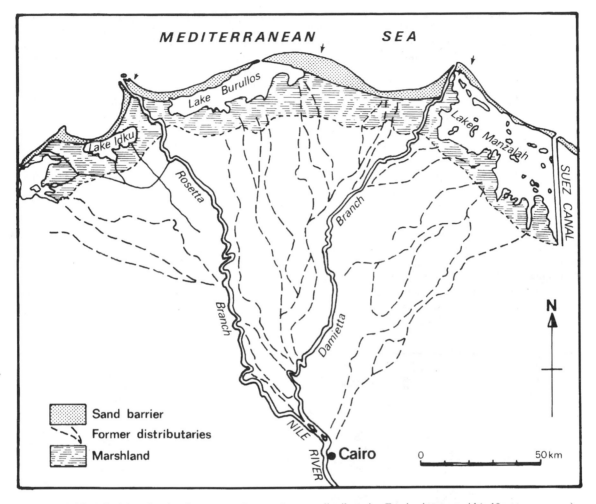

Figure 10.1   The Nile delta, showing the pattern of past and present distributaries. Erosion is now rapid (>10 metres per year) on the arrowed shore sectors.

tidal Ord estuary (Wright et al. 1973). As deposition proceeds, natural levees bordering the river channel are prolonged seaward, and these various bars and shoals move forward in front of them. However, where wave energy increases these river-mouth deposits are reshaped into smooth arcuate shore-parallel swash bars that are driven shoreward and eventually incorporated in beaches on the delta coast.

On deltas where the main river branches into distributaries, progradation is by means of sedimentation at and around the mouths of these channels, particularly during river floods. Waves and currents can spread the sediments alongshore, sorting them into sandy beaches and spits and separating finer sediment to settle in backing lagoons and swamps or be dispersed seaward.

As tide range increases, tides penetrate further upstream, impeding or reversing river discharge and causing overbank flow, crevassing, and splay deposition. Deltaic river channels subject to such tidal oscillations are essentially estuaries (Hart 1995), with features similar to those described in the previous chapter.

THE MISSISSIPPI DELTA

Much attention has been given to the large, complex delta built by the Mississippi, which has a catchment of 3.3 million square kilometres and delivers about 240 billion (10^9) kilogrammes of sediment to the river mouth each year (Coleman et al. 1998). The sediment yield to the delta consists of fine sand, silt and clay (the clay fraction being about 70 per cent), and during Holocene times the volume of fluvial sediment deposited to form the 28 500 square kilometre delta was about 2800 cubic kilometres.

Numerous oil borings on the Mississippi delta have shown that it is underlain by a great thickness of Holocene sediment, occupying a crustal depression, the subsidence of which is partly due to isostatic adjustments of the earth's crust beneath the accumulating sedimentary load (p. 30). A long history of subsidence can be deduced from this stratigraphic evidence, for the deposits on river terraces bordering the Mississippi valley can be traced down into the wedge of sediment beneath the delta (Russell 1967).

The delta has formed in several stages during this subsidence, and consists of seven successive partly overlapping subdelta lobes. As Figure 10.2 shows, one of these grew southward between 5500 and 4400 years ago, and others have been added subsequently in different positions to the east and south-east. As each

lobe grew, channel gradients in the river were reduced, and the outflow of water and sediment diminished. Bordering levees were then breached during floods to form crevasses, which became shorter and steeper outflow channels. Some of these were later sealed by deposition but others persisted to form distributaries, at the mouths of which sedimentation formed a new subdelta lobe, producing eventually the branched pattern shown in Figure 10.2. The most recent outgrowth is to the south-east, where elongated sedimentary jetties (digitate or bird's foot delta) are extending seaward beside diverging distributary channels. Meanwhile subsidence has continued, and the older lobes, no longer maintained by sedimentation, are sinking beneath the sea in coastal embayments of irregular configuration. As submergence proceeds, wave action has sorted the sandy fraction from the more widely dispersed silt and clay and shaped it into beaches, spits and barrier island chains, such as the Iles Dernières to the east, which commemorate the margins of the partly submerged fourth subdelta lobe.

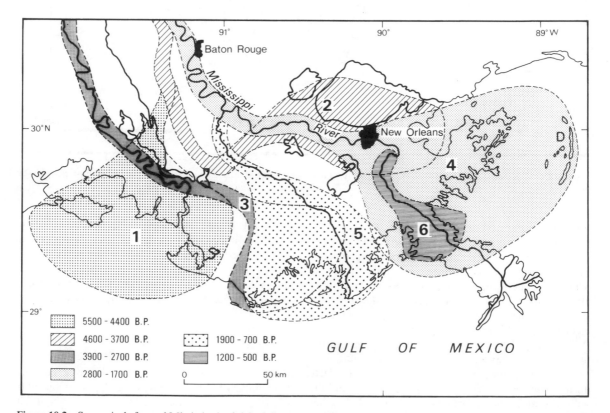

Figure 10.2 Successively formed Mississippi subdelta lobes, now partly submerged and dissected as the result of subsidence. The most recent delta growth has been south-eastward from New Orleans. D – Iles Dernières.

The delta has thus grown out into the Gulf of Mexico, a microtidal and generally low wave energy environment, the nearshore area being a broad shallow plain up to 20 kilometres wide.

Wave action is weak, so that sediment delivered to river mouths during recurrent Mississippi floods is deposited to form parallel subaqueous ridges alongside channels that continue out across the sea floor. As sedimentation proceeds these ridges are augmented to become emerged jetties of sand and silt, prolonging the natural levees on either side of the river mouths. This growth of a digitate delta has been aided by colonizing reedswamp (known as *roseau* in Louisiana), within which sediment is trapped (Figure 10.3), much in the manner described from coastal lagoons, notably the Mitchell River silt jetties in the Gippsland Lakes (p. 243).

Confinement of the lower Mississippi between artificial levees during the past 250 years has intensified its outflow to the Gulf and increased the sediment supply to the river mouths. The modern delta coastline is highly indented and marshy, with only minor sandy beaches and spits, and off the river mouths deposition of shoals has resulted in local upwellings known as mudlumps (diapirs), which are offshore extrusions of prodelta clay, squeezed up as the result of nearby sediment loading.

Although the Mississippi delta has grown on a low wave energy coast, it is subject to occasional hurricanes. In 1957, for example, Hurricane Audrey submerged the Mississippi delta when it raised the sea level by up to 3.6 metres. Under these conditions, large waves sweep sandy sediment inland from the shore, and deposit it as cheniers, low ridges on the deltaic plain (p. 145).

The features of the Mississippi delta are of great interest, but they are not representative of the world's deltas, most of which are of simpler configuration.

DELTA OUTLINES

The size and shape of a delta depends partly on the pattern and rate of sediment yield, on the configuration of the coast and the nearshore bathymetry of the water body (open sea, gulf, lagoon, lake) into which it has grown, and partly on the effects of waves and currents on the accumulating sediments. Small rivers have built

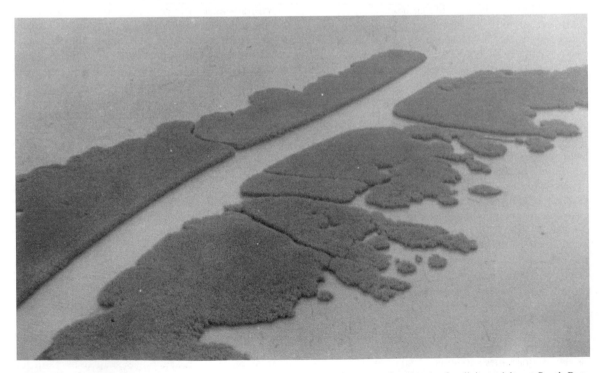

Figure 10.3 Reedswamp growing in shallow water traps sediment and initiates the growth of a digitate delta at South Pass, Mississippi delta.

deltas at the heads of estuaries or the sheltered shores of coastal lagoons, or on low wave energy shores of tideless seas, but on coasts dominated by strong wave action and tidal scour protruding deltas have been built only by large rivers draining catchments that have yielded abundant sediment, especially during floods. A gravelly intertidal delta has been built by recurrent floods off the mouth of the River Lyn in north Devon, and was enlarged during the Lynmouth flood in 1952 (Figure 4.1). Deltas have been built on arid coasts where intermittent streams flowing from a wadi have deposited cones of sediment extending into the sea as arcuate salients, as at Ghabour, south of Hurghada on the Gulf of Suez coast in Egypt.

Wave-dominated deltas

The shape of a protruding delta is related to the effects of wave action and accompanying currents, which tend to disperse sediment and smooth the coastline (Wright 1985). Wave action is reduced where the nearshore water is shallow, as on deltas (such as the Mississippi, where the 10 metre depth contour is about 17 kilometres offshore) where fluvial deposition has built a broad shallow submarine profile.

It is possible to classify deltas in terms of outlines related to rates of fluvial sediment supply and incident wave energy. The branching digitate outline of the Mississippi delta is found where wave energy is usually low, and fluvial sediment supply abundant (Figure 10.4). Similar long, narrow projections built by rapid deposition of large quantities of fine-grained sediment during floods from the Cimanuk and Solo Rivers in Indonesia extend out into the shallow Java Sea, where wave energy is low and protruding silt jetties produced by river-mouth deposition can persist (Hoekstra 1993). The Volga delta has numerous distributaries opening to an irregular marshy shore which prograded rapidly as the level of the Caspian Sea fell between 1930 and 1977, and has been partly submerged by the ensuing sea level rise (Figure 3.9). These deltas can be classified as river-dominated deltas, as distinct from wave-dominated and tide-dominated deltas.

Where wave action is somewhat stronger (and the nearshore sea floor less shallow) deltas develop smoother, cuspate outlines. The Danube delta (Figure 10.5) (10 metre depth, five kilometres offshore) has several distributaries, which have waxed and waned as large quantities of fluvial sediment have been delivered to their mouths, but although progradation has been rapid, digitate outgrowths have not formed.

Instead, wave action has shaped newly-deposited material into cuspate salients, sorting sand from silt and clay to form beaches and spits. The Ebro delta on the Mediterranean coast of Spain (10 metre depth, 3.5 kilometres offshore) has a cuspate outline and a generally sandy coastline, with bordering beaches prolonged as trailing spits by waves that come in from the east and break obliquely as they pass along the northern and southern coasts (Figure 10.4) (Guillen and Palanques 1997). The Burdekin river in north-eastern Australia divides into distributaries in its delta region, and a series of spits have been built northward by longshore drift resulting from the prevalence of waves generated by the south-east trade winds, deflecting distributary outlets and culminating in a long recurved spit built northward to Cape Bowling Green. A similar pattern is seen on the Godavari delta in eastern India, shaped by south-easterly waves which have drifted sand from the river mouth northward to the Kakinada spit, and on the Kelantan delta in Malaysia, where the large recurved Tumpat spit has also grown northward from the river mouth. Although it has formed on a coast exposed to high wave energy the Nandi delta in Fiji has a cuspate outline shaped by wave refraction around an outlying coral reef.

With stronger incident wave action, delta outlines become lobate, as on the Niger delta in West Africa (Figure 10.4), a large delta containing 1400 cubic kilometres of Holocene sediment, fringed by a sandy barrier. High energy ocean swell from the South Atlantic is diminished across a shelving sea floor (10 metre depth, 9 kilometres offshore) to waves that are sufficient to suppress delta outgrowth and distribute sandy sediment alongshore. The delta has several channels diverging across a marshy flood-plain and a mangrove zone to outlets through the barrier, each of which has an offshore sand bar (Allen 1970). The Nile delta (Figure 10.1) is also rounded and lobate, with a fringing sandy beach, because the nearshore sea floor has a concave profile (10 metre depth, 3 kilometres offshore), which allows moderate Mediterranean wave action to reach the shore.

Still stronger oceanic waves, reaching the coast across a narrow nearshore slope (10 metre depth, 2 kilometres offshore) have blunted the delta of the São Francisco in Brazil (Figure 10.4). Sediment delivered to the river mouth is quickly dispersed and sorted by waves, the sandy fraction being deposited on bordering beaches so that the delta coast has prograded by the addition of sandy beach ridges shaped by refracted ocean swell (Wright and Coleman 1973). Finally, in a high wave energy environment on a coast where the

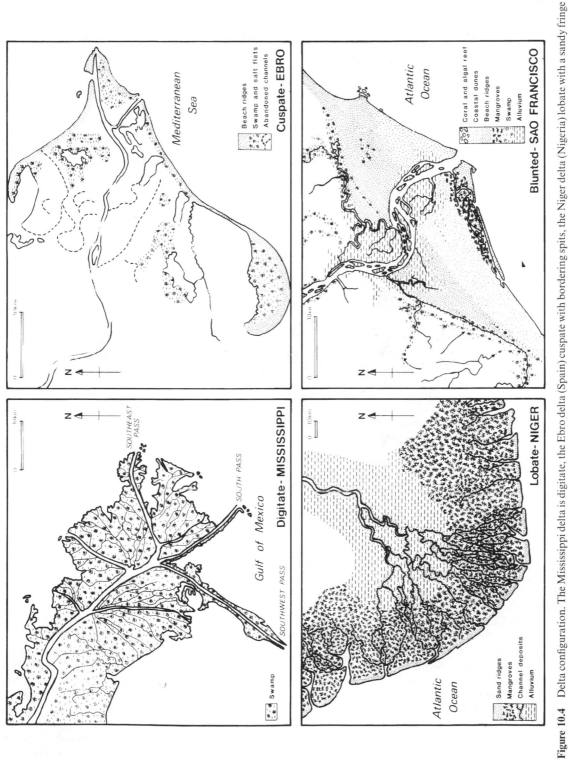

Figure 10.4   Delta configuration. The Mississippi delta is digitate, the Ebro delta (Spain) cuspate with bordering spits, the Niger delta (Nigeria) lobate with a sandy fringe and the São Francisco delta (Brazil) blunted by high wave energy.

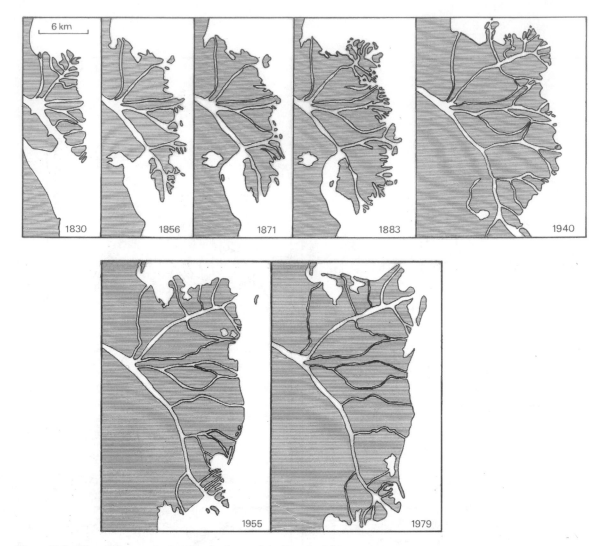

Figure 10.5 Historical stages in the growth of part of the Danube delta.

nearshore profile is steep (10 metre depth, 1 kilometre offshore), the Senegal River opens on a sandy Atlantic Ocean coastline that is straight rather than protruding and has a blocked delta, the river mouth having been deflected southward by longshore spit growth (p. 173).

Deltas are poorly developed on high wave energy coasts because sediment delivered by rivers has been dispersed by strong wave action, the coarse fraction (sand and gravel) being deposited in beach and barrier formations along the coast and the finer material being carried away in suspension by the sea or settling on the sea floor. However, on the stormy west coast of South Island, New Zealand, rivers such as the Haast have brought down sufficient gravelly material from glaciated mountain catchments to build slightly protruding, blunt deltas.

Protruding deltas are rare on the coasts of Britain and north-west Europe, partly because of high wave energy and strong tides, and partly because fluvial sediment yields have been generally low. Most of the rivers draining to the Atlantic and Pacific coasts of the United States flow into estuaries and lagoons, and have not yet built deltas that protrude into the sea. In Australia protruding deltas have generally not developed because of high wave energy and meagre fluvial discharge (Wright et al. 1980). An exception is the De Grey delta on the north-west coast, where wave energy is moderate because the continental shelf is broad and the coastal waters macrotidal. Most Australian rivers do not yet provide sufficient sediment to fill valley mouths drowned by Late Quaternary marine submergence. The Murray–Darling system has a catchment basin of more than a million square kilometres, but in its lower reaches the Murray flows through a semi-arid region and loses so much water by evaporation that it has not delivered sufficient sediment to fill the lagoons (Lakes Albert and Alexandrina) at its mouth (Figure 9.10). Some rivers such as the Snowy and the Shoalhaven have brought down more sediment and reclaimed their drowned valleys as depositional plains, but as in Senegal the building of protruding deltas has been prevented by strong ocean swell arriving through relatively deep water. On the Queensland coast the growth of protruding deltas such as that of the Burdekin has been possible because ocean swell is excluded by the Great Barrier Reef.

Tide-dominated deltas

Although deltas are best developed on microtidal coasts (as around the Mediterranean Sea) they also occur where the tide range is large, as on the Irrawaddy delta (tide range 5.5 metres) and the Ganges delta where the tides rise and fall 4.5 metres. These are on low wave energy coasts where broad shallow areas offshore diminish incoming waves. Deltas on macrotidal coasts show distinctive features, notably funnel-shaped estuarine river mouths produced by the ebb and flow of tidal currents, and associated linear river-mouth bars (p. 201).

The Mahakan delta in Kalimantan, Indonesia, is an example of a protruding delta with several distributaries opening to wide mouths on a mesotidal coast where wave action is moderate (Allen et al. 1979). Large quantities (about eight million cubic metres per year) of fine-grained sediment are delivered by the river to its distributaries and deposited at and around their mouths, but much of the delta plain is occupied by intertidal mangrove swamps passing landward into swamp forests. Because of tidal currents the delta outline is irregular, rather than cuspate.

On the west coast of peninsular Malaysia, the Klang delta, near Kuala Lumpur, has been shaped by tidal processes as fluvially-supplied sediment was delivered to the coast (Coleman et al. 1970). The delta has been built into the Strait of Melaka, here about

50 kilometres wide. There is low to moderate wave energy, chiefly from the north-west, and the mean spring tide range at Port Klang, beside one of the river mouths, is 4.2 metres, producing strong currents flowing south-eastward as the tide rises and north-westward (augmented by a monsoon-generated northward current) as it falls. These currents disperse sediment mainly north-westward, where it has been deposited to form the large trailing intertidal Angsa Bank, with parallel mud-crested ridges and current-scoured sandy troughs. Tidal currents have been more important than waves in shaping the Klang delta, much of which is submerged at the highest tides. There are extensive tidal flats with mangrove islands and interconnecting tidal channels, and rapid peat accumulation has contributed to aggradation in these areas. The southern shores of the delta are eroding, but the seaward spread of mangroves has prograded parts of the western and northern shores at rates of up to six metres per year. Tidal channels between mangrove islands meander and migrate laterally, one bank building up as the other is undercut, and some become sealed off as the mangroves spread.

The Ord River, opening into Cambridge Gulf on the north coast of Australia, has an extremely variable seasonal flow, and discharges an average of 22 billion kilogrammes of sediment annually (Wright et al. 1973), but the large delta occupies the head and fringes of the funnel-shaped (estuarine) gulf and a great deal more infilling will be necessary to form a coastal protrusion. Meanwhile, the fluvially supplied sediment is widely dispersed by strong tidal currents (often more than three metres per second), mean spring tide range being of the order of six metres. Off the mouth of Cambridge Gulf there are sub-parallel tidally scoured channels between submerged linear sand ridges (King Shoals). The gulf-head delta has extensive sandflats and mudflats exposed at low tide, and there are several long inlets that may have been former outlets of the Ord River. Wave action from the Timor Sea is weak, and the coastal morphology is almost entirely tide-dominated.

The Red River delta in Vietnam has both wave-dominated and tide-dominated shores in a mesotidal environment where wave energy diminishes northward along the coast (Mathers and Zalasiewicz 1999). The wave-dominated southern part has beach ridges built of sand reworked from deposits swept into the sea by occasional major floods interspersed with muddy tidal lagoon deposits, while the tide-dominated northern part has numerous inlets and tidal creeks fringed by mangroves

Effects of nearshore currents

Nearshore currents can move fine-grained sediment supplied by rivers along the coast. The Amazon River has filled a former gulf with a large swampy deltaic plain, but growth of a protruding delta has been prevented by the Guyana current, which flows north-west past the mouth of the river and sweeps fine-grained sediment along the coast past Surinam to Trinidad (Gibbs 1970). Deposition of this sediment has formed extensive shore mudflats, and sand sorted from the fine-grained sediment by waves forms low beaches and cheniers (p. 145). On the coast of Papua muddy water from the Purari River is swept westward by waves arriving from the south-east and deposited in mudflats and mangrove swamps (Thom and Wright 1983).

Effects of cold climate

Arctic deltas, such as those of the Ob, Lena and Yenisey in Russia, and the Mackenzie, Yukon and Colville in North America, show features related to cold climate weathering and erosion processes (Walker 1998). Frost action produces ice wedge polygons and ice-heaved soil-capped conical hills up to 50 metres high and 400 metres in diameter, known as pingos. Fluvial sediments in cold regions are generally coarse, with much sand and gravel. Sand deposits on river bars are frozen and snow-covered in winter and saturated by river flooding in the spring thaw, but in summer they dry out and are built up by the wind into river-bank dunes (including barchans) which may become sparsely vegetated (p. 181). There are numerous lakes and peaty swamps on the delta plain, modified by interactions between river floodwaters and coastal and sea ice (Hill et al. 1994). Some of the sediment swept down by spring floods is deposited on the frozen sea and carried away when the sea ice disintegrates into dispersing floes in summer. In Alaska the Point Barrow coast has a number of deltaic salients built by former glacifluvial deposition and now being re-shaped by wave action.

DELTA EVOLUTION

Stages in the evolution of a delta may be traced from the time when the sea attained its present level relative to the land, and sedimentation began to fill a valley-mouth inlet through to the complete filling of that inlet and the formation of a depositional landform protruding into the sea. As has been noted (p. 257), the Ord delta in northern Australia is still at the stage of a gulf-head delta, but many rivers have filled the former valley-mouth inlets and now protrude in various forms.

The outlines of such a delta change as the result of deposition and erosion along its coasts. Deltas continue to prograde as long as the supply of sediment, mainly from the river, exceeds its removal by wave and current action. Progradation can be spectacular in shallow seas, especially on parts of tropical deltas, where progradation of over 200 metres per year has been recorded on some Javanese deltas (Figure 10.6) (Bird and Ongkosongo 1980).

There is historical evidence of delta growth, especially around the Mediterranean Sea over the past 2000 years. The port of Ostia, near Rome, became silted as the result of the growth of the Tiber delta, and is now three kilometres inland. In Turkey the ancient port of Ephesus, built where the River Kayster opened into the Bay of Anatolia, became stranded by delta growth and the spread of marshland, particularly after a breakwater was built across the mouth of the bay in 150 BC, and is now 24 kilometres from the sea (Kraft et al. 1988).

Delta progradation can be accelerated by increased sediment yield due to deforestation, overgrazing or unwise cultivation leading to rapid soil erosion, or mining activities in the river catchment. The Mahakan delta in Kalimantan expanded rapidly following forest clearance in its catchment. The rapid growth of the George River delta into a Tasmanian coastal lagoon was the result of increased fluvial sediment yield caused by tin mining upstream (Bird, J.F. 2000) (p. 230). Sluicing for tin has augmented the sandy loads of several rivers in Malaysia, where the Pahang delta has been enlarged by sand washed down from a mined catchment. In Cornwall, the River Fal has built a small delta into a sheltered arm of the Carrick Roads ria (p. 230). It grew rapidly during the phase when the river brought down large quantities of kaolin from the china clay quarries in its upper catchment, but growth ceased and delta shore erosion began after this supply of sediment was curtailed by conservation works in the mining area upstream (Figure 10.7) (Bird 1998).

The Jaba delta on the island of Bougainville began to grow in 1972 after tailings from copper mines in the catchment flowed down the Jaba River to the coast at the rate of 26 million tonnes per year. By 1977 an arcuate salient had formed on the shores of Empress Augusta Bay, but surveys by Wright et al. (1980) showed that delta growth was then curtailed by moderate south-easterly wave action (and a 1975

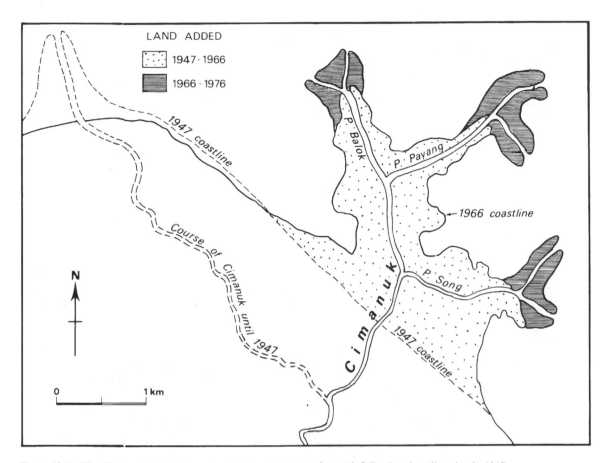

Figure 10.6 The Cimanuk delta, Indonesia, showing the extent of growth following river diversion in 1947.

tsunami), which spread sandy sediment northward to prograde beaches along the coast, leaving a small wave-dominated lobate delta at the river mouth.

The Colorado delta in Texas prograded rapidly after log jams were removed from the river channel in 1929, forming several distributaries and extending across a lagoon at Matagorda Bay towards a sandy coastal barrier, but when an artificial outlet was cut in 1932 through the barrier to the sea, fluvial sediment, carried through to the Gulf of Mexico, was dispersed by wave action, so that a protruding delta failed to develop (Wadsworth 1966).

Delta growth is slower into deep water, or where a submarine canyon exists offshore, as off the mouth of the River Ganges. Reference has been made to the subsidence and erosion of earlier subdelta lobes on either side of the modern Mississippi delta, and similar features are seen on a smaller scale on the Rhône delta in southern France. Subdelta lobes built by the Petit Rhône and the Vieux Rhône have been cut back by marine erosion following the decay of these distributaries and the shifting of the river to a south-eastern outlet. Sediment eroded from these former subdelta lobes has been reworked and sorted by marine erosion, producing sand that has been carried westward by the dominant south-easterly waves to be built into depositional forelands at Pointe de Beauduc and Pointe de l'Espiguette respectively (Suanez and Provansal 1998).

Erosion of deltaic coastlines may follow diminished sediment yield from the river (p. 148). The delta of the Argentina River on the Ligurian coast in Italy grew for as long as the river delivered sediment, but extensive headwater soil erosion has laid bare rocky outcrops, and sediment yield has diminished, so that the delta is now eroding (Bird and Fabbri 1993). More often, fluvial sediment yield has been reduced by dam construction upstream. Erosion on the shores of the Rhône delta accelerated following dam construction

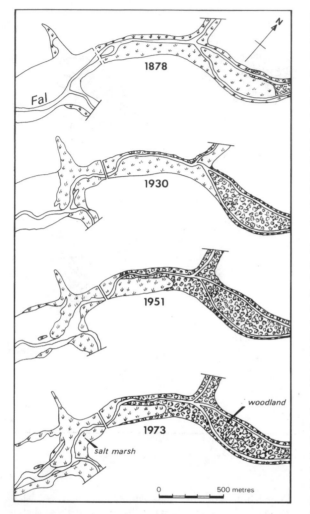

Figure 10.7 The growth of the Fal delta into Carrick Roads, Cornwall, during a phase when the river was delivering a sediment load augmented by china clay waste from its upper catchment. Control of china clay pollution has been followed by delta shore erosion, so that by 1998 the delta had returned to approximately its 1930 outline.

that reduced fluvial sediment yield from 11 to less than five million tonnes annually (Paskoff 1994) and in Ghana the building of the Akosombo Dam on the Volta River greatly reduced discharge and sediment yield to the coast, resulting in an increase in the rate of coastline recession from two to three metres per year to eight to ten metres per year (Ly 1980).

Erosion has been prevalent on the shores of the Nile delta since the beginning of the twentieth century, and in recent decades it has been locally rapid, up to 120

metres per year near the Rosetta mouth (Figure 10.1) (Sestini 1992). This coastline previously prograded as the result of the delivery of Nile sediment to the coast by floods, but the building of dams, especially the Aswan High Dam in 1964 depleted the supply of Nile sediment to the delta coastline and so accelerated erosion (Stanley and Warne 1998). Fluorescent tracers were used by Badr and Lotfy (1999) to show net eastward longshore drift of sand along the eroding coastline of 1.48 to 3.21 million cubic metres per year, and losses offshore of 0.39 to 0.44 million cubic metres per year. Some of the drifting sand has been washed into the mouths of the Rosetta and Damietta distributaries and the Burullus lagoon outlet (Badr and Lotfy 1999).

Natural diversion of a river mouth is generally followed by erosion of the former delta and the building of a new one. An example is the Yellow River in China, where there has been erosion of the delta abandoned after the river changed its outlet during an 1855 flood and growth of another delta at the new outlet northward into the Gulf of Bo Hai. (Chen Jiyu et al. 1985). Similar changes followed the diversion of the Hwang Ho River in China in 1852, the Rio Sinu in Colombia in 1942, the Ceyhan in Turkey in 1935 and the Medjerda in Tunisia in 1973.

The modern Po delta has grown since a new outlet was cut in the late sixteenth century to divert the river away from the Venice region, where it threatened to fill the lagoon and deprive the city of its natural defensive moat. The Po branches into distributaries, and the seaward fringes of intervening marshy deltaic islands are lined by sandy barriers. In recent decades, delta enlargement has ceased because of a diminishing sediment yield due to river impoundments and soil conservation works upstream (Fabbri 1985, Cencini 1998).

When the fluvial sediment supply to a delta is reduced or cut off, and erosion of the delta coastline follows, there is usually an interval between the halting of the fluvial sediment supply and the onset of erosion because the change from a convex, aggrading sea floor profile to a concave, eroding sea floor profile begins offshore, and is transmitted landward. As the concave profile intersects the deltaic coastline, erosion begins, or is accelerated (Figure 10.8).

DELTA SHORES

The advance or retreat of deltaic coastlines can be traced from successive maps and air photographs, as in Figure 10.5, which shows stages in the growth of

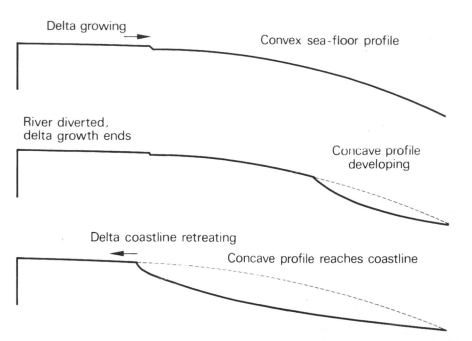

Figure 10.8 A growing delta usually has a convex nearshore sea floor profile. When sea level rises and the delta coastline is submerged and eroded a concave profile is initiated offshore by erosion. This migrates shoreward until it intersects the coastline, which then recedes more rapidly.

the Danube delta since 1830. Nossin (1965) used cartographic records to reconstruct the evolution of the north Padang delta in Malaya since the early seventeenth century. Historical changes on the Ganges–Brahmaputra delta front were determined from successive charts and modern Landsat imagery by Allison (1998), who found that progradation had added about seven square kilometres per year to the delta coastline since 1792 while the submerged delta front had expanded eastward and been reduced by erosion on the western side.

Beaches and spits border many deltaic coastlines, supplied with sand or gravel by rivers, derived from erosion and sorting of delta shore sediments by wave action, or washed in (often with shells and other marine organisms) from nearshore shallows. Beaches, spits and barriers built by marine processes on the shores of a delta often enclose lagoons and swamps (as in the Kizilirmak delta in northern Turkey), and are sometimes themselves incorporated as the delta grows larger. In northern Italy the Tagliamento delta has a series of symmetrical parallel sandy beach ridges marking stages in the progradation of a cuspate delta coastline. Reference has been made to the formation of barrier islands from beaches fringing subdelta lobes of the Mississippi that have subsided (p. 252), and similar features are seen on the Apalachicola delta in Florida (Donoghue and White 1995).

Cheniers are long, narrow, low-lying strips of sand that have been deposited by wave action during occasional high tides or storm surges on deltas and coastal plains, and marked out by contrasts in vegetation (p. 145). They are found on the Mississippi delta and in Surinam on the coast of South America, where several sand ridges have been emplaced on broad swampy deltaic plains adjacent to the Amazon and the Orinoco (Psuty 1985).

Where wave energy is low on delta shores, or in bays and lagoons behind spits and barriers along these shores, there are freshwater swamps and salt marshes, and in low latitudes mangrove swamps, fronted by mudflats. The shores of the Mississippi, and Mahakan deltas have sectors of salt marsh or mangroves which form irregular, sometimes crenulate, mid-tide shorelines that prograde as fine-grained sediment is deposited. If the delta coast becomes exposed to stronger wave action (e.g. if a sheltering spit is submerged or truncated, or if there is a relative sea level rise, perhaps due to delta subsidence) progradation ceases, and these swampy shores may be eroded. Alternatively, higher wave energy may deliver sandy sediment to form new fringing beaches.

11

Coral and Algal Reefs

INTRODUCTION

Reefs built in the sea or on the shore by organisms are termed biogenic. By far the most extensive are coral reefs, which incorporate algae and other organisms, but there are also algal reefs and reefs built by oysters, bryozoans, serpulids, mussels and tubeworms. Coral reefs are coastal landforms where they fringe the coastline, where they have emerged to form limestone islands (usually also reef-fringed), and where they are surmounted by islands of coralline sediment.

CORAL REEFS

Coral growth is confined to warm seas, where the mean temperature of the coldest month does not fall below 18°C, and the warmest month does not exceed 34°C. Reefs built by coral and associated organisms thus occur extensively in tropical waters, particularly between latitudes 30°N and 30°S in the western parts of the Pacific, Indian, and Atlantic oceans (Davis 1928, Guilcher 1988). Their distribution is too intricate to be shown adequately on a textbook map, but is portrayed in the Russian *Atlas Okeanov* (1974, 1977). In the Pacific coral reefs extend north to Japan and Hawaii, in the Atlantic to Florida and the Bermuda Islands, and in the Indian Ocean to the Red Sea and the Arabian Gulf. They are well developed in the Caribbean Sea, the Philippines and Indonesia, and around the coasts of Australia, particularly off the east coast of Queensland, where the Great Barrier Reef extends from Torres Strait in the north to the Bunker and Capricorn reefs in the south (latitude 24°S). In the Coral Sea reefs extend farther south, to the Middleton and Elizabeth atolls (29°, 30°S) and the west coast of Lord Howe Island (32°S). There are scattered reefs off the northern coast of Australia, and off the west coast of the continent they extend as far south as Houtman Abrolhos, in latitude 28° to 29°S

(Fairbridge 1967). In the Indian Ocean they extend south to Mozambique and Natal, and in the Atlantic Ocean to the Brazilian Abrolhos.

The global distribution of coral reefs is related to the dispersal of free-floating planktonic coral larvae by ocean currents to areas warm enough for coral growth. The paucity of reefs in the eastern Atlantic and Pacific (there are no reefs on the Galapagos Islands or on Easter Island) results from the westward flow of equatorial currents away from these areas (as well as the upwelling of cold water along the coast) whereas the richness of the reefs off north-eastern Australia (including the Great Barrier Reef) is related to an abundant larval supply in currents arriving from the Pacific and the Coral Sea. There is relatively poor development of coral reefs off the west coast of Australia, where reefs running parallel to the mainland coast are submerged ridges of dune calcarenite (p. 191), with only a veneer and fringe of living corals and calcareous algae. The flow of ocean currents across the Indian Ocean results in more extensive coral reefs to the west, and in a similar way the poor development of reefs in the eastern Atlantic gives place to more extensive reefs in the Caribbean. Sandstone reefs similar to those off the west coast of Australia are found along the north-east coast of Brazil, where the supply of coral larvae has been meagre, both the northward summer current and the southward winter current arriving from ocean areas where there is little or no development of coral reefs (Guilcher 1988).

ORIGIN OF CORAL REEFS

Coral reefs bordering the coast are termed fringing reefs and those that lie offshore and parallel to the coast are barrier reefs. There are also patch reefs, which are isolated coral reef platforms of various shapes and sizes (Hopley 1994). Atolls are oceanic reefs that encircle a lagoon.

Coral reefs are built by polyps, small coralline organisms that extract calcium carbonate from sea water and grow by accretion into a variety of branching skeletal structures, forming a coral garden. These structures can form a habitat for calcareous algae, which grow with the coral, as well as foraminifera, molluscs and other shelly organisms. They have a similar effect to mangroves in that they diminish current flow and promote sedimentation, so that fragments of shells, corals and algae (such as the sand-producing *Halimeda*) are deposited in the spaces between the coral garden structures; with precipitated carbonates they form a solid calcareous reef limestone (Jones and Endean 1977). The seaward slopes of coral reefs are often very steep (up to 50°), descending to aprons of reef-derived sand and gravel that decline more gradually to the sea floor. While coral is essential for reef building, it generally forms only a small proportion of a solid reef structure as seen in sections or quarries in emerged coral reefs.

Corals require a firm sea floor substrate (usually rocky) and coral larvae cannot establish on mud or on mobile sand and gravel, or where sea floor sedimentation is proceeding rapidly. An adequate supply of sunlight is essential for algal photosynthesis, and growth of coral is best in clear, warm water. Intensity of sunlight diminishes downwards into the sea, and although live corals have been found in exceptionally clear water at depths as great as 100 metres, the maximum depth at which reefs are being built is rarely more than 50 metres. In coastal waters turbidity due to land-derived sediment (chiefly silt and clay) in suspension reduces penetration by sunlight and impedes growth of reef-building organisms. Off river mouths this cloudiness and the blanketing effects of deposits of inorganic sediment often prevent the growth of reef-building organisms, and break the continuity of fringing reefs. Corals can dispose of small accessions of sediment, but are choked by continued deposition of large quantities of detritus, or by heavy loads of sediment dumped suddenly from discharging river floods. In Indonesia coral reefs are missing from sectors of the coast that receive lava flows (as on Anak Krakatau) or ash deposits from volcanic eruptions (as at Parangtritis, on the south coast of Java, where rivers supply sand erupted from the Merapi volcano). In general, coral reefs are not found on coasts where sandy beaches are extensive.

Corals are marine organisms, found where sea salinity is within the range of 27 to 38 ppt, and most luxuriant in 34 to 36 ppt. Dilution by fresh water discharged by rivers impedes coral growth and contributes to the persistence of gaps in bordering reefs. On the other hand excessive salinity may explain why reefs are absent from certain parts of the coast such as Hamelin Pool, in Shark Bay, Western Australia, where salinity is up to 48 ppt in summer. Where ecological conditions are suitable, corals begin to grow with associated algae, to form a coral garden and eventually a solid reef (Fairbridge 1967).

Oceanic coral reefs have been forming since at least Eocene times, and borings have shown that they extend to great depths: on Bikini Atoll the base of the coral rests upon volcanic rocks 1400 metres below sea level, well below the limits at which reef-building corals now grow.

RATES OF GROWTH

Where ecological conditions are favourable, and there is an adequate supply of mineral nutrients in the sea water, corals grow up towards the sea surface. There are differences between the growth rates of individual organisms, which can be quite rapid, the branches of some staghorn corals (*Acropora* spp.) extending by up to 20 centimetres per year, and the reef formation as a whole. Measurements of mean upward growth rates of coral reefs are generally in the range 0.4 to 0.7 millimetres per year (Hopley and Kinsey 1988), with up to one centimetre per year in favourable conditions (Buddemeier and Smith 1988). Studies of reef growth during the Late Quaternary marine transgression, when the sea rose at an average rate of about a metre per century, indicate average rates of upward growth of up to eight millimetres per year (Davies 1983). This enabled a variety of growing corals in the reef framework to be within a few metres of sea level when the transgression slackened about 6000 years ago, and then to extend upward and outward to form existing reefs.

Corals grow upward until they reach a level where they are briefly exposed at low tide (Figure 11.1). As they cannot survive prolonged exposure to the atmosphere they then die off, but associated algae, other organisms and sediments may continue to build the solid reef platform up to this level. Algae (chiefly *Porolithon*) may build a slightly higher reef crest, awash at high tide, and on ocean coasts where there is strong wave action a seaward ramp may be formed. Reefs built in more sheltered waters, such as the Java Sea in Indonesia, are flatter and without algal ramparts. It is possible that some reef platforms have been built up to a slightly higher Holocene sea level and then

Figure 11.1   Reef platform exposed at low tide on the Great Barrier Reef, Australia, consisting mainly of dead coral.

planed off by marine erosion following subsequent emergence (Hopley 1982).

Coral growth continues on the steep bordering slopes of reefs declining below low tide level. Optimum coral growth occurs where sea water is being circulated sufficiently to prevent clogging by silt deposition, to maintain a uniformly high temperature, to renew the supply of plankton and other nutrients on which the coral polyps feed and to maintain the supply of oxygen, particularly at night when it is no longer replenished by algal photosynthesis and the concentration of dissolved carbon dioxide tends to rise in reef waters. A high concentration of dissolved carbon dioxide impedes coral growth, and could become corrosive.

Where the slopes of reefs are exposed to strong wave action, as on the seaward flank of a barrier reef or the outward slopes of an atoll, and there is adequate circulation of sea water for vigorous coral growth, waves may break off, or inhibit the formation of, the more intricate skeletal corals, so that the seaward slopes consist of more compact coral growth. On the leeward side, in more sheltered waters, a greater variety

of growth forms coral gardens. The breaking of skeletal corals by storm waves generates large quantities of coralline sand and gravel, much of which is banked up as sedimentary aprons on the lower slopes of reefs or laid down on lagoon floors, while some is thrown up by wave action on to reef platforms (p. 272).

Coral growth is also limited by predatory organisms, such as the crown-of-thorns seastar (*Acanthaster planci*), which in the past few decades has grown in plague proportions on Indian and Pacific Ocean reefs, notably on parts of the Great Barrier Reef. The seastar feeds on, and thus kills, corals. The Great Barrier Reef outbreak began near Cairns in the 1960s, possibly as the result of excessive collecting of triton shells diminishing predation on the seastar, which has subsequently spread northward and southward. The impact has been severe on branching *Acropora* corals, broken sticks of which form abundant gravelly debris on and around the affected reefs, locally augmenting gravel deposits on reef platforms. Although many consider the outbreaks to be a result of human impacts, stratigraphic evidence of seastar remains in coral lagoon sediments shows that there have been previous outbreaks of this kind.

Coral reef ecosystems have certainly been modified by human activities, especially during recent decades. Marine pollution has occurred in many coral reef areas, and some reefs have been quarried for sand, gravel and building stone, or damaged by the cutting of boat access channels, or by boat anchors. Coral reefs have also been damaged by human activities such as weapons testing and fishing with explosives. Such activities generate sediment turbidity, which impedes coral growth in neighbouring areas. Around Sulawesi, in Indonesia, coral reefs have been impoverished by increased sedimentation resulting from soil erosion, the greater turbidity having made them less vigorous, and reduced the number of species, especially near coastal towns. Excessive nutrients from eroding soils, agricultural fertilizers and sewage pollution have caused eutrophication, which is detrimental for coral growth, and can lead to the killing of corals by the growth of other organisms (Guilcher 1988). There have been reports of corals damaged by bleaching, possibly as a consequence of higher sea temperatures associated with the El Niño Southern Oscillation in the Pacific Ocean, or as a result of increasing ultra-violet radiation due to atmospheric ozone depletion (Brown 1990).

FRINGING REEFS

Fringing reefs have been built upwards and outwards in the shallow seas that border continent or island shores (Figure 11.2). They consist of a reef platform at low tide level, similar in many ways to the low tide shore platforms found on cliffed limestone coasts (p. 88), except that fringing reefs have been built up, rather than cut down, to this level. Some fringing reefs are really shore platforms cut into emerged older coral reefs, but with a veneer of modern coral, as on the shores of Mbudya Island, in Tanzania.

Fringing reefs are usually thin, resting on a rocky foundation, and are generally widened by the growth of coral along their seaward margins. As coral growth is inhibited by fresh water and active sedimentation fringing reefs are not found near river mouths, but are well developed around offshore islands and on coastal salients, extending round headlands. They originated on shores that had ecologically suitable conditions for colonization by reef-building coral communities when the Late Quaternary marine transgression brought the sea close to its present level, establishing the general outlines of the present coast, and they have grown seaward during the ensuing stillstand of sea level. Some fringing reefs may embody the relics of Pleistocene fringing reefs which developed at an earlier stage close to the present shore, and were dissected during low sea level phases.

Where conditions have become adverse for the growth of coral reefs (e.g. because of the diversion of a river mouth towards them) they are dead and disintegrating. Incipient fringing reefs are rare, presumably because ecological transitions from adverse to suitable conditions for coral establishment have been exceptional during the past few centuries. Scattered

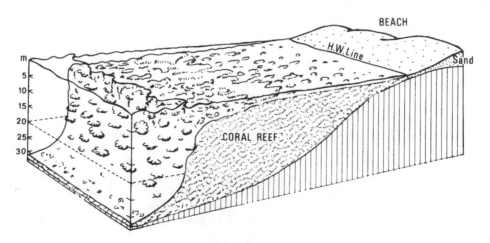

Figure 11.2   The structure of a fringing reef. Some fringing reefs have raised seaward rims backed by a moat (boat channel) which remains flooded at low tide.

corals occupy bouldery shores in north-east Queensland, as along the coast of the Macalister Range, and coral gardens have formed on shores around the volcanic islands of Krakatau, Indonesia since the explosive eruption of 1883 (Bird and Ongkosongo 1980).

Some fringing reefs, especially in East Africa, are not shore platforms of the kind shown in Figure 9.14, but have a broad higher outer segment backed landward by a shallow depression known as a boat channel, typically 100–200 metres wide and up to three metres deep. At low tide the outer segment is exposed and the boat channel becomes a lagoon, while at high tide the whole area is submerged, and waves wash over and through to the coastline, where there is often a beach (or low beach ridges) consisting largely or wholly of coralline sand and gravel. There is greater vigour of coral and algal growth at the outer margins of such a reef, but the boat channel may indicate a relative rise of sea level since the fringing reef was initiated (Bird and Guilcher 1982).

At Yule Point on the north-east coast of Queensland the reef platform adjoining the coast is strewn with sand, partly of terrigenous origin, and there are patchy mangroves bordering the shore near the high tide line. Live coral is confined to the outer edge, which is briefly exposed at low tide. This fringing reef originated as a nearshore reef which became attached to the mainland as the result of progradation of the adjacent coast by sand deposition (Bird 1971b).

Fringing reefs are well developed around high islands in the Pacific as in Tahiti, New Guinea and Indonesia, and off the north-eastern and northern coasts of Australia, as on Lizard Island, Hayman Island and Snapper Island. They also border promontories on the north coast of Australia, notably on the shores of Arnhem Land and between Port Hedland and North West Cape. They are missing from the swampy shores of the Gulf of Carpentaria and other northern gulfs which are receiving deposits of land-derived sediment, where they give place to intertidal sandflats of the kind described by Russell and Russell (1966).

BARRIER REEFS

Barrier reefs have been built offshore and roughly parallel to the coastline (Guilcher 1988). Geological studies have shown that many barrier reefs were initiated as fringing reefs in Eocene times, and have grown upward during the Tertiary and Quaternary (Figure 11.3). In 1835 Charles Darwin observed barrier reefs and atolls during his voyage in the *Beagle*, proposed the subsidence theory, which suggested that barrier reefs (and atolls) were the outcome of upward growth from ancient fringing reefs bordering continental margins and islands that have subsided beneath the oceans. There is no doubt that oceanic reefs have grown upward as their foundations subsided, but there have also been major oscillations of sea level, especially in the Quaternary (p. 28), and that barrier reefs have periodically emerged and been dissected by erosion, then submerged again to be rebuilt as coral growth revived (Figure 11.3).

Reefs that subsided or were submerged too rapidly for corals to maintain their upward growth are now submerged below the depth at which reef-building organisms could revive them. Submerged barrier reefs have been detected by soundings in the Pacific Ocean, particularly off the south-east coast of New Guinea, in the Fiji archipelago and the Coral Sea. The patchy development of reefs off northern Australia may be due to excessive subsidence in the Timor Sea, where a submerged barrier reef off the Sahul Shelf indicates that sea floor subsidence has been too rapid for reef growth to be maintained (Davis 1928).

Existing barrier reefs attained their present form during the later stages of the Late Quaternary marine transgression and the ensuing sea level stillstand. The largest and most famous is the Great Barrier Reef, which extends for about 2000 kilometres from north to south off the Queensland coast and the following description is based on a detailed study by Hopley (1982). It commemorates the alignment of a Queensland coastline that has subsided tectonically. In the north it consists of a chain of elongated or crescentic reefs with intervening gaps, generally about a kilometre wide, lying about 130 kilometres off the tip of Cape York peninsula, 50 kilometres off Port Stewart, less than 15 kilometres off Cape Melville, and about 56 kilometres seaward from Cooktown. Off Cairns there is a broad transverse passage known as Trinity Opening, which allows ocean waves to penetrate to the mainland coast (swell reaches the shore between Buchan Point and Yule Point), but the outer barrier reef reappears farther south, diverging from the mainland coast until it lies about 100 kilometres off Rockhampton. At the southern end it breaks up into the scattered reef platforms of the Capricorn and Bunker reefs, with Lady Elliot Island the southernmost reef, just south of latitude 24°S. The southern limit of reef building in these coastal waters may be determined by the temperature of the sea, but coral growth has also

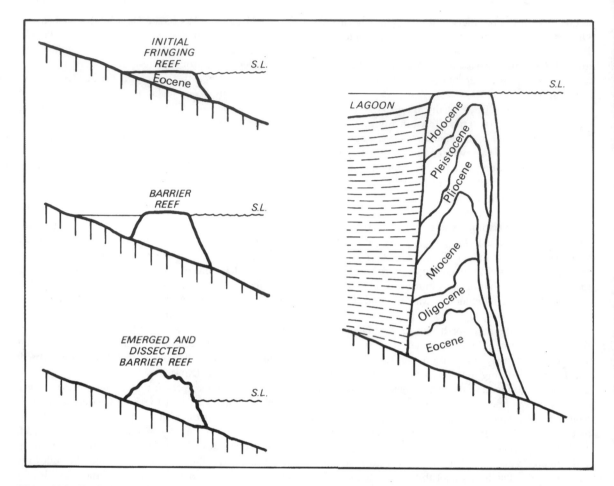

Figure 11.3 Evolution of a barrier reef from a fringing reef that grew upward as sea level rose, relative to the land. Barrier reef stratigraphy shows that there were successive phases of upward growth, interrupted by emergence and dissection in phases of low sea level and followed by the revival of corals to widen and raise the reef structure. The modern barrier reef has formed in Holocene times (during and since the Late Quaternary marine transgression), and encloses the dissected remnants of Pleistocene and earlier reefs.

been inhibited by the drifting of sand northward across the sea floor from Fraser Island.

The Great Barrier Reef plunges steeply (40°–50°) on its seaward side in to water more than 1800 metres deep in the northern section, shallowing to about 180 metres south of Trinity Opening. Coral is growing to a depth of about 45 metres on this seaward slope, which is marked by patterns of grooves and spurs (buttresses) aligned at right angles to the reef edge. These serrations are common on reef margins exposed to strong wave action, the grooves showing evidence of scouring by swash and backwash, and abrasion by waves armed with reef debris, while the spurs are crowned by a rich compact coral and algal growth. The pattern is similar

to that of other rhythmic shore forms, such as beach cusps (p. 125).

At low tide the upper parts of the Great Barrier Reef emerge as a chain of elongated platforms up to 25 kilometres long and up to a kilometre wide, typically with an outer algal rim (rampart) on which the surf breaks, and a backing slope declining gently (5°–10°) into the calmer waters of the lagoon. Yonge Reef, north of Cooktown, is a typical outer barrier segment, crescentic in form and recurved at the northern and southern ends bordering gaps in the reef. These gaps may commemorate the sites of river outlets at low sea level stages, but they persist because the rising and falling tides produce strong scouring currents which prevent them from being sealed off by coral growth and sedimentation. In one section, south-east of Cape Melville, there is a double barrier, with parallel reefs separated by a lagoon about eight kilometres wide. The inner barrier is less regular in form than the chain of outer barrier reefs, and may be a relic of an earlier barrier reef outflanked by the growth of a younger reef to seaward.

The lagoon between the Great Barrier Reef and the mainland coast of Queensland is generally between 18 and 45 metres deep, with a rather featureless floor formed by the deposition of land-derived sediment carried in by rivers and reef-derived sediment washed in by the sea. Lagoon floor sediments just behind the barrier reef show a high proportion of calcareous organic material (typically 80 per cent to 90 per cent carbonates) diminishing as terrigenous (land-derived) sediment increases towards the shore. The width of the lagoon behind a barrier reef is related to the pre-existing sea floor topography and the rate and pattern of upward reef growth during the Late Quaternary marine transgression. Coral is growing within the lagoon off the Queensland coast to depths of about 12 metres, and columns of reef limestone have grown up from the lagoon floor to form patch reefs a little above low tide level, notably in the Steamer Channel north of Cairns. The shapes of patch reefs are related to waves generated by the prevailing south-easterly trade winds in Queensland coastal waters. Several have a horseshoe form, with arms trailing north-westward. Cairns Reef has this form, and a more advanced stage is represented by Pickersgill Reef, the arms of which curve round and almost enclose a shallow lagoon.

Other major barrier reefs (Davis 1928) include those that run parallel to the coasts of New Caledonia for over 600 kilometres, enclosing lagoons up to 12 kilometres wide and up to 100 metres deep, the Great Sunda Reef south-east of Kalimantan; and the Great Sea Reef, 260 kilometres long, to the north of Fiji. In the Caribbean the Belize barrier reef, extending 220 kilometres along the coast, has three offshore atolls and over 1000 sand cays and mangrove islands. The Belize barrier reef differs from Pacific barrier reefs in having extensive coral gardens, with delicate corals growing upward, and fewer solid coral reef platforms with dead corals on the surface exposed at low tide.

Coasts bordered by barrier reefs show evidence of Late Quaternary submergence by the sea in the form of inlets, embayments and drowned valley mouths. The general absence of cliffing on such coasts is evidence that the barrier reefs grew up as the continental shelf subsided, so that the coast was consistently protected from the action of strong ocean waves. There is thus a contrast between the reef-protected Queensland coast, which has numerous promontories and high islands off-shore with only limited cliffing and the New South Wales coast, farther south and beyond the protection of the barrier reef, where headlands are more strongly cliffed.

Where coastlines bordered by barrier reefs have fringing reefs as well (as in north-east Queensland) the latter are secondary forms, initiated along the coast only after the Late Quaternary marine transgression, which allowed the barrier reef to build to its present level, came to an end.

ATOLLS

Atolls are reefs built up to sea level, typically circular or ovoid in plan, more or less continuous and surrounding a lagoon. They originated as fringing reefs around islands that were lowered by crustal subsidence during Tertiary and Quaternary times, the fringing reef growing up to become a barrier reef enclosing a lagoon (an almost-atoll), and eventually an atoll, with the central island lost from view as submergence continued. Like barrier reefs, atolls were exposed to subaerial karstic weathering during low sea level phases of the Pleistocene, and rebuilt when sea level rose again (McLean and Woodroffe 1994).

Atolls are of three kinds (Guilcher 1988). There are oceanic atolls, which have localized (generally volcanic) foundations at depths exceeding 550 metres, shelf atolls which rise from the continental shelf and have foundations at depths of less than 550 metres, and compound atolls where the ring-shaped reef surrounds or encloses relics of earlier atolls.

Oceanic atolls are common in the west Pacific and are present in the Coral Sea and the northern Tasman Sea. Typical features of an oceanic atoll are shown in Figure 11.4. The reef platform on the windward side is often wider, with an algal rampart in the breaker zone, and sometimes one or more low islands of coral debris, while outlets to the ocean are generally on the leeward side. The lagoon floor is a smooth depositional surface, from which pinnacles and ridges of live coral may protrude, but sand and gravel fans are formed when reef-derived sediment is washed in through entrances or over the bordering reef (Kench 1998).

Atolls are well developed in the Indonesian region, especially in the Flores Sea, and are also found off the north coast of Australia, where barrier reefs have not formed. Seringapatam, 460 kilometres north of Broome, is an example, rising abruptly from a depth

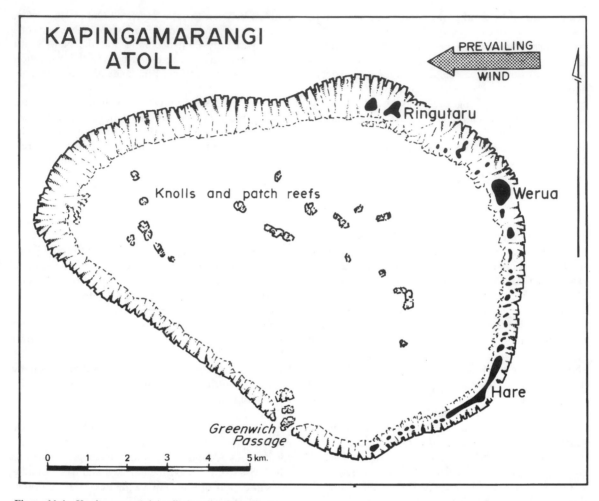

Figure 11.4   Kapingamarangi Atoll, Caroline Islands.

of almost 550 metres near the outer edge of a broad, sloping continental shelf. The enclosing reef is about 900 metres wide, and the lagoon, 9.6 kilometres long and 6.4 kilometres wide, has an average depth of 35 metres. Scott Reefs, not far away, consist of a similar enclosed atoll and a second atoll that is incomplete, with the superficial form of a horseshoe reef. Water more than 180 metres deep surrounds and separates the two reefs, but the enclosed and partly enclosed lagoons have floors at a depth of 35–45 metres (Fairbridge 1967).

Compound atolls are found in the Houtman Abrolhos, a group of reef platforms and patches with remnants of emerged Pleistocene reefs rising sharply from a depth of 55 metres on the outer part of the continental shelf off the west coast of Australia.

These southernmost (29°S) coral reefs in the Indian Ocean show vigorous coral and algal growth in water warmed by the south-flowing Leeuwin current. Pelsart Island is 11 kilometres long, a few hundred metres wide, and up to two metres high, an emerged Pleistocene coralline reef on the eastern limb of the V-shaped Pelsart Group, essentially a tilted atoll. The reefs are penetrated by deep sinkholes, produced by subaerial solution processes when they were dissected during Pleistocene low sea level phases (Guilcher 1988).

In the Pacific Ocean there are a number of sea mounts, generally extinct volcanoes which rise steeply from the deep ocean floor, several of which have been planed off to form flat-topped guyots, platforms at depths of between 550 and 900 metres. Some bear fossil corals of Cretaceous age, but had evidently subsided tectonically below the limit of coral reef growth during Tertiary and Quaternary times. Those that remained within the range of coral growth have been built up as atolls, but some have failed to maintain upward growth and form drowned atolls, notably in the Caroline Islands (Davis 1928).

Small-scale atoll-like reefs up to a kilometre in diameter and enclosing lagoons less than 20 metres deep are known as faros in equatorial regions (especially in the Maldive Islands, 2°S to 7°N), where they may have originated from coral reefs that lived through the Last Glacial low sea level phase in these low latitudes. During Pleistocene low sea level phases ocean temperatures fell by up to 6°, so that coral growth could then have continued only in the warmer parts of the oceans, probably within latitudes 10°–15°N and S. (Guilcher 1988).

Still smaller are microatolls, intertidal ring-shaped organic reef structures typically one to six metres in diameter, with slightly raised rims of living coral, built up to about low neap tide level, and thus standing above the general level of a reef platform (Stoddart and Scoffin 1979).

EMERGED CORAL REEFS

Coral reef platforms are built up to a level just above low tide, and emerged reefs, consisting of dead coral and associated reef organisms standing above low tide level, can result from tectonic uplift or from a lowering of sea level. Emerged reefs have been reported from several sites in the Pacific, notably in the Society, Tuamoto and Cook Islands, many with solution notches indicating phases of stability separated by tectonic uplift or sea level fall. An emerged reef forms part of the Pelsart Group in Houtman Abrolhos, off the coast of Western Australia, which was evidently built when the sea stood three to eight metres above its present level during a Late Pleistocene stillstand. It is bordered by modern reef platforms, which have grown up to just above low tide level (Fairbridge 1967).

Similar emerged reefs have been found farther south, on Rottnest Island, beyond the present range of reef-building. Emerged reefs extending up to three metres above present low tide level have been reported at various other places around the Australian coast, including the shores of Melville Island, islands in Torres Strait, and Raine Island off the Queensland coast. They have generally been regarded as reef platforms built up to a slightly higher sea level, then laid bare as a consequence of an ensuing fall in sea level, and

they may correspond with emerged shore platforms cut in dune calcarenite and other formations around the Australian coast (p. 33).

Near Tokyo uplifted early Holocene coral formations are found well to the north of living coastal reefs, implying that the limits of Holocene coral growth in the north Pacific then extended further north, and while the emerged Holocene reef that borders Peel Island, in Moreton Bay (27°30′S), south of the present limit of coral growth in Queensland coastal waters, may indicate a similar southward extension of reef growth (Hopley 1982).

Examples of emerged fringing reefs that have been raised by tectonic movements are found on the mountainous slopes of the Huon peninsula, in north-eastern New Guinea, where they form a stairway of reef terraces, each representing a phase of stillstand between episodes of uplift (p. 34). They are now being rapidly consumed by subaerial processes, notably solution by rainwater. The reef terraces diverge on either side of an axis of upwarping, on which they attain a maximum elevation of about 750 metres in the vicinity of the deeply incised Tewai gorge (Chappell 1974). The structure of the fringing reef terraces can be seen on exposures along the sides of this gorge. A similar sequence of uplifted and tilted fringing coral reefs is seen on the island of Barbados.

Islands formed by the emergence of coral reefs include Christmas Island, 320 kilometres south of Java, which stands 390 metres high and has been uplifted tectonically to such an extent that underlying Eocene marine limestones are exposed. The Loyalty Islands, north-east of New Caledonia, include uplifted atolls such as Maré, where the former lagoon floor is an interior plain surrounded by an even crested rim of reef limestone 30 metres high, the steep outer coastline showing 15 notched terraces indicative of stages in its uplift, and Uvéa, which has been intermittently tilted during upheaval so that the terraced eastern rim has emerged while the western rim, submerged, has a newer loop of developing reefs. Rennell Island is another emerged atoll with high ridges of coral limestone enclosing a lake that is linked to the sea by way of a cavern (Guilcher 1988).

Emerged coral reefs are attacked by rainwater and sea spray solution, and develop a pitted and hollowed karstic topography known as makatea. Their coasts show cliffs with basal notches, fronted by low tide shore platforms similar to those on other limestone and dune calcarenite coasts (p. 88). In the Bismarck Archipelago cliffs at the edge of emerged reefs show two notches, one formed when the sea stood about

1.5 metres above present mean tide level, where the modern notch has a pitted basal slope descending to the modern fringing reef (cf. p. 34).

The Isle of Pines, south of New Caledonia, is fringed by emerged coral reefs that are cliffed, with typical notch-and-visor profiles, and bordered by shore platforms cut to present low tide level and extending seaward by the growth of a modern fringing reef at the same level.

ISLANDS ON CORAL REEF PLATFORMS

Fragments of coral broken off during occasional storms are cast up on to the reef platform as coralline sand and gravel, and occasionally larger boulders. Blocks measuring several cubic metres thrown up during a storm surge may persist on a reef platform for decades, even centuries. They develop features typical of limestone coasts (p. 78), being gradually undermined by notches formed by solution and becoming rugged

and pitted as they are dissolved by rain water and sea spray.

Low islands (known as cays, and also as cayes or keys) have been built on reef platforms by the accumulation of sand, shingle, and boulders formed from reef debris eroded by wave action and thrown up on the platform (Figure 11.5). These carbonate sediments include large lumps of broken reef limestone, broken cylindrical sticks of staghorn coral, algae such as the sand-sized *Halimeda*, discoidal shingle and the gritty sands into which coralline reef material disintegrates.

Cays are found on the Belize barrier reef in the Caribbean, where the term caye or key indicates an island of coralline sand and gravel deposited on a coral reef (Stoddart 1965). Off the Queensland coast there are cays on coral platforms in the Bunker and Capricorn reefs, at Fife Island north of Port Stewart and Green Island, off Cairns (Steers 1937). Waves generated by the prevailing south-east winds have washed sediment across the reef platform at high tide to build the cays generally near the north-west (leeward) edge.

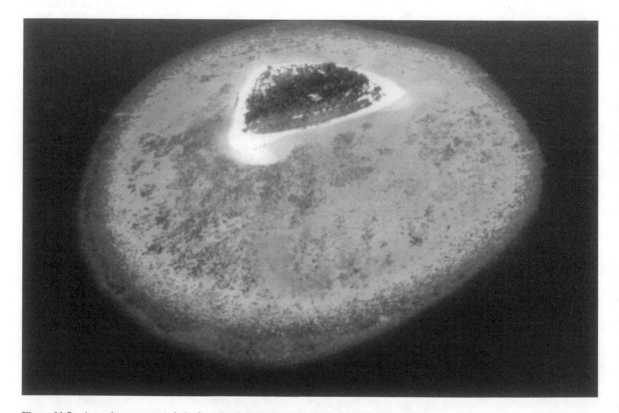

Figure 11.5   A sand cay on a reef platform in the Maldive Islands, Indian Ocean.

Waves refracted around the reef platform converge on the lee side in such a way as to prevent reef debris from being swept over the lee edge of the platform. Generally cays have steeper and narrower beaches on their windward than on their leeward shores.

Algal sand is plentiful on coral cays. A cay is initiated as a sandbank or heap of coral shingle awash at high tide, but as sediment accumulates, and is built above high tide level, it is colonized by grasses and shrubs, then coconut trees, palms (*Pisonia* and *Pandanus* spp.), and *Casuarinas*.

Most cays consist mainly of wave-deposited sediment, but Heron Island, off south-east Queensland, has low dunes of coralline sand built by the south-east trade winds. On reefs more exposed to strong wave action the cays are gravelly, with a predominance of coarse shingle over sand, as on Lady Musgrave Island, near the southern end of the Great Barrier Reef, and on the Gili Islands north-east of Lombok in Indonesia.

Low wooded islands are more depositional islands found on reef platforms, as off north-east Queensland (Figure 11.6). Low Isles, off Port Douglas, is an example

(Stoddart et al. 1978). It consists of a sand cay that formed on the leeward (north-western) side, then shingle ridges which were thrown up as ramparts on the windward side, built from debris eroded from the edge of the reef, with mangroves (mainly *Rhizophora*) in the intervening depression. Changes occur as the result of sand accretion during relatively calm weather and erosion (sometimes with the addition of coral shingle) during storms. Some low wooded islands have several successively built parallel shingle ridges, the older ridges being dark in colour and largely cemented as conglomerate while the younger are white or cream, and unconsolidated. Addition of freshly-formed coralline debris occurs during tropical cyclones, widening the existing rampart or adding a new shingle ridge.

Cays and low wooded islands are related to wave energy. On the outer reefs high wave energy has prevented the formation of such islands, but the inner reefs, partly protected from ocean waves, have moderate wave energy which leads to sand cay formation by refracted waves generated by local winds. The reef platforms off north-east Queensland are in deep water with a relatively broad fetch over which the south-

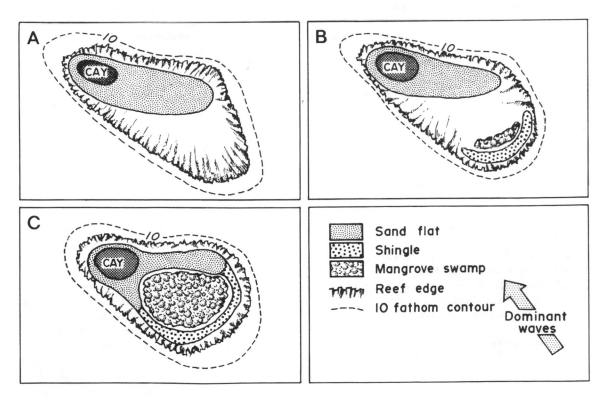

Figure 11.6 Stages in the formation of a cay (A) and low wooded island (B, C) on a reef platform.

easterly trade winds generate waves strong enough to build shingle ramparts on the windward side and cays on the leeward side. Cays and low wooded islands occur in similar conditions of wave energy variation in the reefs off British Honduras.

Cays change in configuration as erosion and deposition modify their shores. Between high and low tide levels cay beaches may be cemented by secondary deposition of calcium carbonate in the zone of repeated wetting and drying, sandy beaches forming the compact layered sandstone known as beach rock (p. 114), whereas shingle becomes a lithified beach conglomerate. The greater resistance of these formations impedes erosion of cay shores, but waves generated during tropical cyclones may expose beach rock on the cay shores as they modify the outline of the cay. It may be swept away altogether, leaving only patterns of eroded beach rock or shingle and conglomerate to indicate the coastlines of vanished cays. Remnants of beach rock on a reef platform can thus be taken as an indication of the former presence of a cay when there is no longer any depositional island.

In 1961 Hurricane Hattie swept westward across the atolls and barrier reef off British Honduras towards the coast south of Belize, accompanied by gusts exceeding 320 kilometres per hour. A 72 kilometre wide storm surge raised sea level briefly by up to 4.5 metres. Cays that had a dense vegetation cover were less modified than those on which the natural vegetation had been cleared and replaced by coconut plantations. The former were eroded on the windward side, with shingle and coral debris being piled up against the vegetation, but the latter were more severely damaged, and seven were swept away completely (Stoddart 1965). The sandy cay of Makuluva, off Suva in Fiji, has migrated eastward across the coral reef platform, the windward (south-west) coast having been cut back, so that lighthouse foundations and water tanks stand offshore in front of a sandy cliff capped by dune woodland, while the leeward coast has a broad prograding beach on to which grasses and shrubs are spreading.

Cays in the Great Barrier Reef region have shown a prevalence of erosion on their coastlines in recent decades, confirmed by extensive outlying exposures of beach rock, so that they have been diminishing in size, perhaps as the result of a rising sea level (p. 280). In many respects the cays are similar to the spits and barriers built by coastal deposition as described in Chapter 6, their distinctive features largely a response to the special environment of the reef platforms on which they have been built.

ALGAL AND OTHER BIOGENIC REEFS

The commonest non-coralline reefs are those built by algae on rocky shores or in the intertidal zone at the base of a cliff. Calcareous algae withdraw carbonates from sea water and build solid reefs or encrustations. Some algae also have sediment-trapping structures. On the coasts of the Caribbean, the Mediterranean and southern Australia calcareous algae such as *Lithophyllum*, and *Porolithon* form solid ripples and rims on shore platforms, particularly those cut in dune calcarenite. On some shore platforms algae produce rimmed terracettes that descend from a zone of higher water level, maintained by frequent large waves, seaward or sideways, and sometimes landward (Guilcher 1988).

Reefs built largely by algae occur on subtropical coasts close to, and beyond, the limits of coral growth, as in Florida and Brazil, around the Mediterranean, and in the hypersaline waters of Shark Bay, Australia. In the absence of corals, or where coral growth is poor, algae can construct reefs, or at least form reefs surmounting submerged ridges of sandstone or dune calcarenite. Algal reefs of this kind are extensive in the Cape Verde Islands, off the west coast of Africa (Davis 1928).

The growth of calcareous algae and associated marine invertebrates can form a protruding shelf, known as a trottoir, a small constructional shore platform projecting from a cliff at about high tide level, notably on the coasts of the Mediterranean (Peres 1968). Occasionally waves break off the outer edge. Similar features have been seen in the Pacific, as in the Mariana Islands.

In south-eastern Australia calcareous structures built by tubeworms (*Galeolaria*) form ledges on rocky shores and harbour walls, or cushions around pier supports. Their upper limits are just above mid-tide level (Bird 1988).

On the shores of the Gulf and Atlantic coasts of the United States oyster reefs built by *Crassostrea virginica* are extensive, and there are also small mounds built by bryozoans, serpulids, and mussels on rocky or sandy foundations. Vermetid gastropods have built reefs in the Mediterranean and Atlantic (Safriel 1975), and form biogenic sandrocks deposited along the seaward margins of mangroves on the low wave energy coast of western Florida. They form in turbid water as tubular concretionary structures that become aggregated as minor reefs.

12

Future Coasts

INTRODUCTION

The changes that will take place on the world's coastline over the coming century cannot be predicted simply by extrapolating the gains, losses and modifications that occurred during the twentieth century, because of the likelihood of climatic changes and a world-wide sea level rise. In Chapter 3 it was noted that sea level has been relatively stable on much of the world's coastline during the past 6000 years, apart from minor oscillations, but there are sectors where the coastal land has continued to rise or fall within this period: some coasts are responding to a slow sea level rise (as on the Gulf and Atlantic seaboards of the United States), others to a lowering of sea level, notably where the land is being uplifted as the result of deglaciation and isostatic recovery (as in Scandinavia and parts of northern Canada). Later chapters showed that coastal landforms have been changing as the result of erosion and accretion in response to coastal processes with the sea at or near its present level, and that there have been additional responses to relative sea level changes. The present chapter considers probable responses to forecast climatic changes and sea level rise.

GREENHOUSE EFFECT AND SEA LEVEL RISE

Global climatic changes are likely to occur (and may already have started) as the result of the human-induced accumulation in the earth's atmosphere of such gases as carbon dioxide, nitrous oxide, and methane, produced mainly by industry and agriculture. In particular, the burning of fossil fuels (coal, oil and natural gas) is returning to the atmosphere carbon dioxide that was withdrawn from earlier atmospheres by plant photosynthesis and retained in swamp forests that became fossil fuel deposits in the geological past (Titus 1988).

Atmospheric monitoring initiated during the International Geophysical Year in 1957 has shown that concentrations of carbon dioxide and other gases in the atmosphere have been increasing. The carbon dioxide concentration, for example, increased from 315 parts per million (ppm) in 1958 to more than 350 ppm in 1998. Such an increase will enhance the natural greenhouse effect, whereby the atmosphere intercepts some of the solar radiation reflected into space from the earth's surface, and so maintains global temperatures at a higher level than would otherwise prevail. It is expected that the mean temperature of the lower atmosphere will increase by between 1.5°C and 4.5°C over the coming century. Such human-induced global warming will lead to expansion of the oceans (the steric effect) and some melting of the world's snowfields, ice sheets and glaciers, resulting in a world-wide rise of sea level. As has been noted, there is evidence from tide gauge records suggesting that the sea has been rising at the rate of between one and two millimetres per year around much of the world's coastline (Pirazzoli 1996), but as the majority of tide gauges are located in port areas and may show local anomalies the pattern and scale of a contemporary marine transgression require confirmation.

Calculations by the Intergovernmental Panel on Climatic Change indicated that global sea level will probably rise 10 to 15 centimetres by the year 2030, accelerating to between 30 and 80 centimetres by the end of the twenty-first century (Wigley and Raper 1993). Such a sea level rise will initiate or accelerate coastal changes around the world (Bird 1993c). An obvious outcome will be that submerging coastlines, presently confined mainly to sectors where the land has been subsiding (Figure 3.10), will become more extensive, and that emerging coastlines will become rarer. A rising sea level will reduce the effects of land emergence, slowing down the advance of coastlines such as those bordering the Gulf of Bothnia. An accelerating sea level rise will eventually equal, then exceed, the rate of land uplift, and in due course all of the world's coastline will be submerging as the result of sea level rise.

Depletion of the world's upper atmospheric ozone layer, which intercepts much of the ultraviolet radiation arriving from the sun, is thought to be due to the effects of chlorine monoxide produced from chlorofluorocarbon (CFC) emissions, generated by aerosol propellants, refrigerators and various industrial processes. In addition to depleting ozone, CFCs also contribute to the enhancement of the greenhouse effect, global warming and sea level rise, but ozone depletion leads to increased ultraviolet radiation, which has adverse effects on the growth and health of plants and animals, and is harmful to humans.

GENERAL EFFECTS OF A RISING SEA LEVEL

If global sea level rises in the manner predicted there will obviously be extensive marine submergence of low-lying coastal areas. High and low tide lines will advance landward, and at least part of the present intertidal zone will become completely submerged. On steep hard rock coasts where there is little or no marine erosion in progress and on solid artificial structures such as vertical sea walls, a sea level rise will simply raise the high and low tide lines and the coastline will remain in its present position.

It is possible that there will be a slight increase in tide ranges around the world's coastline as the oceans deepen, the rise that actually occurs being modified as tidal amplitude is adapted to the changing coastal and nearshore configuration. On many coasts the extent to which the high tide line moves landward will be augmented by an increase in erosion as nearshore waters deepen and larger and more destructive waves break upon the shore. As sea level rises erosion will begin on coasts that are at present stable, and accelerate on coasts that were already receding. On coasts that had been prograding, the seaward advance of the land will be curbed, and erosion may begin (Carter and Devoy 1987). This has been seen on the coasts of the Caspian Sea, where the proportion eroding in 1977 after several years of falling sea level and widespread progradation was 10 per cent, but has increased to over 40 per cent in the ensuing sea level rise (Figure 3.9).

Erosion will increase still further where the climatic changes that accompany the rising sea level lead to more frequent and severe storms, generating surges that penetrate further inland than they do now. Coasts already subject to recurrent storm surges (e.g. the hurricane-prone Gulf and Atlantic coasts of the United States) will have more frequent and extensive marine flooding as submergence proceeds, and more severe erosion and structural damage where larger waves reach the coast through deepening nearshore waters. These effects could be more damaging than the direct impact of a sea level rise (Daniels 1992).

The extent of coastal erosion will also depend on how the nearshore sea floor is modified by the rising sea as the coastline moves landward. Although there will generally be deepening of nearshore waters and a consequent increase in wave energy approaching the coastline, some coasts may receive an augmented sediment supply, derived from increasing fluvial or alongshore sources, and this could maintain, or even diminish, the nearshore profile as the sea rises. On such coasts wave energy will not intensify, and there may be little, if any, coastline erosion; there may even be some progradation.

Attempts have been made to predict the effects of a sea level rise on coastlines (Tooley and Jelgersma 1992, Warrick et al. 1993, Bird 1993c) but they remain speculative because of limited monitoring of recent and continuing changes and the imprecision of available models. A major difficulty is that the prediction is an accelerating sea level rise, with no early prospect of stabilization. If the sea continues to rise, coastal erosion will accelerate and become more widespread as any compensating sedimentation declines. The rising sea will encounter, re-shape and in due course submerge the 'raised beaches' and other emerged shore features that formed during Pleistocene interglacial phases (p. 33). Revival at higher levels of features similar to those now seen on the world's coastline, with some parts eroding while others are stable or prograding, must await the establishment of a new sea level stillstand. This could be when the greenhouse effect has been brought under control, or when all the world's glaciers, ice sheets and snowfields have melted and the water contained in them has flowed back into the oceans, producing a global sea level rise of more than 60 metres (p. 31).

EFFECTS OF A CHANGING CLIMATE

A global sea level rise will be accompanied by other responses to the increase in atmospheric temperature. As well as raising temperatures in coastal regions, global warming will cause a migration of climatic zones, tropical sectors expanding as temperate sectors migrate poleward and the domain of Arctic coasts shrinks. Tropical cyclones may become more frequent and severe, extending into higher latitudes and bringing storm surges and torrential downpours to coasts that now lie outside their range. Coastal climates will also be modified by changes in ocean currents, as already shown by the El Niño Southern Oscillation, which has led to heavy rain and river flooding in China, Ecuador and Peru and droughts in Australia.

As warming proceeds, some regions are expected to receive more rainfall while others become drier (Houghton et al. 1990). Where rainfall increases there will be more frequent and persistent river flooding, and the water table rise will be added to that caused by the rising sea level. Some low-lying parts of coastal plains will become permanent swamps or lagoons, the salinity of which will depend on interactions between increasing marine incursion (and perhaps an upwards movement of subterranean salt) and any offsetting effects of augmented rainfall and freshwater runoff. With increased rainfall, coastal vegetation may become more luxuriant; with drier conditions it is likely to be depleted. Coastal regions that become drier will also have more extensive marine salinity penetration into both surface and underground water. Coastal lagoons will become more brackish, and some will dry out as saline flats. Dessication will also reduce the vegetation cover and allow wind action to be more erosive, especially in coastal dune areas, whereas more vigorous and extensive vegetation will tend to stabilize dunes that are now bare and drifting. Vegetation may also be impoverished by the effects of increasing ultraviolet radiation resulting from atmospheric ozone depletion, and thus less able to trap sediment on dunes or in salt marshes and mangrove swamps. On the other hand, increasing atmospheric carbon dioxide may enhance photosynthesis and plant growth, and produce vegetation that is more effective in trapping dune sand or intertidal sediment.

Most marine plants and animals have specific latitudinal ranges, usually depending on maximum or minimum sea temperatures, which will rise as global warming proceeds. Mangroves, now largely confined to tropical coasts, will extend their range poleward along coasts whereby there are suitable habitats, and corals will extend northward and southward beyond their present latitudinal limits, providing there are suitable substrates and available coral larvae. Many temperate salt marsh species will also migrate poleward, whereas the distribution of kelp, a plant

restricted to cooler waters, will contract to higher latitudes. Global warming is also likely to affect coastal ecosystems by way of thermal stress, a factor already noted where coral bleaching has occurred (p. 264).

EFFECTS ON CLIFFS AND SHORE PLATFORMS

Existing cliffs and rocky shores are largely a consequence of the relatively stable sea level around much of the world's coastline during the past 6000 years. If sea level rises, nearshore waters will deepen, submerging shore platforms and rocky shores, and the deeper water will allow larger waves to reach the coast and attack the base of cliffs and bluffs, accelerating their erosion (Figure 12.1A). On some coasts the rock outcrops are so resistant that the high and low tide lines simply move up the existing cliff face, but elsewhere increasing erosion will prompt demands for coastal defence works, notably the extension and elaboration of sea walls.

Where the coast consists of vegetated bluffs, a rising sea level will increase basal wave erosion and slumping, and the bluffs will develop into retreating cliffs. Existing cliffs will generally become more unstable, and recede more rapidly. Cliff erosion accelerated on the coasts of the Great Lakes in North America during phases of rising water level (Carter and Guy 1988), and a rising sea level resulting from land subsidence following oil extraction around Long Beach, California, has intensified wave attack, accelerating the retreat of soft clay cliffs at Huntington (Bird 1993c). Clayton (1989) estimated that in Britain, cliffs that are already retreating a metre per year will show an accelerated retreat of 0.35 metres per year for every millimetre rise in sea level. On the volcanic island of Nii-jima, off the Japanese coast south of Tokyo, Sunamura (1992) calculated that cliffs cut in poorly consolidated volcanic gravel and ash, now retreating at 1.2 metres per year, will recede at between 2.3 and 2.9 metres per year with a sea level rise of a metre by the year 2100, or more if the accompanying climatic change increases storminess in coastal waters. It will be difficult to recognize an acceleration of cliff retreat, at least in the early stages of rising sea level, unless existing rates of recession have been measured with sufficient accuracy to provide a basis for comparison. Such monitoring is presently only available for very limited parts of the world's cliffed coastline (Bird 1985a).

Shore platforms that have developed in front of cliffs or bluffs will be submerged for longer periods, and

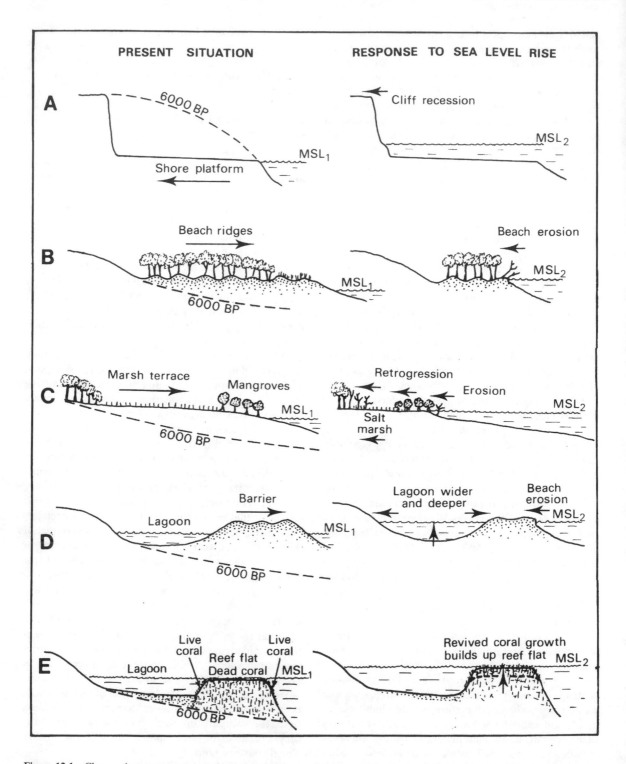

Figure 12.1   Changes in response to a sea level rise, A – on a cliffed coast with a shore platform, B – on beach ridges, C – on a marsh terrace, D – on a barrier-fringed lagoon and E – on a coral reef.

become permanently inundated when the sea level rise exceeds present tide range. Waves that now reach the base of the cliff only at high tide will attack more consistently, thereby accelerating cliff erosion. Again, the eroded material will usually be carried away along-shore, but if it remains as a beach in front of the cliff the shore platform may be protected from further erosion. This will also be the case where submerging shore plat-forms acquire a mantle of accretionary growths of near-shore plant and animal communities. Cementation of sandy ripples on the surface of a shore platform cut in Pleistocene dune calcarenite on the coast near Sorrento, in south-eastern Australia, has been cited as a possible consequence of a slight relative sea level rise (Bird 1982), as have algal encrustations on formerly eroding shore platforms near Port Hedland in Western Australia, and at Nyali in Kenya (Bird and Guilcher 1982). Such changes may at least partly offset the effects of nearshore deepening and intensification of wave attack on backing cliffs.

Cliff-base notches formed by abrasion or solution will be enlarged upward by a rising sea level. The formation of a new notch at a higher level would require a rapid sea level rise followed by a new stillstand.

On soft rock formations waves have shaped the nearshore sea floor into a concave profile that declines seaward from the cliff base, and as the cliffs recede this nearshore sea floor is lowered in such a way as to move the concave profile landward. This has occurred where the cliffs and sea floor are cut into Pleistocene glacial drift on the east coast of England, as at Holderness. A sea level rise is likely to accelerate landward migration of the concave sea floor profiles as the cliff base retreats. Bray and Hooke (1997) have considered adapting the Bruun Rule (p. 147) to the prediction of soft-cliff retreat on the south coast of England.

Cliffs cut in soft rock formations on coasts sheltered from strong wave action are often dominated by gulleying, slumping and other features resulting from subaerial weathering and the effects of runoff and see-page, rather than by marine processes. A vertical cliff profile may be formed briefly after storm waves have removed basal talus and downwashed deposits, but subaerial processes soon restore the degraded profile. On such cliffs a rising sea level is likely to increase marine erosion, reducing and eventually suppressing the features developed by subaerial processes, as under-cutting of the cliff base maintains a steeper or vertical receding cliff, but an increase in rainfall could maintain subaerial erosion by runoff.

A sea level rise is likely to increase the frequency of coastal landslides and produce new and more extensive slumping along coasts, especially where the rock formations dip seaward. The frequency of slumping will accelerate where a cliff is cut back through horizontal or landward-dipping strata into an area where the rocks dip seaward. Coastal landslides will become more frequent if rainfall increases, and less if the coastal climate becomes drier.

A by-product of increased landsliding will be an augmented supply of sediment to adjacent and down-drift beaches, perhaps offsetting the effects of a sea level rise on these beaches.

Rates of cliff recession are influenced by the avail-ability of rocky debris, including beach material, which can be mobilized by wave action and used as ammuni-tion for cliff-base abrasion. Cliffs fringed by a narrow beach have retreated more rapidly than cliffs on the same formation where the beach is wide and high (so that waves do not reach the cliff base), or where there is no beach material (so that the cliff base is attacked only by the hydraulic action of breaking waves). Debris eroded from cliffs is generally carried away alongshore by wave action, but if during a sea level rise this sediment is retained as a persistent talus apron of sand and gravel, or a protective beach in front of the cliff, or as bars and terraces in the nearshore zone, wave energy will diminish, and cliff retreat will slow down and perhaps be halted. On the other hand the depletion and narrowing of a beach along the base of a cliff will result in more vigorous wave abrasion and accelerated cliff retreat, but if the beach is completely removed, wave abrasion will diminish, and with purely hydraulic action cliff retreat could decelerate.

Rates of cliff retreat as sea level rises may be estimated by comparing recession rates on similar rock formations with differing exposure. The retreat of a cliff fronted by a seaward-sloping shore platform and subject to only brief episodes of basal wave attack at high tide can be compared with a cliff cut in similar material, where wave attack is stronger and more persistent, because nearshore water is deeper. Such a contrast can be found where a resistant formation dips alongshore, so that it protects a cliff from sustained wave attack on one sector, but not on the next. On the coast of Kimmeridge Bay, in southern England, cliffs cut in Jurassic shale and limestone are being cut back more rapidly behind shores where the sea floor declines steeply than where the cliff base is protected by a flat structural shore platform. After a sea level rise has submerged the structural platform the rate of cliff recession will increase, while that of the unprotected sectors will also accelerate as the water deepens.

There have been few measurements of cliff recession rates on subsiding coasts, but in southern England, which is believed to have been subsiding at about two millimetres a year, surveys by Brunsden and Jones (1980) indicated that the slumping cliffs of Lyme Bay have been retreating at an average rate of 40 metres per century. If sea level rise accelerates, these cliffs could steepen with recession proceeding too quickly for mass movement to maintain the existing topography, or the present jumbled coastal morphology could persist as the coastline retreats. The second possibility is more likely if the sea level rise is accompanied by a climatic change that increases rainfall, seepage and runoff on this coast.

EFFECTS ON BEACHES, SPITS AND BARRIERS

Reference has been made to the evidence from tide gauge records that a world-wide sea level rise of about one to two millimetres per year has been taking place during the past few decades, offset on some coasts by equal or greater land uplift, and complicated by geophysical factors that have raised or lowered the surface of the oceans (Pirazzoli 1996). The beach erosion problem (p. 147) will intensify if, as is predicted, a global sea level rise increases in the next few decades. A sea level rise will generally result in a deepening of nearshore water, so that larger waves break upon the shore, initiating erosion on beaches or accelerating it where it is already taking place. Beach erosion will become even more extensive and severe than it is now.

On beaches that had been prograding the seaward advance will be curbed, and erosion may begin (Figure 12.1B). Where beach-ridge plains and coastal barriers have shown Holocene progradation a sea level rise is likely to initiate or accelerate erosion along their seaward margins (Figure 6.9). Coastal barriers that are already transgressive (migrating landward) will continue to migrate as sea level rises, and some barriers that have remained stationary, or have prograded during the Holocene stillstand may become transgressive as the result of erosion along their seaward margins, with washovers and landward drifting of dunes. This has happened on the coasts of the Caspian Sea during the marine transgression that began in 1977. In the south-west, where the nearshore gradient is very gentle (about 1 : 1000) the rising sea level has led to beach erosion accompanied by the formation of transgressive barriers in front of shallow lagoons that are moving diachronously landward (Leontiev and Veliev 1990).

On coasts where the beach fringe is narrow, backed by high ground, beaches will soon disappear unless the sea level rise increases nearby cliff erosion and generates additional sediment to maintain them. There will still be beaches that persist as they retreat through beach ridge plains or coastal barriers of sand and shingle. Beaches that front salt marshes or mangrove swamps are likely to be eroded and overwashed, with the sea invading the land that lies behind them. Where sea walls have been built to halt coastline recession, beaches fronting them will be depleted or removed altogether by scour due to reflection of incident waves.

Beaches will be maintained where there is a continuing supply of sand and shingle, or where the supply is increased as the result of accelerated nearby cliff erosion, greater sediment yield from rivers because of heavier or more effective rainfall, catchment devegetation or disturbance by tectonic uplift or volcanic activity. If the nearshore profile is maintained or made shallower by accretion or tectonic uplift, or if the longshore sediment supply continues at a sufficient level (with sediment supplied by rivers or erosion of cliffs and shore outcrops, especially where these supplies increase as the result of a wetter or stormier climate), beaches may persist, or even be prograded, as sea level continues to rise. In most cases the high tide line will move landward as the result of submergence and accompanying erosion as the nearshore profile also migrates landward.

The effects of a sea level rise can already be seen on beaches where there has been submergence as a result of land subsidence in recent decades, as on the Atlantic seaboard of the United States (Leatherman 1990), on the Frisian Islands in the southern North Sea (Eitner 1996) and the north-western Adriatic. Submergence has already contributed to beach profile erosion on these coasts, much as predicted by the Bruun Rule (p. 147). According to this rule the extent of recession of the transverse beach profile will be 50 to 100 times the dimensions of the rise in sea level, so that a one metre rise would cause the beach to retreat by 50 to 100 metres. Since many seaside resort beaches are no more than 30 metres wide, the implication is that these will have disappeared by the time the sea has risen 15 to 30 centimetres (perhaps by the year 2050), unless they are artificially replaced.

The Bruun Rule states that on beaches which had attained a profile of equilibrium (p. 126) the response to a sea level rise will be erosion of the upper beach and withdrawal of sediment from the beach to the adjacent sea floor in such a way as to restore the previous transverse profile in relation to the higher sea level (Figure

5.31). There has been much discussion of its applicability to precise forecasting of the extent of coastline retreat after a sea level rise (SCOR Working Group 1991). A pre-condition of the Bruun Rule is that the beaches are initially in equilibrium (Pilkey et al. 1993), which has been interpreted in various ways. Beach erosion is already widespread and few of the world's sandy beaches are presently in equilibrium.

In order to apply the Bruun Rule it is necessary to determine a seaward boundary of the profile that will be restored at a higher level. Bruun (1988) suggested that the boundary should be at the line where predominantly coarser nearshore sediment gives place to generally finer offshore material, where the water has become too deep for waves to move material of beach calibre, but this requires detailed sedimentological surveys of the sea floor, which were not necessarily available before a sea level rise began. On many coasts there are gradual transitions, and commonly the nearshore topography is variable, with sand bars, rocky outcrops, biogenic structures such as coral reefs, or a muddy substrate offshore.

Restoration of the transverse profile can only be completed after the sea has become stable at a higher level, coastline recession coming to an end as a new equilibrium is attained. The Bruun Rule is likely to underestimate the extent of beach recession while sea level is actually rising, and is difficult to apply to a prediction of an accelerating sea level rise without any indication of the level at which it will eventually stabilize. The Bruun Rule takes no account of changes in processes or rates of sediment supply and removal resulting from the climatic variations that may accompany a sea level rise, such as increased rainfall and run-off, stronger winds and stormier seas.

The Bruun Rule deals only with sediment interchange between the beach face and the nearshore sea floor, and omits gains or losses as the result of longshore drift, and their effects on changing beach profiles. It is thus more likely to apply to swash-dominated rather than drift-dominated beaches. Many beaches also lose sediment blown to backshore dunes, washed into tidal inlets, or swept over barriers or spits into lagoons and swamps.

It is difficult to predict the proportions of sediment that will be lost seaward, alongshore and landward from an eroding beach: these proportions may or may not maintain budgets of the kind shown in Figure 5.37 on an already-submerging coast.

If climatic changes accompanying a sea level rise lead to increased rainfall, beaches may receive more sand or gravel washed down coastal slopes, and changes in coastal wind regimes may increase sand blown from hinterland dunes to beaches, or modify incident wave regimes to increase longshore drift. An abundant supply of sediment from alongshore could maintain or prograde a beach during a phase of rising sea level, either directly, or by shallowing the nearshore zone so that shoreward drifting ensues.

Where the transverse beach and nearshore gradient is low, the effects of a rising sea level are complicated because storm waves will steepen and re-shape the upper beach profile by throwing sand or gravel up above high tide level to form a barrier in front of a low-lying area which becomes submerged as a lagoon. The barrier and lagoon are then driven landward as sea level continues to rise. As it does so, the steepened beach on the seaward side may maintain its profile as sediment is lost, partly by landward overwash and partly by losses seaward (Ignatov et al. 1993). Analysis of transgressive Caspian coast barriers as possible models for responses to accelerating sea level rise elsewhere showed that the Bruun Rule should be extended to a third dimension, and should include longshore drift, washovers and the effects of wind action on beaches (Kaplin and Selivanov 1995).

Where beaches are bordered by nearshore sand bars, a sea level rise may be accompanied by upward growth and landward movement of these bars as the beach is cut back, providing there is a sufficient supply of suitable sediment available to maintain the nearshore profile. The Bruun Rule could be extended to predict that the landward retreat of the outer limit of the sand bars will be equal to the extent of beach recession, so that the overall profile is preserved (Dubois 1992). A complication is that sand bars usually consist of finer material than is present on the beach face, so that erosion and seaward movement of beach sand may not provide sediment of suitable calibre for their maintenance, at least in their existing form. On the shores of Lake Michigan, Dubois (1977) found that beach face erosion as the lake level rose was matched by accretion on the landward side of the nearshore bar, which widened as its outer slope remained unchanged.

There are variations in the applicability of the Bruun Rule in differing wave energy environments. On the generally low to moderate wave energy coasts of New Jersey and Maryland, Everts (1985) found that the Bruun Rule overestimated measured beach recession, but on the Pacific coast, where wave energy is greater, erosion was between two and four times greater than predicted. Prediction of the extent of beach recession as sea level rises thus faces a number of difficulties. As long as the sea is rising beach-fringed coastlines will continue to recede as erosion accompanies submergence, and if the sea rises at an increasing rate the beach erosion will accelerate. A century hence most beach-fringed coastlines will have retreated substantially, and (in the absence of protective structures or artificial nourishment) they will be eroding more rapidly. Prediction of the position of the coastline would be possible if the sea level rise were followed by a stillstand, which could occur after the human modifications of the atmosphere that enhance the greenhouse effect were brought under control.

Evidence from past marine transgressions

The prediction that beach erosion will be initiated or accelerated by a rising sea level has to be reconciled with evidence that in the geological past some marine transgressions were accompanied by shoreward drifting of sea floor sediment, and that beaches formed and prograded on coastlines as sea level rise slackened and came to an end. This was the case on many coasts during the Late Quaternary marine transgression, which brought the sea up to, or close to, its present level about 6000 years ago. Some of the beaches that were then formed have since prograded as Holocene beach ridge plains and barriers, although where sea level is still rising (as on the Gulf and Atlantic coasts of the United States), many beaches are now the seaward fringes of barriers migrating intermittently landward.

It should be noted that the Late Quaternary marine transgression advanced across a land surface that had previously emerged from beneath the sea during a marine regression that occurred about 80 000 years ago (p. 38). This surface had been strewn with Pleistocene beach deposits left stranded as the emergence took place and to these were added fluvial and aeolian deposits, as well as glacial and periglacial deposits in high latitudes. There were also unconsolidated materials where shelf rock outcrops had been subaerially weathered during the prolonged low sea level phase. This shoaly topography was the source from which sand and gravel deposits were derived and swept landward to form beaches and barriers, many of which subsequently prograded as the marine transgression came to an end.

The future of beaches

Some existing coastal lowlands have features similar to those of continental shelves in the Late Pleistocene,

notably where there are extensive coastal dunes, but many do not have sediment mantles that would provide beach sediment for a rising sea, and of course many have been developed and urbanized. It is unlikely that many existing beaches will be maintained or prograded during a phase of rising sea level, but as sea level rises across existing coastal plains, especially those with dunes, there may eventually be the formation of beaches by shoreward drifting of sediment up to the new coastline, along the contour on which submergence comes to an end. By then present-day beaches will have been submerged, buried or destroyed, so that the short-term response to a rising sea level will be the onset or acceleration of erosion on existing beaches.

EFFECTS ON COASTAL DUNES

Most coastal dunes lie behind beaches, and as sea level rises, beach erosion will lead to increased backshore cliffing of dunes by storm waves, which will occur more frequently where the climate becomes more boisterous. As coastal dune fringes are cut back, more blowouts will be initiated, and some of these may grow into large transgressive dunes as sand is excavated and blown landward. A rising sea level will thus accentuate the development of transgressive dune formations (Van der Muelen 1990, Carter 1991).

If climatic changes accompanying a sea level rise result in drier and windier conditions, coastal dunes that are at present stable and vegetated may become unstable as the vegetation cover is weakened and sand mobilized, and dunes that are already active will become more mobile. On the other hand a wetter or calmer climate could facilitate dune stabilization, even where the coastal fringe is being cut back.

EFFECTS ON SALT MARSHES, MANGROVES AND INTERTIDAL AREAS

A rising sea level will submerge existing intertidal areas, including sandflats, mudflats, salt marshes and mangroves. As the nearshore water deepens, stronger wave action will initiate or accelerate erosion on microcliffs along the seaward margins of salt marsh and mangrove terraces (Phillips 1986). Tidal creeks which intersect salt marshes and mangrove areas will widen and deepen, and extend headward as they are submerged. As submergence proceeds, the retreat of the seaward margin will be matched by a diachronous transgression of the landward margins of salt marshes or mangroves on to the hinterland at a rate related to the transverse gradient. Where the hinterland is low-lying, new intertidal areas will form to landward: salt marshes or mangroves will migrate to displace fresh-water or terrigenous vegetation communities, or invade backing salt flats. The vegetation zones will thus move landward to maintain their position in relation to the shifting intertidal zone (Figure 12.1C). Some may widen, others become narrower or coalescent, in relation to variations in the transverse profile (Titus 1988). Landward migration of salt marshes and mangroves will be impeded where the hinterland rises steeply, the vegetation zones being compressed (squeezed) as the sea level rises. The salt marsh or mangrove fringe will disappear completely on the many sectors of coast where these communities are backed by a sea wall. Similar changes will occur on coasts bordered by fresh-water swamps, such as the reedswamp bordering parts of the Baltic coast (p. 213).

These changes will be countered in areas where sedimentation continues (or is augmented by climatic changes leading to increased fluvial sediment yields) at a sufficient rate for the depositional terrace to be maintained by vertical accretion as sea level rises (Redfield 1972). The seaward margin will then remain in place, and existing vegetation patterns could persist (Pethick 1981), with variations in species zoning and distribution related to the transverse profile (Reed 1990). There will be landward migration of the inner salt marsh and mangrove communities, providing there are suitable low-lying habitats. The outcome will be a widening of the aggrading salt marsh or mangrove terrace.

There is also a possibility that as sea level rises, some of the sediment stirred from bordering mudflats by strengthening wave action will be washed up into the salt marsh or mangrove fringe. This has happened in the Lagoon of Venice, where the salt marshes have continued to aggrade while their area was reduced by erosion (p. 237). Coasts with wide bordering mudflats, such as the Bay of Saint Michel in France and Bridgwater Bay in western England, may well maintain their salt marshes as the result of shoreward drifting of muddy sediment as the sea level rises, and a similar response could occur where wide mudflats front mangroves, as in the gulfs of northern Australia, but the extent to which this will happen is still a matter for debate. In the Gulf of Papua, Pernetta and Osborne (1988) decided that fluvial deposition on the Purari delta would provide sufficient sediment to maintain the mangrove ecosystem as sea level rose, whereas in the tidally dominated estuaries of the adjacent Kikori coast, where sedimentation is slower, mangroves would

retreat landward and become reduced in area. Similar conclusions have been reached in northern Australia (Woodroffe 1995).

Variations in the growth rates of salt marsh plants, in sedimentation and peat accretion, in tidal regimes and species response, all complicate attempts to predict exactly how salt marshes and mangroves will respond to a rising sea level (Reed 1995). Where intertidal areas are enriched with nutrients derived from eroding sediments or enhanced runoff from the hinterland it is possible that invigorated growth of salt marshes or mangroves could match a sea level rise by building up the depositional terrace with accumulating peat. In the absence of sustaining sediment accretion, it is thought that mangroves could maintain themselves on their accumulating peat with the sea level rising at up to nine centimetres per century, but that they would be impeded by a faster submergence, and collapse when the rise exceeded 12 centimetres per century (Ellison and Stoddart 1991).

Evidence of the response of salt marshes and mangroves to a rising sea level can be found on coasts that are already submerging, as on the Atlantic seaboard of the United States (Bird 1993c). In Chesapeake Bay some salt marsh islands have diminished in area and others have already disappeared, while salt marsh plants are invading backing meadow land (Kearney and Stevenson 1991). Reference has been made to the submergence and erosion of salt marshes on the Essex coast, where land subsidence is also proceeding (p. 208).

Intertidal areas, comprising sandflats, mudflats and rocky shores, will be modified as sea level rises. The outer part of the existing intertidal zone will become permanently submerged, and as backing salt marshes and mangroves are eroded and coastal lowland fringes cut back, areas previously occupied by this vegetation will become mudflats or sandflats, and underlying rocky areas may be exposed. Again, landward migration of this kind will not be possible where sea walls have been built along the coast, so that the area of sandflats, mudflats and rocky shores will be reduced in width, and eventually submerged by the rising sea. Seagrasses and marine algae will tend to migrate shoreward, their inner limits spreading on to the sandy and muddy substrates that form as beaches are submerged and salt marshes and mangroves are cut back, their outer margins dying away as the water becomes too deep. As in salt marsh and mangrove terraces, changes in the area of sandflats, mudflats and rocky shores as sea level rises will depend on the extent to which submergence is offset by continuing sediment accretion. The sediment supply is more likely to be maintained in the vicinity of river mouths and may increase where fluvial sediment yields are augmented by larger or more intensive rainfall and runoff; it could also be maintained where there are shallow sea areas from which sediment can drift shoreward as submergence proceeds.

EFFECTS ON ESTUARIES AND LAGOONS

As sea level rises, estuaries and lagoons will generally widen and deepen, and may transgress inland. Tides will penetrate farther upstream, tide ranges may increase, and there will be changes in patterns of shoal deposition (Shennan and Sproxton 1991). The discharge of river floods will be impeded by the rising sea, so that flooding becomes more extensive and persistent, and a higher proportion of fluvial sediment will be retained within the submerging estuaries, instead of being delivered to the sea floor or adjacent coasts. There may be contrasts within an estuary, as between the windward and leeward shores of Chesapeake Bay (Stevenson and Kearney 1996). However, if accompanying climatic changes increase rainfall in coastal regions, or in the hinterland, some of these changes may be at least partly offset by greater fluvial discharge and sediment supply to the coast.

An analysis of the effects of a sea level rise averaging six millimetres per year in the Blackwater estuary in Essex, using a formula that relates channel width (w) to discharge (Q) ($w = aQ^{0.21}$, where a is a constant) indicated that (in the absence of preventive structures such as shore walls) the augmented discharge would increase channel width between high and low water mean spring tide, by between 400 and 500 metres at the mouth, diminishing to zero 10 kilometres upstream (Pethick 1998). It is difficult to apply this approach in estuaries where channels split around islands.

Coastal lagoons will generally be enlarged and deepened as sea level rises (Figure 12.1D), with submergence and erosion of their shores and fringing swamp areas, and widening and deepening of tidal entrances, increasing the inflow of sea water during rising tides and drought periods. Erosion of enclosing barriers may lead to breaching of new lagoon entrances, and continuing submergence may eventually remove the coastal barriers and reopen the lagoons as marine inlets and embayments. On the other hand, new lagoons may be formed by sea water incursion into low-lying areas behind dune fringes on coastal plains, or where depressions are flooded as the water table rises to form seasonal or permanent lakes and swamps.

As sea level rises, the currents that flow through existing tidal entrances and gaps between barrier islands, as on the Dutch and German North Sea coasts, may be strengthened, augmenting the inflow of water and sediment. In parts of the Wadden Sea increased sediment inflow has been building up intertidal and nearshore sandflats and mudflats, and maintaining vertical accretion on salt marshes, even though coastal subsidence is in progress.

There will be much variation in the nature and extent of changes in coastal lagoons as sea level rises, depending on their existing configuration and dynamics, and the extent to which they have been modified by human activities. The Lagoon of Venice (Figure 9.8), for example, has been maintained partly by continuing subsidence in the north-west Adriatic region, and partly by the diversion of rivers such as the Po and Brenta, which had been carrying sediment into it. Recent changes in the Lagoon of Venice have been documented from successive air photographs, beginning with those taken from Airship Parseval in 1913 (Cavazzoni 1983).

The deepening and enlargement of estuaries and coastal lagoons may be countered where there is a supply of sediment, arriving at a sufficient rate to offset the effects of submergence. The Holocene stratigraphy of the New Jersey coast shows that sea level rise (due to land subsidence) during the past 2500 years was accompanied by the upward growth and expansion of salt marshes so that coastal lagoons backing sandy barrier islands were reduced in area (Psuty 1986). Tropical estuaries may be modified as sediment yields change in response to greater or lesser rainfall and runoff, and variations in catchment vegetation related to climate (Wolankski and Chappell 1996). Rapid sedimentation could prevent the enlargement of coastal lagoons as sea level rises, but submergence will postpone infilling of the residual lagoon basin.

EFFECTS ON DELTAIC COASTS

A rising sea level will cause submergence and erosion of low-lying deltas and coastal plains, especially where there is little or no compensating sediment accretion. As sea level rises, progradation of most deltas will be curbed, and erosion will become more extensive and more rapid. Deltaic coastlines that are already receding because of submergence and erosion will show accelerated retreat, but progradation will continue around the mouths of rivers that continue to supply sufficient sediment to the coast, including those where the sediment yield has been increased as the result of human activities such as deforestation, farming and mining.

Most large deltas already show isostatic subsidence under the weight of accumulating sediment, and submergence of areas no longer being maintained by sedimentation. This is obvious on the Mississippi delta, where former subdelta lobes no longer maintained by sedimentation have subsided beneath the sea (Figure 10.2). Coastal subsidence has curtailed the seaward growth of large deltas such as those of the Rhine, Guadalquivir, Colorado and Amazon, and a rising sea level will increase the effects of subsidence, extending submergence and intensifying erosion.

The extent to which these effects will be countered, if sea level rise is accompanied by increasing water and sediment yields from rivers, depends on changes within the river catchments. Deposition of delta sediments will accelerate, if rainfall and runoff from the river catchments increases, as a consequence of climatic changes associated with global warming. It has been estimated that the sediment yield from Javanese rivers will increase by up to 43 per cent as the result of greater runoff from their catchments during the coming century (Parry et al. 1991). Under these circumstances sedimentation may build some deltas upward to maintain their area as sea level rises, and sedimentation may continue to aggrade natural levees alongside river channels and extend them seaward. Submerging deltas that are now lobate or arcuate in form may become digitate, like the modern Mississippi delta, with intervening areas of submerging swamp and deepening sea. It should be noted that an increase in water and sediment yields from rivers because of greater runoff could be partly offset if warmer and wetter conditions increase the extent and luxuriance of catchment vegetation

EFFECTS ON CORAL AND ALGAL REEFS

Corals and algae on the surface of intertidal reef platforms will be stimulated by a rising sea level, leading to a revival of upward reef growth (Figure 12.1E), initiated by the expansion and dispersal of presently sparse and scattered living corals on reef platforms. If global sea level has indeed been rising during the past century at 1.0 to 1.2 millimetres per year, and the present sea level is therefore 10 to 12 centimetres higher than it was in the 1890s (p. 45), it could be expected that coral reefs would be among the first features to show indications of such a sea level rise, but there have not yet been reports of the general spread of living corals or

accelerating coral growth. Few reefs have been mapped and monitored with sufficient accuracy for such a change to be measured, but there is likely to be a widespread revival of coral growth on reef platforms as sea level continues to rise. That reef platforms are quickly responsive to sea level changes is shown by the renewed upward growth of coral on reefs in Houtman Abrolhos, Western Australia, as a result of a relative rise of sea level.

Revival of coral growth will be strongly influenced by ecological factors which influence the ability of coral species to recolonize submerging reef platforms (Stoddart 1990). Global warming will modify the distribution of corals by increasing sea temperature and salinity in areas that become more arid. Recent reports of widespread coral bleaching (Brown 1990) may be a consequence of higher temperatures in tropical seas during the El Niño events in 1982–83, 1987 and 1996–98, when coral mortality was extensive on Pacific Ocean reefs. This suggests that increasing sea temperatures are likely to impede coral growth and reef aggradation.

Climatic changes that result in higher rainfall and greater discharges of fresh water will also impede coral growth, especially where runoff from the hinterland produces more extensive turbidity in coastal waters around coral reefs.

The response of coral reefs to a rising sea level will depend on the rate at which the sea rises. A slowly rising sea should stimulate the revival of coral growth on reef platforms, but an accelerating sea level rise may lead to the drowning and death of some corals, and eventually the submergence of inert reef formations. Neumann and Macintyre (1985) pictured the relationship between rates of sea level rise and rates of upward reef growth in terms of keep-up, catch-up, and give-up reefs, noting that there will be variations in response related to ecological composition. As measurements of mean upward growth rates of existing coral reefs are up to one centimetre per year (p. 264) it is likely that reef formations will grow upward to match sea level rise as long as it is within this range. Theoretically, coral reefs could also expand northward and southward beyond their present latitudinal limits as sea temperatures rise, but this is likely to be a slow response because of impeding factors resulting from human impacts. On the other hand, artificial reefs may be nurtured to enhance the role of corals in absorbing carbon dioxide to offset the greenhouse effect.

A slackening sea level rise could permit some surviving corals to grow up, and others to recolonize as the reef shallows. Reefs submerged during a period of rising sea level could catch up if the rate of submergence then diminished, and especially if a new stillstand were to ensue.

The impacts of a sea level rise on existing coral reefs can also be considered in terms of stratigraphic evidence of what happened to coral reefs during the Late Quaternary marine transgression. It should be borne in mind, however, that the ecology and geomorphology of existing reef formations differ in various ways from those when Late Quaternary reef revival commenced on submerging coastlines and pre-existing dissected reef limestones, and a much cooler climate had started to ameliorate. The most notable differences are the impacts of human activities on existing coral reefs during the past few centuries, which may have made the world's coral reefs much less capable than they were under the natural conditions of the Late Quaternary to respond to a sea level rise (Yap 1989).

Spencer (1995) reviewed the literature on coral reef growth and found that reefs were likely to keep up with a sea level rise of less than about one centimetre per year, to be growing upward by one to two centimetres per year, and to be drowned when the sea level rise exceeded two centimetres per year. However, the actual response will vary with accompanying changes in sea temperature and salinity, the incidence and effects of tropical cyclones and biophysical constraints (including human impacts). There will also be variations depending on the nature of existing reefs, coral gardens of Caribbean type being more easily maintained as sea level rises, than solid Indo–Pacific reef platforms, which may grow upward as coral gardens rather than solid structures.

Measurements can be made on reefs that are subsiding at known rates. On tilting coral reefs and atolls, such as Uvéa in the Loyalty Islands the submerged portion shows rapid, if patchy, upward growth of corals, with an inner zone of slow submergence where upward growth is being maintained, passing laterally to a central zone of moderate submergence where the corals are growing, but are failing to keep up with the rising sea, and an outer zone of more rapid submergence where the corals have died and the reefs are drowned (Figure 12.2). The inner submerging portion of the Uvéa atoll shows keep-up growth of coral gardens, rich in corals and accumulating sediment, but forming fragile structures rather than a solid reef platform. Keep-up reefs are thus likely to be coral garden structures that cannot be walked over. As existing Pacific and Indian Ocean reef platforms are submerged by the rising sea, reviving coral growth is likely to form coral gardens similar to those found in the Caribbean, or in shallow water to the lee of existing reefs, rather than consolidated reef platforms, which require a sea level stillstand for their completion, or even a phase of growth to a higher Holocene sea level followed by down-cutting and planation of the reef limestone.

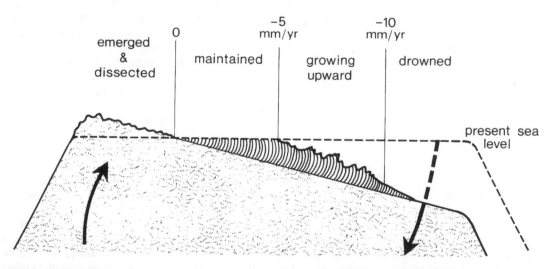

Figure 12.2   Tilting reefs show the response of corals to a sea level rise, but there is much variation in limiting rates of sea level rise (Spencer 1995). This example is based on Uvéa, in the Loyalty Islands, New Caledonia, where coral growth of up to 10 mm/yr was measured on a reef that has been tilting. Observations here indicate that where sea level rises at up to 5 mm/yr corals are likely to maintain the reef, where the rise is between 5 and 10 mm/yr they will continue to grow upward but will fail to maintain the reef, and where the rise exceeds 10 mm/yr they will be drowned.

Islands on coral reefs will be modified by a rising sea level. Many cays and low wooded islands will be eroded by larger waves approaching through deepening waters, and may disappear, overwashed by storm surges, especially if there is an increasing frequency and severity of tropical cyclones as sea and atmospheric temperatures rise. On the other hand, there is a possibility that where submergence is slow enough reviving coral growth on the surrounding reef platforms will at least partly offset erosion of cay shores by impeding wave attack. Low islands may even be enlarged by accretion of coralline material derived by stronger wave action from the growing reef gardens.

Appendix

Classification of Coastal Landforms

TYPES OF CLASSIFICATION

Classification is the grouping of similar features in classes that are contrasted with other dissimilar features. A simple classification is implicit in the topics identified in chapter headings and sub-headings used in textbooks such as this one, as a basis for discussion and explanation. Ideally a geomorphological classification should elucidate relationships between landforms and processes, and lead towards an explanation of coastal evolution and coastline changes.

Attempts to classify coastal landforms (including shores and shoreline features) have not been entirely successful. This is partly because of a widespread belief in the superiority of genetic classifications, based on the origin of the landforms, over purely descriptive classifications of landforms as observed: cliffed coasts, delta coasts, mangrove coasts, and so on. The difficulty is that a genetic classification can only be applied satisfactorily when the mode of origin of coastal landforms is known, but use of particular types or associations of landforms as indicators of particular modes of origin has led to errors. Schwartz (1971) observed that some landforms, such as barrier islands, can evolve in more than one way, an example of multicausality that has also been found to apply to such features as cuspate forelands, foredunes, and shore platforms. Examples of various kinds of classification follow.

ATLANTIC AND PACIFIC TYPE COASTS

An early coastal classification was put forward by Suess (1906), who distinguished Atlantic coasts, which run across the general trend of geological structures, from Pacific coasts, which run parallel to structural trends. The former are characteristic of the Atlantic shores of Britain and Europe; the latter of the Pacific coasts of North and South America. The Atlantic type is also exemplified by the north–west coast of Australia

between Broome and Darwin, and the Pacific type by the coast of New South Wales, which is parallel to the north–south strike of the rocks of the Eastern Highlands. Cliffed coastlines that transgress geological structures are sometimes termed discordant, whereas those that follow the strike of a particular geological formation are termed concordant.

CLASSIFICATION AND PLATE TECTONICS

An updated version of the Suess scheme was proposed by Inman and Nordstrom (1971), who devised a geophysical classification based on plate tectonics, recognizing that the earth's crust is a pattern of plates separated by zones of spreading and zones of convergence, the plate margins moving at rates of up to 15 centimetres per year. They contrasted subduction coasts where one plate is passing beneath another, with trailing-edge coasts on a diverging plate margin and marginal sea coasts on the lee side of island arcs, and described characteristic features of each of these (Table A.1). It was a broad-scale classification, dealing with first-order (continental) features (ca. 1000 km long × 100 km wide × 10 km high) whereas coastal geomorphologists are primarily concerned with second order features (ca. 100 km long × 10 km wide × 1 km high) such as major embayments or deltas, and smaller shore features of third (10 km × 1 km × 100 m), fourth (1 km × 100 m × 10 m) or even higher order. Current ripples on sandy shores can be seventh order features (1 m × 10 cm × 1 cm).

COASTS OF SUBMERGENCE AND EMERGENCE

In 1899 Gulliver distinguished coasts formed by submergence from coasts formed by emergence. This was

Table A.1 Coastal Classification based on Inman and Nordstrom (1971).

1. *Subduction coasts* (plate-edge coasts) occur where oceanic plates are subsiding beneath the edges of continental plates, which are buckled and uplifted as mountain chains as around the Pacific Basin and the Mediterranean. There are narrow continental shelves, frequent earthquakes and active vulcanicity. The coasts are typically steep, with geological structures parallel to the coast (cf. Suess' Pacific coasts). Subduction coasts also occur along island arcs, where a dense oceanic plate is subsiding beneath a less dense continental plate some distance offshore, producing a deep linear trench, often with associated volcanoes, as in the Aleutian Islands.
2. *Trailing edge coasts* (plate-embedded coasts) on the sides of continents that are moving away from the rising and spreading mid-ocean zones. They are backed by plateaux or hilly country, are tectonically more stable than collision coasts, and often have transverse geological structures (cf. Suess' Atlantic coasts). They can be divided into:
    a. Neo-trailing edge coasts where new rifts have recently opened and separation is so far small. Typically these are steep coasts with little or no continental shelf, as on the coasts bordering the Gulf of California and the Red Sea.
    b. Afro-trailing edge coasts where both margins are trailing. Predominantly plateaux and plains, with relatively narrow continental shelves as in Greenland and Africa.
    c. Amero-trailing edge coasts, where the other side of the continent has a collision coast. Typically broad coastal plains and wide continental shelves, rather stable, without active volcanoes, e.g. the east coast of the United States.
3. *Marginal sea coasts* on the protected side of island arcs, e.g. east Asian coasts.

developed into a genetic classification by Johnson (1919), who described coastlines (he used the American term shorelines) of submergence, coastlines of emergence, neutral coastlines (with forms due neither to submergence nor emergence, but to deposition, e.g. delta coastlines, alluvial plain coastlines, glacial outwash coastlines and volcanic coastlines) and compound coastlines (with an origin combining two or more of the preceding categories). Most coasts fall into the compound category, because they show evidence of both emergence, following high sea levels in interglacial phases of the Pleistocene, and submergence, due to the Late Quaternary marine transgression. Because of this recent marine transgression, coastlines of submergence predominate, but examples of emerged coastlines are found in regions that are rising because of isostatic recovery following deglaciation, as in Scandinavia or in regions subject to uplift associated with tectonic deformation, as in parts of Indonesia and New Guinea. A classification could be made of coastlines having had similar histories of changing relative sea level in Quaternary times.

CLASSIFICATION BASED ON CLIMATE

In 1936 Aufrère proposed a coastal classification based on climate. Categories include coasts with a permanent ice cover (no marine processes), coasts with a seasonal ice cover (seasonal marine processes and abundant sediment from glacial sources), temperate humid coasts (as in Europe), tropical humid coasts (with abundant fluvial sediment in deltas and coastal plains), arid coasts (without rivers; marine sediments dominant) and semi-arid coasts (some river features; sebkhas). Cold coasts are a distinct category, subject to the effects of ice and snow and to freezing and thawing of rocks, beaches and the nearshore sea area. They pass in lower latitudes to the periglacial zone, which extends to Patagonia, northern Scandinavia, north and north-east Russia, Alaska and eastern Canada and the sub-Antarctic islands.

The global distribution of coastal climates (Bird 1997b) shows sector variations related to latitude and wind regime. In general, lateral transitions are gradual, but there are some rapid transitions from humid tropical to arid within comparatively short distances in Ecuador and Colombia, in West Africa and northern Madagascar.

Elucidation of the climatic factor in coastal geomorphology requires comparative studies of coastal sectors that are similar geologically, but differ in climate. Reference has been made (p. 89) to contrasts in the features of limestone coasts in tropical, temperate and cold environments (Guilcher 1958), and on the eastern seaboard of Australia the humid tropical sector in north Queensland shows distinctive features related to its climatic environment differing from those seen on other parts of the Australian coast (Bird and Hopley 1969). On the other hand, comparisons of coasts of similar aspect in the same latitude (e.g. the Uruguay–southern Brazil, South African and south-east Australian coasts between 28°S and 35°S) may show contrasts unrelated to climate, or reveal contrasts in climatic history.

CLASSIFICATION BASED ON COASTAL PROCESSES

Various attempts have been made to describe and explain the pattern of coastal processes effective around the world's coastline. Variations in wave climate were discussed by Davies (1980), who distinguished swell and storm wave environments, trade and monsoon influences and tropical cyclone influences, produced world maps of these, and discussed the distribution of high, moderate and low wave energy coasts (p. 16). He also presented maps of the global distribution of tidal types (semidiurnal, mixed and diurnal) and, perhaps more usefully in terms of geomorphology, tide ranges divided into microtidal, mesotidal and macrotidal (to which may be added megatidal) categories (p. 18).

Hansom (1988) distinguished wave-dominated from tide-dominated coasts, and added a third (more doubtful) category of wind-dominated coasts.

INITIAL AND SUBSEQUENT COASTS

A distinction can be made between initial and subsequent coast forms, initial forms being those that existed when the present relative levels of land and sea were established and marine processes began work (on most coasts about 6000 years ago) and subsequent forms that have since developed as the result of marine action (Gulliver 1899). Shepard devised a classification on these lines, making a distinction between primary coasts shaped largely by non-marine agencies and secondary coasts that owe their present form to marine action: his 1976 version is given in Table A.2. It was essentially a genetic classification, with descriptive detail inserted to clarify the subdivisions, and it recognized that, because of the world-wide Late Quaternary marine transgression, the sea has not long been at its present level relative to the land, so that many coasts have been little modified by marine processes. Snead (1982) published a collection of 247 annotated photographs illustrating the various categories of Shepard's classification.

Shepard's aim was to devise a classification that would prove useful in diagnosing the origin and history of coastlines from a study of charts and air photographs, but it is dangerous to assume that the origin and history of a coast can be deduced from such evidence without field investigation. A straight coast (p. 2) may be produced by deposition, faulting or emergence of a featureless sea floor or submergence

of a coastal plain; an indented coast by submergence of an undulating or dissected land margin, emergence of an irregular sea floor, differential marine erosion of hard and soft outcrops along the coast or transverse tectonic deformation (folding and faulting) of the land margin. It is doubtful whether configuration is a reliable indication of coastal evolution.

STABLE AND MOBILE COASTS

Cotton (1952) made a distinction between coasts of stable and mobile regions, stable regions being those that have escaped the Quaternary tectonic movements that have affected mobile regions, especially around the Pacific rim, where they still continue. On the coasts of stable regions he separated those dominated by features produced by Late Quaternary marine submergence from those dominated by (mainly Pleistocene) features inherited from earlier emergence. On the coasts of mobile regions he separated those where the effects of Late Quaternary marine submergence have not been counteracted by recent uplift of the land from those where recent uplift of the land has led to emergence.

STEEP AND FLAT COASTS

De Martonne (1909) used a morphological distinction between steep and flat coasts as a basis for classification, suggesting a number of sub-types, some descriptive (estuary coasts, skerry coasts), others genetic (fault coasts, glacially-sculptured coasts). Ottmann (1965) followed a similar approach, recognizing five classes (A to D rocky coasts, and E coasts that have advanced as the result of deposition) with some subdivisions (Table A.3).

GEOLOGY IN COASTAL CLASSIFICATION

Classification of rocky coasts on the basis of geology and structure was advocated by Russell (1967). In the course of his very extensive coastal travels he had noted a striking similarity in features developed on crystalline rocks, irrespective of their climatic and ecological environments: granites that outcrop on parts of the coasts of Scandinavia, south-west Australia, South Africa and Brazil all show similar domed surfaces related to large-scale spalling and conspicuous joint-control. Limestones (including chalk and coral), basalts and sandstones each show distinctive kinds of coastal

Table A.2   Classification of Coastal Landforms (based on Shepard 1976).

I. PRIMARY COASTS, with configuration resulting from non-marine processes:

A. *Land erosion coasts*, shaped by erosion of the land surface and subsequently drowned by sea-level rise or sinking of the land:

1. Ria coasts (drowned river valley mouths): relatively shallow estuaries usually with V-shaped cross-sections, deepening seaward – Chesapeake Bay (USA), Carrick Roads (Cornwall) and the Galician ria coast (NW Spain). May be dendritic (branching river valleys in homogeneous or horizontal strata) or trellised (river valleys incised into inclined beds of unequal resistance).

2. Drowned glaciated valley mouths: deeply indented with many islands; including fiords and glacial troughs, deep (often > 100 metres), with U-shaped cross-sections but shallower towards the mouths – fiord coasts in Norway and on the SW coast of South Island, New Zealand, Gulf of St Lawrence.

3. Drowned karst topography: embayments with oval depressions representing sinkholes, common in limestone area – Dalmatian coast, Yugoslavia.

B. *Subaerial deposition coasts*, prograded by land-derived deposits:

1. River deposition coasts:
   a. Deltaic coasts: usually in the form of one or more lobes protruding into the sea. May be digitate (Mississippi), lobate (Rhône) or cuspate (Tiber).
   b. Compound delta coasts: coastal plain formed by confluent deltas – as on the north coast of Alaska from Point Barrow to the Mackenzie delta.
   c. Compound alluvial fan coasts: at base of mountains, usually straightened by marine erosion – east coast of South Island, New Zealand.

2. Glacial deposition coasts:
   a. Partially submerged moraines, hummocky topography, usually straightened by marine deposition and/or erosion – parts of Long Island, New York, USA.
   b. Partially submerged drumlins, usually with oval islands, elongated in direction of ice movement – Boston Bay, Massachusetts; Clew Bay, western Eire.
   c. Glacial outwash plain coasts, where there has been glacifluvial deposition in front of large glaciers – seaward margin of Malaspina Glacier, Alaska.

3. Wind deposition coasts:
   a. Dune prograded coast: where dunes have spilled across the shore: rare and localized – part of south side of San Miguel Island, California, and the southern shores of Corner Inlet, Victoria.
   b. Dune coasts: where dunes back a beach – Coos Bay, Oregon; Ninety Mile Beach, Australia.
   c. Fossil dune coasts: where consolidated dunes (e.g. dune calcarenites) form coastal cliffs – on the Eyre Peninsula in South Australia.

4. Landslide coasts: where masses fallen from a cliffed coast project into the sea – Humbug Mountain, Oregon; Black Ven in Lyme Bay, Dorset.

C. *Volcanic coasts*
   1. Lava-flow coasts: lobate protrusions of lava into the sea, usually with submarine slopes continuing steeply seaward – south coast of Hawaii and many other volcanic islands.
   2. Tephra coasts: where deposition of fragmental volcanic material has formed roughly convex prograded coasts – San Benedicto Island west of Mexico; Surtsey off Iceland.
   3. Volcanic collapse or explosion coasts: concave bays on a volcano – Hanauma Bay, Hawaii; Krakatau, Indonesia.

D. *Coasts shaped by tectonic movements*
   1. Fault coasts: straight or gently arcuate fault scarps forming the land margin – NW San Clemente Island, California.
   2. Fold coasts: where the coast has been shaped by recent folding – Wellington district, New Zealand.
   3. Sedimentary extrusions:
      a. Salt domes: oval-shaped islands – southern Arabian Gulf.
      b. Mudlumps: small, short-lived islets resulting from upthrust of mud off river mouths – Mississippi delta.

E. *Ice coasts*: where glacier fronts extend into the sea: much of the Antarctic coast and locally in Greenland.

II. SECONDARY COASTS, with configuration resulting mainly from marine agencies or marine organisms:

A. *Wave erosion coasts*:
   1. Wave-straightened cliffs: contrast with fault coast in having gently sloping sea floors and shore platforms – Atlantic shores of Cape Cod, chalk coasts bordering the English Channel.
      a. Cut in homogeneous materials – the Sussex chalk coast.
      b. Hogback strike coasts: where hard layers of folded rocks strike parallel to the coast so that erosion forms a straight coastline – Lulworth coast, Dorset.
      c. Fault line coast: formed by erosion along a fault line that has juxtaposed harder and softer formations – Polurrian Cove, Cornwall, cliff along line of Lizard Fault (Bird 1998, Figure 52).
      d. Elevated shore platform coasts: seaward edge of emerged shore platform – Montague Island, Alaska (emerged during 1964 earthquake)
      e. Submerged shore platform coasts: backed by plunging cliffs – Jervis Bay, south-eastern Australia.
   2. Made irregular by wave erosion – differ from ria coasts (drowned river valleys) in that the embayments do not extend deeply into the land, and are not necessarily at river mouths:

a. dip coasts: where alternating hard and soft rock out-crops intersect the coast at an angle, and the softer sectors have been cut out as coves or bays – east and west coasts of the Isle of Wight.

b. heterogeneous formation coasts: where wave action has cut back the weaker sectors, forming irregularities – Lizard peninsula, south-west England.

c. joint-guided coasts – cliffed coast of Port Campbell, Australia.

B. *Marine deposition coasts*: coasts prograded by wave or current deposition:

1. Barrier coasts – separated from mainland by lagoons or swamps: Gulf and Atlantic coasts, USA.
   a. Barrier beaches: single ridge – Chesil Beach, Dorset, England.
   b. Barrier islands: multiple ridges, dunes and overwash flats – Padre Island, Texas.
   c. Barrier spits: connected to the mainland – Orfordness, East Anglia.
   d. Bay barriers: barriers across the mouths of bays – Slapton Ley, Devon.
   e. Overwash fans: lagoonward extensions of barriers deposited by storm surges – shingle fans in The Fleet lagoon, behind Chesil Beach.

2. Cuspate forelands: large projecting sand (or shingle) points with or without enclosed lagoons – Cape Hatteras and Dungeness.

3. Beach ridge plains: differ from barriers in having no backing lagoon – Disaster Bay, New South Wales.

4. Mud flats or salt marshes: on deltaic or other low coasts where very gentle offshore slope stops large waves from breaking near the coast – Bridgwater Bay, Somerset.

C. *Coasts built by organisms*

1. Coral reef coasts: formed by corals and algae, mainly in the tropics, with beaches later built above sea level by storm waves:
   a. Fringing reef coasts: as in south-east Papua.
   b. Barrier reef coasts: separated from the mainland coast by a lagoon – Great Barrier Reef, Australia.
   c. Atolls: coral reefs enclosing a lagoon.
   d. Elevated reef coasts: where fringing reefs have been uplifted to form one or more coastal terraces – Huon peninsula, New Guinea.

2. Serpulid reef coasts: small sectors of beach coast built out by the cementing of serpulid worm tubes on to rocks or beaches – found in Bermuda.

3. Oyster reef coasts: built by oyster shells, sometimes with storm-piled beaches of shells – locally on the shores of Atchafalaya Bay, Gulf Coast USA.

4. Mangrove coasts: inlets, estuaries and sheltered embayments, mainly in the tropics, where the shore is fringed by mangroves – south-west Florida, Port Darwin, Australia.

5. Salt marsh coasts: inlets, estuaries and sheltered embayments, mainly in the temperate zone, where the shore is fringed by salt marshes – Chesapeake Bay, USA; Bay of St Michel, France. See also B4 (above).

6. Freshwater swamp coasts: inlets, estuaries and sheltered embayments bordering sea areas of low salinity – Baltic coasts of Sweden, Finland and Estonia.

D. *Artificial or man-made coasts*
   Extensive in the Netherlands, where reclaimed lowlands are bordered by lengthy sea walls along the coast, and in Hong Kong, Singapore, Malaysia and the head of Tokyo Bay.

---

landforms, with contrasts related to structure, wave energy, and weathering regimes, while unconsolidated formations show another range of characteristic landform types. Bedrock coasts are commoner in cold, arid, and temperate regions than in the humid tropics, where depositional mantles and aprons are extensive, and the limited sectors of coastal rock outcrop are often profoundly weathered.

ADVANCING AND RECEDING COASTS

A coastline may advance because of coastal emergence and/or progradation by deposition, or retreat because of coastal submergence and/or retrogradation by erosion (Figure 1.2). Valentin (1952) used this distinction as the basis for a system of coastal classification that could be shown on a world map (scale 1 : 50 000 000) of coastal configurations, a classification that was partly genetic and partly descriptive. Examples of each type of coast shown on Valentin's world map are given in Table A.4.

Valentin's classification took account of changes in the relative levels of land and sea, and was based on observed evidence of the gain or loss of land. He also considered changes in progress, which can be expressed as an interaction of vertical movements (emergence and submergence) and horizontal movements (erosion and deposition), shown diagrammatically in Figure 1.2. On this diagram the line ZOZ′ indicates coasts that are neither advancing nor retreating, either because emergence is balanced by erosion (ZO) or because submergence is offset by deposition (OZ′): the point O represents a static coast on which no changes of any kind are taking place. Changes are most marked towards A, where emergence accompanied by deposition leads to rapid advance of the coast, and towards R, where erosion accompanying submergence results in

Table A.3   Coastal Classification based on Ottmann (1965).

A   Steep coasts which plunge to oceanic depths – French Alpes-Maritimes.
   A1 – terraced or faulted steep coasts – California; Algeria.
   A2 – steep coasts of volcanic islands and (emerged) atolls – Krakatau, Indonesia.

B   Vertical cliffs with shore platforms at sea level – Normandy.

C   Steep coasts which plunge to submerged platforms – Marseille.
   C1 – with steep-sided inlets – calanques of Provence.
   C2 – with fiords – western Patagonia.

D   Partially submerged coasts without notable cliffing – parts of Brittany:
   D1 – ria coasts – south-east Corsica.
   D2 – dissected coasts – Dalmatian coast.
   D3 – glacial drift coasts – parts of Denmark; Puget Sound, north-western USA.

E   Low depositional coasts behind gently-shelving sea floors:
   E1 – estuaries – Loire.
   E2 – deltas – Orinoco.
   E3 – marshlands and mangroves – Amazon.
   E4 – dunes – Les Landes, south-west France.
   E5 – with barrier reefs – north-east Brazil.

Table A.4   Classification based on Valentin (1952).

A   Coasts that have advanced:
   1. due to emergence:
      a. emerged sea-floor coasts – the southern shores of Hudson Bay, Canada, where the land is still emerging as the result of isostatic recovery following deglaciation of the Canadian Shield.

   2. due to organic deposition:
      b. phytogenic (formed by vegetation): mangrove coasts – in sheltered embayments especially in the tropics, e.g. Gulf of Papua.
      c. zoogenic (formed by fauna): coral coasts – fringing reefs on the Queensland coast, bordered by the Great Barrier Reef.

   3. due to inorganic deposition:
      d. marine deposition where tides are weak: lagoon barrier and dune-ridge coasts – SE coast of Australia and the Gulf and Atlantic coasts of the United States.
      e. marine deposition where tides are strong: tideflat and barrier-island coasts – Dutch, German and Danish North Sea coasts.
      f. fluvial deposition: delta coasts, such as the Mississippi and the Nile, and the confluent deltas of northern Java.

B   Coasts that have retreated:
   1. due to submergence of glaciated landforms:
      g. confined glacial erosion: fiord-skerry coasts – Norway and British Columbia; the west coast of South Island, New Zealand has fiords, but skerries (fringing islands) are rare. Port Davey, in Tasmania, shown as a fiord by Valentin, is actually a ria (drowned river valley).
      h. unconfined glacial erosion: fiard-skerry coasts – Baltic shores of Sweden and Finland and the Arctic coasts of Canada.
      i. glacial deposition morainic coasts – Baltic Germany and Poland.

   2. due to submergence of fluvially-eroded landforms:
      j. on young fold structures: embayed upland coasts – the west coast of the United States, Chile, and around the Mediterranean, notably in Greece and Turkey.
      k. on old fold structures: ria coasts – SW Ireland, SW England, Brittany, and Galicia (NW Spain).
      l. on horizontal structures: embayed plateau coasts – bordering the Red Sea, the west coast of India, and the Argentine coast.

   3. due to marine erosion:
      m. cliffed coasts – extensive on the coasts bordering the English Channel and the North Sea, and along the south and south-east coasts of Australia.

rapid retreat. Rapid erosion may lead to recession of an emerging coast (X), while rapid deposition may prograde a coast that is being submerged (Y).

Bloom (1965) elaborated Valentin's scheme by adding a time-axis passing through O. On the three-dimensional diagram thus produced, it is possible to portray the course of evolution of a particular coast where relationships between emergence and submergence, erosion and deposition, have varied through time. Bloom gave an example from the Connecticut coast, where radiocarbon dates from buried peat horizons have yielded a chronology of relative changes of land and sea level in Holocene times. At some stages the sea gained on the land during submergence, even though deposition continued; at other stages deposition was sufficiently rapid to prograde the land during continuing submergence. Now there is widespread erosion on the seaward margins of salt marshes, possibly because of resumed submergence. Each condition gives different points on Valentin's diagram, and Bloom's three dimensional scheme offers a means of portraying coastal evolution as a wavy line drawn from a selected origin on the time axis and linking successive positions on the intersected planes.

Valentin attempted to measure present-day emergence and submergence from tide gauge records, accepting the evidence for a world-wide eustatic rise of sea level in progress at a rate of about one to two millimetres per year. He considered that where mean tide levels are rising more rapidly than this, submergence is accelerated by tectonic subsidence; where they are rising less rapidly, and where they are falling, submergence is partly offset by tectonic uplift and the coast is emerging. Tide gauge records are not available for many sections of coast, and this kind of analysis remains somewhat speculative. Valentin's map showing present-day coastal evolution distinguished between emerging, prograding, submerging, and retrograding coasts, and gave figures for changes of level where possible.

Valentin's scheme was an advance on earlier classifications made in terms of theoretical marine cycles, initiated by submergence or emergence as propounded by Johnson in terms of the Davisian cycle concept and reproduced in early textbooks. The cyclic hypothesis has outlived its usefulness (even as a teaching device) and can be seen in retrospect to have hindered the progress of coastal research. The essential idea was that a coast formed by submergence, initially indented, would pass through stages of youth as promontories were cut back and sediment derived from them built laterally into spits and barriers enclosing embayments, sub-maturity as the barriers retreated and coastal erosion drove back the headlands, and maturity when all traces of the original indentations were destroyed, and the barriers came to rest as beaches along a smoothed, cliffed coast. It is quite possible that some coasts have evolved in this way, but the hypothetical sequence cannot be accepted as normal or typical. The objections are that newly-submerged coasts are not necessarily indented; that crenulation by marine erosion can be a prolonged phase, not necessarily followed by this smoothing and simplification of coastal outlines; that where spits and barriers have formed the sediment may not have been derived from erosion of intervening cliffed headlands; that spits and barriers, formed on a submerged coast, are not always driven landward, but will prograde seaward if sediment supply is abundant; that the theoretical scheme is not cyclic, returning to an initial form, but a postulated sequence, the general validity of which has never been demonstrated; and that such idealized theoretical schemes of coastal evolution prompt facile interpretation of observed coastal features, which may appear to fit the theoretical sequence, and are accepted as evidence of it without proper investigation of their

geomorphological evolution. Similar objections apply to the hypothetical sequence of evolution of coasts following emergence, and neither scheme can be applied unless the coast has become stable, following submergence or emergence.

The advantage of non-cyclic classifications, such as those proposed by Valentin and Bloom, is that they pose problems and stimulate further research instead of fitting observed features into presupposed evolutionary sequences. The liberation of coastal geomorphology from premature assumptions of cyclic development was taken a stage further by McGill (1958) in compiling a map of the coastal landforms of the world.

COMPOSITE CLASSIFICATIONS

McGill's map (scale 1 : 25 000 000), published by the American Geographical Society in 1958, is an example of a composite classification. It showed the major landforms of the coastal fringe, 5–10 miles (8–16 kilometres) wide, with selected shore features and certain other information relevant to analysis of coastal evolution. Major coastal landforms were classified in terms of their lowland or upland hinterlands, giving the categories listed in Table A.5. This information is supplemented by indications of selected shore features, constructional or destructional, in the backshore, foreshore, and offshore zones, classified by the agent responsible: sea, wind, coral, or vegetation. This enables such features as backshore dunes, tidal flats, sedimentary and coral barriers, and 'tidal woodlands' (usually mangrove swamps) to be indicated, and some variations in coastal ecology to be addressed. Finally, the map showed such relevant features as the limits of present-day glaciation and the maximum extent of ice during the Last Glacial phase; the approximate limit of permafrost; the average limit of polar sea ice, with spring maximum and autumn minimum of pack ice; the approximate limit of reef coral (taken as the 20°C isotherm of sea water in the coldest month); tide ranges larger than 10 feet (about three metres), and larger than 20 feet (about six metres), both generally and where localized in embayments; certain characteristics of coastal embayments, their form and orientation to the coastal margin; coastal areas affected by isostatic recovery following deglaciation; and the location of karst topography on limestone coasts.

McGill's map provided a useful aid to research, prompting further inquiry and comparative study of coastal landforms as a means of isolating and measur-

Table A.5   Coastal Classification by McGill (1958).

I. Lowlands:
  A. Constructional (depositional) plains:
    1. dominantly flat-layered structure:
      a. existing ice
      b. plains of glacial deposition
      c. plains of fluvial deposition (including deltas)
      d. dune plains
      e. coral flats
      f. lava plains

  B. Destructional (erosional) plains:
    1. flat-layered structure:
      a. sedimentary plains formed by glaciation
      b. sedimentary plains formed by fluvial erosion
      c. volcanic plains formed by glaciation
      d. volcanic plains formed by fluvial erosion
    2. complex structure:
      a. formed by glaciation
      b. formed by fluvial erosion

II. Uplands:
  A. Plateaux:
    1. flat-layered structure:
      a. glaciated sedimentary plateau
      b. fluvially-eroded sedimentary plateau
      c. glaciated volcanic plateau
      d. fluvially-eroded volcanic plateau

  B. Hills:
    1. flat-layered structure:
      a. glaciated sedimentary plateau
      b. fluvially-eroded sedimentary plateau
      c. glaciated volcanic plateau
      d. fluvially-eroded volcanic plateau
    2. complex structure:
      a. formed by glaciation
      b. formed by fluvial erosion

  C. Mountains: subdivisions as for hills

  D. Constructional (depositional) uplands:
      a. ice plateau
      b. dune hills
      c. elevated coral flats
      d. lava plateau
      e. volcanoes (active and inactive).

ing the factors at work. It stimulated further investigation of coastal landforms, and the making of more detailed morphological maps of coastal features. The American Geographical Society also published 1:50 000 000 maps of the world showing coastal climates (Bailey 1958) and coastal vegetation (Axelrod 1958), which give a valuable background to the analysis of coastal geomorphology, and these could also be developed on a local scale for research on particular coasts.

Examples of local classification include Alexander's (1966) use of symbols for detailed mapping of the coast, illustrated with reference to Tanzania. Wiseman et al. (1973) divided the Alaskan Arctic coasts into 157 eight-kilometre segments for morphometric analysis, and by grouping segments with similar landform suites generated 29 provinces ranging from 15.6 to 159.6 kilometres in coastline length and separated by well-defined boundaries. This concept is analogous to the land systems approach in terrain studies, and can be applied hierarchically, with recurrent associations of minor compartments grouped into longer coastal sectors.

CHANGING COASTLINES

Classification of coastlines should take account of the fact that changes are occurring at varying rates. Some changes are cyclic (e.g. in response to alternations of stormy and calm weather); others gradual, and others sudden; while some coastlines have remained stable over periods of up to about 6000 years (e.g. hard rock cliffs that have not been modified since the Late Quaternary marine transgression brought the sea up to about its present level, relative to the land).

Steers (1964) gave examples of coastline changes mapped over specified periods on several sectors of the coastline of England and Wales, and similar evidence from the United States coasts was illustrated by Shepard and Wanless (1971). Between 1972 and 1984 the IGU Commission on the Coastal Environment carried out surveys and reviewed published work in order to assess global patterns of coastline change over the preceding century (Bird 1985a). It was found that the extent of land lost because of erosion or submergence exceeded the extent of land gained by accretion, emergence or reclamation.

The most rapid erosion, more than 100 metres per year, has occurred on cliffs cut in soft sediments, while recession of up to a kilometre per year has been recorded on ice coasts (melting glaciers) (p. 58). Longshore growth of spits has in places exceeded 100 metres per year (p. 166). The coastlines of the Caspian Sea advanced by up to 50 kilometres when sea level fell between 1930 and 1977, when the sea began to rise again, and coastlines retreated (p. 45).

Gains or losses of at least 10 metres per year have occurred on some beach-fringed and deltaic coastlines, but this has been exceptional, and has usually lasted only for a few years. Landslides have resulted in the sudden advance of the coastline by tens of metres, followed by more gradual erosion of the deposited material (p. 66). Beaches receiving increased sediment loads from rivers draining catchments modified by deforestation, earthquakes, volcanic activity, or mining have prograded at rates of several metres per year (p. 138). Coastlines at and near river mouths have shown rapid erosion after the building of dams upstream curtailed fluvial sediment supply (p. 148).

Short-term changes have occurred during storm surges or as the result of earthquakes, and have been rapid on cliffs cut in soft rock, on the shores of deltas, and along beaches, but few coastlines have continued to change at rates of more than a metre per year. The coasts of the Gulf of Bothnia have shown an annual advance of up to a metre per year as the result of emergence of the sea floor due to continuing isostatic uplift of the land. Where rapid estuarine silting has occurred, the seaward spread of salt marshes and mangrove swamps has advanced intertidal shorelines at rates of several metres per year. On the other hand, rapid erosion and disintegration has resulted in the loss of tens of hectares of salt marsh on the shores of the Lagoon of Venice, where sea level has been rising, partly as the result of coastal land subsidence (p. 237).

ARTIFICIAL COASTLINES

Long sectors of coastline have become artificial during recent decades, partly as the result of engineering works designed to combat erosion and partly as a consequence of embanking or infilling to extend coastal land. On developed coasts (Nordstrom 1994) the proliferation and extension of anti-erosion works, notably sea walls and boulder ramparts, has resulted in large proportions of artificial coastline: 85 per cent in Belgium, 51 per cent in Japan, 38 per cent in England and 21 per cent in South Korea (Walker 1988). Coastal land has been artificially extended on a large scale in Singapore, Hong Kong, Tokyo Bay in Japan, western Malaysia and the Netherlands (Bird 1985a). The category of artificial coastlines is increasing rapidly, and there is the dismal prospect that much of the world's coastline will eventually become artificial as attempts are made to halt submergence and erosion (Figure A.1). Problems of the kind illustrated in Figure A.2 will become more widespread as sea level rises.

PROGRESS IN COASTAL GEOMORPHOLOGY

Attempts to classify coastal landforms have raised a number of themes that require further study. It is evident that there are many unsolved problems in coastal geomorphology, and in concluding this book it may be useful to draw attention to some of them, in the hope that the next generation of coastal researchers will deal with them, and so advance the subject.

1. Coastal landforms have been strongly influenced by the history of Late Quaternary land and sea level changes, which has varied from one sector of coast to another, and requires more precise mapping. On some coasts sectors only a few kilometres apart may have had very different land and sea level histories.

2. The patterns and rates of change on coastlines require further mapping and documentation over specific periods, particularly the past century, and explanation in terms of processes: much more information is needed on coastlines that have been little studied (i.e. outside North America, Britain and Europe, and Australasia).

3. It is necessary to explain why particular landforms (e.g. cliff and shore morphology, notably seaward-sloping and subhorizontal shore platforms) occur on some sectors of the coast and not on other, essentially similar sectors (e.g. slope-over-wall profiles, extensive on the Atlantic coast of Europe, are poorly developed in equivalent latitudes on the Pacific coast of North America).

4. More attention should be given to the documentation and analysis of variations in transverse shore and nearshore profiles, particularly on beaches and coasts that have shore platforms, with a critical review of the idea that these tend towards some kind of equilibrium.

5. There is much variation in the nature and composition of beach materials, and more detailed research is needed into their origins and mode of delivery to shore sectors, with particular attention to sediment supply (past and present) from the sea floor.

6. Artificially nourished beaches should be regarded as field experiments and monitored to elucidate process–response relationships (a procedure that can improve subsequent beach nourishment projects).

Figure A.1   An artificial coastline formed after a sea wall and groynes failed to retain the beach at Litorale di Pellestrina, near Venice on the coast of Italy, and limestone blocks were dumped on the beach.

Figure A.2   Desperate remedies. A boulder bank has been built on the shingle beach at Fairlight Cove, Sussex, in an attempt to halt cliff recession which has destroyed several houses. As sea level rises, this kind of problem will become more widespread.

7. Analysis of the variability of coastal environments is necessary before research findings on one sector of coast can be applied to other sectors (e.g. the work of Californian engineers on beaches as rivers of sand is not widely applicable on the world's coastline).

8. The effects of a rising sea level on coastlines, especially beaches, can be studied on subsiding coasts, but more precise models are needed, including a critical review of the Bruun Rule and its adaptation to the three-dimensional coastal environment.

9. Attempts to extrapolate evidence of past changes (e.g. from the stratigraphy of sediments deposited during a marine transgression) to present and future changes should take account of the environments in which such changes occurred, and of preceding environmental conditions.

10. Substantial sectors of the world's coastline remain to be mapped and documented in detail.

References

Aagard, T. (1991) Multiple-bar morphodynamics and its relations to low frequency edge waves. *Journal of Coastal Research*, **7**: 801–813.

Adam, P. (1990) *Salt marsh ecology*. Cambridge University Press, Cambridge.

Admiralty Tide Tables (1998) 4 volumes. Hydrographer of the Navy, London.

Ager, D.V. (1980) *The Geology of Europe*. John Wiley & Sons, New York.

Ahnert, F. (1960) Estuarine meanders in the Chesapeake Bay area. *Geographical Review*, **50**: 390–401.

Alexander, C.S. (1966) A method of descriptive shore classification and mapping as applied to the north-east coast of Tanganyika. *Annals, Association American Geographers*, **56**: 128–140.

Allen, G.P., Laurier, D. and Thouvenin, J. (1979) *Etude sedimentologique du delta de la Mahakan*. Notes et Memoires, Total, Paris, 15.

Allen, J.R., Psuty, N.P., Bauer, B.O. and Carter, R.W.G. (1996) A field data assessment of contemporary models of beach cusp formation. *Journal of Coastal Research*, **12**: 622–629.

Allen, J.R.L. (1965) Coastal geomorphology of eastern Nigeria: beach ridge, barrier islands and vegetated tidal flats. *Geologie en Mijnbouw*, **44**: 1–21.

Allen, J.R.L. (1968) *Current Ripples*. North-Holland Publishing Co., Amsterdam.

Allen, J.R.L. (1970) Sediments of the modern Niger delta. In: J.P. Morgan (ed.) *Deltaic sedimentation, modern and ancient*, Society of Economic Paleontologists and Mineralogists, Special Publication **15**, pp. 138–151.

Allen, J.R.L. (1989) Evolution of salt marsh cliffs in muddy and sandy systems: a qualitative comparison of British west coast estuaries. *Earth Surface Processes and Landforms*, **14**: 85–92.

Allen, J.R.L. (1996) Shoreline movement and vertical textural patterns in salt marsh deposits. *Proceedings of the Geologists' Association*, **107**: 15–23.

Allen, J.R.L. and Pye, K. (eds) (1991). *Saltmarshes: morphodynamics, conservation and engineering problems*, Cambridge University Press, Cambridge.

Allison, M.A. (1998) Historical changes on the Ganges–Brahmaputra delta front. *Journal of Coastal Research*, **14**: 1269–1275.

Allison, R.J. and Brunsden, D. (1990) Some mudslide movement patterns. *Earth Surface Processes and Landforms*, **15**: 297–312.

Alveirinho Dias, J.M. and Neal, W.J. (1992) Sea cliff retreat in southern Portugal: profiles, processes and problems. *Journal of Coastal Research*, **8**: 641–654.

Atlas Okeanov (1974) Pacific (Tikhii) Ocean. Moscow.

Atlas Okeanov (1977) Atlantic and Indian Oceans. Moscow.

Aufrère, L. (1936) Le rôle du climat dans l'activité morphologique littorale. *Proceedings, 14th International Geographical Congress*, Warsaw, **2**: 189–195.

Augustinus, P.G.E.F. (1995) Geomorphology and sedimentology of mangroves. In: G.M.E. Perillo (ed.) *Geomorphology and Sedimentology of Estuaries*, Elsevier, Amsterdam, pp. 333–357.

Axelrod, D.I. (1958) *Coastal vegetation of the world*. Map issued with Second Coastal Geography Conference (ed. R.J. Russell). Washington, DC.

Badr, A.A. and Lotfy, M.E. (1999) Tracing beach sand movement using fluorescent quartz along the Nile delta promontories, Egypt. *Journal of Coastal Research*, **15**: 261–265.

Bagnold, R.A. (1940) Beach formation by waves: some model experiments in a wave tank. *Journal of the Institution of Civil Engineers*, **15**: 27–52.

Bagnold, R.A. (1941) *The physics of blown sand and desert dunes*. Chapman and Hall, London.

Bailey, H.P. (1958) An analysis of coastal climates, with particular reference to humid mid–latitudes. In: R.J. Russell (ed.) *Proceedings, Second Coastal Geography Conference*, Washington, pp. 23–56.

Baker, G. (1943) Features of a Victorian limestone coastline. *Journal of Geology*, **51**: 359–386.

Baker, G. (1956) Sand drift at Portland Harbour, Victoria. *Proceedings, Royal Society of Victoria*, **68**: 151–198.

Baker, G. (1958) Stripped zones at cliff edges along a high wave current energy coast, Port Campbell, Victoria. *Proceedings, Royal Society of Victoria*, **71**: 175–179.

Bakker, T.W.M., Jungerius, P.D. and Klijn, J.A. (eds) (1990) *European coastal dunes*. Catena Supplement **18**.

Barnes, R.S.K. (1980) *Coastal lagoons, the natural history of a neglected habitat*. Cambridge University Press, Cambridge.

Barrow, G. (1906) *The Geology of the Isles of Scilly*. Memoirs of the Geological Survey, London.

Bascom, W.N. (1951) Relationship between sand size and beach face slope. *Transactions, American Geophysical Union*, **32**: 866–874.

Bascom, W.N. (1954) The control of stream outlets by wave refraction. *Journal of Geology*, **62**: 600–605.

Bauer, B.O., Sherman, D.J., Nordstrom, K.F. and Gares, P.A. (1990) Aeolian transport measurement and prediction across a beach and dune at Castroville, California. In: K.F. Nordstrom, N. Psuty and R.W.G. Carter (eds) *Coastal dunes: form and process*. John Wiley & Sons, Chichester, pp. 39–55.

Belperio, A.P. (1993) Land subsidence and sea level rise in the Port Adelaide estuary: Implications for monitoring the greenhouse effect. *Australian Journal of Earth Sciences*, **40**: 359–368.

Berendsen, H.J.A. (1998) Birds-eye view of the Rhine-Meuse delta. *Journal of Coastal Research*, **14**: 740–753.

Berthois, L. (1951) Façonnement et granulométrie des galets marins au Delec, près Brest. *Revue de Géomorphologie dynamique*, **2**: 259–275.

Bird, E.C.F. (1965) The formation of coastal dunes in the humid tropics: some evidence from North Queensland. *Australian Journal of Science*, **27**: 258–259.

Bird, E.C.F. (1967a) Depositional features in estuaries on the south coast of New South Wales, *Australian Geographical Studies*, **5**: 113–124.

Bird, E.C.F. (1967b) Coastal lagoons of south-eastern Australia. In: J.N. Jennings and J.A. Mabbutt (eds) *Landform Studies from Australia and New Guinea*, pp. 365–385, Australian National University Press, Canberra.

Bird, E.C.F. (1970) Coastal evolution in the Cairns district. *Australian Geographer*, **11**: 327–335.

Bird, E.C.F. (1971a) The origin of beach sediments on the North Queensland coast. *Earth Science Journal*, **5**: 95–105.

Bird, E.C.F. (1971b) The fringing reefs near Yule Point, North Queensland. *Australian Geographical Studies*, **9**: 107–115.

Bird, E.C.F. (1974) Dune stability on Fraser Island. *Queensland Naturalist*, **21**: 15–21.

Bird, E.C.F. (1978a) *The Geomorphology of the Gippsland Lakes Region*. Ministry for Conservation, Victoria, Environmental Studies Series, **186**.

Bird, E.C.F. (1978b) The nature and source of beach materials on the Australian coast. In: J.L. Davies and M.A.J. Williams (eds) *Landform Evolution in Australasia*, Canberra, pp. 144–157.

Bird, E.C.F. (1981) Beach erosion problems at Wewak, Papua New Guinea. *Singapore Journal of Tropical Geography*, **2**: 9–14.

Bird, E.C.F. (1982) Foundations of the Nepean Peninsula. In: C. Hollinshed (ed.) *Lime, Land & Leisure*, Flinders Shire Council, pp. 1–24.

Bird, E.C.F. (1985a) *Coastline changes*. John Wiley & Sons, Chichester.

Bird, E.C.F. (1985b) Recent changes on the Somers–Sandy Point coastline. *Proceedings, Royal Society of Victoria*, **97**: 115–128.

Bird, E.C.F. (1986) Mangroves and intertidal morphology in Westernport Bay. *Marine Geology*, **77**: 327–331.

Bird, E.C.F. (1987) The effects of quarry waste disposal on beaches on the Lizard Peninsula, Cornwall. *Journal of the Trevithick Society*, **14**: 83–92.

Bird, E.C.F. (1988) The tubeworm *Galeolaria caespitosa* as an indicator of sea level rise. *Victorian Naturalist*, **105**: 98–104.

Bird, E.C.F. (1989) The beaches of Lyme Bay. *Proceedings, Dorset Natural History and Archaeological Society*, **111**: 91–97.

Bird, E.C.F. (1991) Changes on artificial beaches in Port Phillip Bay, Australia, *Shore and Beach*, **59**: 19–27.

Bird, E.C.F. (1993a) *The Coast of Victoria*. Melbourne University Press, Melbourne.

Bird, E.C.F. (1993b) Physical setting and geomorphology of coastal lagoons. In: B. Kjerfve (ed.) *Coastal Lagoon Processes*. Elsevier, Amsterdam, pp. 9–39.

Bird, E.C.F. (1993c) *Submerging Coasts: The effects of a rising sea level on coastal environments*. John Wiley & Sons, Chichester.

Bird, E.C.F. (1995) *Geology and scenery of Dorset*. Ex Libris, Bradford on Avon.

Bird, E.C.F. (1996a) *Beach Management*. John Wiley & Sons, Chichester.

Bird, E.C.F. (1996b) Lateral grading of beach sediments: a commentary. *Journal of Coastal Research*, **12**: 774–785.

Bird, E.C.F. (1997a) *The shaping of the Isle of Wight*. Ex Libris, Bradford on Avon.

Bird E.C.F. (1997b) General features of the coastline. In: E. Van der Maarel (ed.) *Dry Coastal Ecosystems*, Elsevier, Amsterdam, pp. 1–10.

Bird, E.C.F. (1998) *The coasts of Cornwall*. Alexander, Fowey.

Bird, E.C.F. and Christiansen, C. (1982) Coastal progradation as a by–product of human activity: an example from Hoed, Denmark. *Geografisk Tidsskrift*, **82**: 1–4.

Bird, E.C.F. and Dent, O.F. (1966) Shore platforms on the south coast of New South Wales. *Australian Geographer*, **10**: 71–80.

Bird, E.C.F. and Fabbri, P. (1987) Archaeological evidence of coastline changes illustrated with reference to Latium, Italy. *Déplacement des lignes de rivage en Mediteranée*, Colloques Internationaux C.N.R.S.: 107–113.

Bird, E.C.F. and Fabbri, P. (1993) Geomorphological and historical changes on the Argentina delta, Ligurian coast, Italy. *GeoJournal*, **29**: 428–439.

Bird, E.C.F. and Green, N. (1992) Induration of ferruginous shore outcrops. *Victorian Naturalist*, **109**: 64–69.

Bird, E.C.F. and Guilcher, A. (1982) Observations préliminaires sur les récifs frangeants actuels du Kenya et sur les formes littorales associées. *Revue de Géomorphologie Dynamique*, **31**: 113–135.

Bird, E.C.F. and Hopley, D. (1969) Geomorphological features on a humid tropical sector of the Australian coast. *Australian Geographical Studies*, **7**: 89–108.

Bird, E.C.F. and Jones, D.J.B. (1988) The origin of foredunes of the coast of Victoria, Australia. *Journal of Coastal Research*, **4**: 181–192.

Bird, E.C.F. and Koike, K. (1986) Man's impacts on sea level changes: a review. *Journal of Coastal Research*, Special Issue **1**: 83–88.

Bird, E.C.F. and May, V.J. (1976) *Shoreline changes in the British Isles during the past century*, Bournemouth College of Technology, Bournemouth.

Bird, E.C.F. and Ongkosongo, O.S.R. (1980) *Environmental changes on the coasts of Indonesia*. United Nations University, Tokyo.

Bird, E.C.F. and Ranwell, D.S. (1964) The physiography of Poole Harbour, Dorset. *Journal of Ecology*, **52**: 355–66.

Bird, E.C.F. and Rosengren, N.J. (1984) The changing coast-line of the Krakatau Islands, Indonesia. *Zeitschrift für Geomorphologie*, **28**: 346–366.

Bird, E.C.F. and Rosengren, N.J. (1987) Coastal cliff management: an example from Black Rock Point, Melbourne, Australia. *Journal of Shoreline Management*, **3**: 39–51.

Bird, E.C.F. and Schwartz, M.L. (eds) (1985) *The World's Coastline*. Van Nostrand Reinhold, New York.

Bird, E.C.F. and Schwartz, M.L. (2000) Shore platforms at Cape Flattery, Washington. *Washington Geology*, **28**: 21–26.

Bird, E.C.F., Dubois, J.P. and Iltis, J.A. (1984) *The Impacts of Opencast Mining on the Rivers and Coasts of New Caledonia*. United Nations University, Tokyo.

Bird, J.F. (2000) The impact of mining waste on the rivers draining into Georges Bay, Northeast Tasmania. In: S. Brizga and B.L. Finlayson (eds) *River management: the Australian experience*, John Wiley & Sons, Chichester, pp. 151–172.

Black K.A., and Rosenberg, M.A. (1992) Natural stability of beaches around a large bay. *Journal of Coastal Research*, **8**: 385–397.

Bloom, A.L. (1965) The explanatory description of coasts. *Zeitschrift für Geomorphologie*, **9**: 422–436.

Bloom, A.L. (1979) *Atlas of Sea Level Curves*. Cornell University, New York.

Bodéré, J.C. (1979) Le rôle essentiel des débacles glacio-volcaniques dans l'évolution recente des côtes sableuses en voie de progradation du sud-est d'Islande. In: A. Guilcher (ed.) *Les Côtes Atlantiques de l'Europe*, University of Western Brittany, Brest, pp. 55–64.

Bodge, K.R. (1992) Representing equilibrium beach profiles with an exponential expression. *Journal of Coastal Research*, **8**: 47–55.

Borowca, R.K. (1990) The Holocene development and present morphology of the Leba dunes, Baltic coast of Poland. In: K.F. Nordstrom, N. Psuty and R.W.G. Carter (eds) *Coastal dunes: form and process*. John Wiley & Sons, Chichester, pp. 289–313.

Borowca, M. and Rotnicki, K. (1994) Intensity, directions and balance of aeolian transport on the beach barrier and the problem of sand nourishment. In: K. Rotnicki (ed.) *Changes of the Polish Coastal Zone*. Quaternary Research Institute, Adam Mickiewicz University, Poznan, pp. 108–115.

Bourman, R.P. (1990) Artificial beach progradation by quarry waste disposal at Rapid Bay, South Australia. *Journal of Coastal Research*, Special Issue **6**: 69–76.

Bourman, R.P. and Harvey, N. (1983) The Murray mouth flood tidal delta, *Australian Geographer*, **15**: 403–406.

Brampton, A.H. (1977) *A computer model for wave refraction.* Hydraulics Research Station, Wallingford, Report **172**.

Bray, M.J. (1997) Episodic shingle supply and the modified development of Chesil Beach, England. *Journal of Coastal Research*, **13**: 1035–1049.

Bray, M.J. and Hooke, J.M. (1997) Prediction of soft-cliff retreat with accelerating sea-level rise. *Journal of Coastal Research*, **13**: 453–467.

Bray, M.J., Carter, D.J. and Hooke, J.M. (1995) Littoral cell definition and budgets for Central South England. *Journal of Coastal Research*, **11**: 381–400.

Bremner, J.M. (1985) Southwest Africa/Namibia. In: E.C.F. Bird and M.L. Schwartz (eds) *The World's Coastline*, Van Nostrand Reinhold, New York, pp. 645–651.

Brown, B.E. (1990) Coral bleaching. *Coral Reefs*, **8**: 153–232.

Brown, M.J.F. (1974) A development sequence: disposal of mining waste on Bougainville, Papua New Guinea. *Geoforum*, **18**: 19–27.

Brunsden, D. and Jones, D.K.C. (1980) Relative time scales in formative events in coastal landslide systems. *Zeitschrift für Geomorphologie*, Supplementband **34**: 1–19.

Brunsden, D. and Prior, D.B. (1984) *Slope instability*. John Wiley & Sons, Chichester.

Bruun, P. (1962) Sea level rise as a cause of shore erosion. *Proceedings, American Society of Civil Engineers, Waterways and Harbour Division*, **88**: 117–130.

Bruun, P. (1988) The Bruun Rule of erosion by sea level rise: a discussion of large-scale two- and three-dimensional usages. *Journal of Coastal Research*, **4**: 627–648.

Bryant, E.A. (1985) Rainfall and beach erosion relationships, Stanwell Park, Australia, 1895–1980: worldwide implications for coastal erosion. *Zeitschrift für Geomorphologie*, Supplementband **57**: 51–65.

Bryant E.A., Young, R.S. and Price, D.M. (1996) Tsunami as a major control on coastal evolution, southeastern Australia. *Journal of Coastal Research*, **12**: 831–840.

Buddemeier, R.W. and Smith, S.V. (1988) Coral reef growth in an era of rapidly rising sea level. *Coral Reefs*, **7**: 51–56.

Byrne, J.V. (1964) An erosional classification for the northern Oregon coast. *Annals, Association of American Geographers*, **54**: 329–335.

Cailleux, A. (1948) Lithologie des dépôts émergés actuels de l'embouchure du Var au Cap d'Antibes. *Bulletin de l'Institut Océanographique de Monaco*, **94**: 1–11.

Cambers, G. (1976) Temporal scales in coastal erosion systems. *Transactions, Institute of British Geographers*, **1**: 246–256.

Carey, A.E. and Oliver, F.W. (1918) *Tidal lands*. Blackie, Edinburgh.

Carlton, J.M. (1974) Land–building and stabilization by mangroves. *Environmental Conservation*, **1**: 285–294.

Carr, A.P. (1962) Cartographic record and historical accuracy. *Geography*, **47**: 135–144.

Carr, A.P. (1965) Shingle spit and river mouth: short-term dynamics. *Transactions, Institute of British Geographers*, **36**: 117–130.

Carter, C.H. and Guy, D.E. (1988) Coastal erosion: processes, timing and magnitude at the bluff toe. *Marine Geology*, **84**: 1–17.

Carter, R.W.G. (1988) *Coastal Environments*. Academic Press, London.

Carter, R.W.G. (1991) Near-future sea-level impacts on coastal dune landscapes. *Landscape Ecology*, **6**: 9–40.

Carter, R.W.G. and Devoy, R.J.N. (eds) (1987) The hydro-dynamic and sedimentological consequences of sea level rise. *Progress in Oceanography*, **18**: 136–154.

Carter R.W.G., Hesp, P.A. and Nordstrom, K.F. (1990) Erosional landforms in coastal dunes. In: K.F. Nordstrom, N. Psuty and R.W.G. Carter (eds) *Coastal dunes: form and process*, John Wiley & Sons, Chichester, pp. 217–250.

Carter, R.W.G. and Wilson, P. (1990) The geomorphological, ecological and pedological development of coastal foredunes at Magilligan Point, Northern Ireland. In: K.F. Nordstrom, N. Psuty and R.W.G. Carter (eds) *Coastal dunes: form and process*, John Wiley & Sons, Chichester, pp. 129–157.

Carter, R.W.G, and Woodroffe, C.D. (eds) (1994) *Coastal evolution. Late Quaternary morphodynamics*. Cambridge University Press, Cambridge.

Cartwright, P.E. (1974) Years of peak astronomical tides. *Nature*, **248**: 656–657.

Castaing, P. and Guilcher, A. (1995) Geomorphology and sedimentology of rias. In: G.M.E. Perillo (ed.) *Geomorphology and Sedimentology of Estuaries*, Elsevier, Amsterdam, pp. 69–111.

Cavazzoni, S. (1983) Recent erosive processes in the Venetian Lagoon. In: E.C.F. Bird and P. Fabbri (eds) *Coastal Problems in the Mediterranean Sea*, Bologna, pp. 19–22.

Cencini, C. (1998) Physical processes and human activities in the Po delta, Italy. *Journal of Coastal Research*, **14**: 774–793.

Chapman, D.M., Geary, M., Roy, P.S. and Thom, B.G. (1982) *Coastal evolution and coastal erosion in New South Wales*. Coastal Council of New South Wales, Sydney.

Chapman, V.J. (1976) *Mangrove Vegetation*. Cramer, Vaduz.

Chappell, J. (1974) Geology of coral terraces, Huon Peninsula, New Guinea. *Bulletin, Geological Society of America*, **85**: 553–570.

Chappell, J. (1983) A revised sea-level record for the last 300 000 years. *Search*, **14**: 99–101.

Chardonnet, J. (1948) Les calanques provençales, origin et divers types. *Annales de Géographie*, **57**: 289–297.

Chen Jiyu, Liu Cangzi and Yu Zhiying (1985) China In: E.C.F. Bird and M.L. Schwartz (eds) *The World's Coastline*. Van Nostrand Reinhold, New York, pp. 813–822.

Christiansen, C., Christoffersen, H., Dalsgard, H. and Nornberg, P. (1981) Coastal and nearshore changes associated with die-back in eel grass (*Zostera marina*). *Sedimentary Geology*, **28**: 163–173.

Chun-Hsin, Chung (1985) The effects of introduced *Spartina* grass on coastal morphology in China. *Zeitschrift für Geomorphologie*, Supplementband **57**: 169–174.

Churchill, D.M. (1959) Late Quaternary eustatic changes in the Swan River district. *Proceedings, Royal Society Western Australia*, **42**: 53–55.

Clayton, K.M. (1989) Sediment input from the Norfolk cliffs, eastern England – a century of coast protection and its effect. *Journal of Coastal Research*, **5**: 433–442.

Clayton, K.M. and Shamoon, N. (1998) A new approach to the relief of Great Britain, II. A classification of rocks based on relative resistance to denudation. *Geomorphology*, **25**: 155–171.

Codignotto, J.O. and Aguirre, M.C. (1993) Coastal evolution, changes in sea level and molluscan fauna in north-eastern Argentina during the Late Quaternary. *Marine Geology*, **110**: 163–175.

Codignotto, J.O., Kokot, R.R., and Marcomini, S.C. (1992) Neotectonism and sea level changes in the coastal zone of Argentina. *Journal of Coastal Research*, **8**: 125–133.

Coleman, J.M. (1981) *Deltas*. Burgess, Minneapolis.

Coleman, J.M., Gagliano, S.M. and Smith, W.G. (1970) Sedimentation in a Malaysian high tide tropical delta. In: J.P. Morgan (ed.) *Deltaic sedimentation: modern and ancient*. Society of Economic Palaeontologists and Mineralogists, Special Publication **15**: 185–197.

Coleman, J.M., Roberts, H.H. and Stone, G.W. (1998) Mississippi River delta: an overview. *Journal of Coastal Research*, **14**: 698–717.

Colombo, G. (1977) Lagoons. In: R.S.K. Barnes (ed.) *The Coastline*. John Wiley & Sons, New York, pp. 63–81.

Cooper, J.A.G. (1994) Lagoons and microtidal coasts. In: R.W.G. Carter and C.D. Woodroffe (eds) *Coastal Evolution. Late Quaternary shoreline morphodynamics*, Cambridge University Press, Cambridge, pp. 219–265.

Cooper, W.S. (1958) *Coastal sand dunes of Oregon and Washington*, Memoirs, Geological Society of America, **72**.

Cooper, W.S. (1967) *Coastal dunes of California*. Memoirs, Geological Society of America, **104**.

Cotton, C.A. (1952) Criteria for the classification of coasts *Proceedings 17th Conference, International Geographical Union*, Washington, D.C., 315–319.

Cotton, C.A. (1956) Rias sensu stricto and sensu lato. *Geographical Journal*, **122**: 360–364.

Cotton, C.A. (1963) Levels of planation of marine benches. *Zeitschrift für Geomorphologie*, **7**: 97–111.

Cotton, C.A. (1974) *Bold Coasts*. Wellington, New Zealand.

Cowell, P.J. and Thom, B.G. (1994) Morphodynamics of coastal evolution. In: R.W.G. Carter and C.D. Woodroffe (eds) *Coastal Evolution. Late Quaternary shoreline morphodynamics.* Cambridge University Press, Cambridge, pp. 33–86.

Craig, A.K. and Psuty, N.P. (1968) The Paracas papers. *Studies in Marine Desert Ecology,* **1**: 44–77.

Cressard, A.P. and Augris, C. (1982) Etudes des phénomènes d'érosion cotière liès a l'extraction de materiaux sur le plateau continentale. *Proceedings, 4th Conference, International Association of Engineering Geology,* **7**: 203–211.

Cruz, O., Coutinho, P.N., Duarte, G.M., Gomes, A. and Muehe, D. (1985) Brazil In: E.C.F. Bird and M.L. Schwartz (eds) *The World's Coastline,* Van Nostrand Reinhold, New York, pp. 85–91.

Curray, J.R. (1960) Sediments and history of Holocene transgression, continental shelf, northwest Gulf of Mexico. *Recent Sediments, Northwest Gulf of Mexico,* American Association of Petroleum Geologists' Symposium, Tulsa, Oklahoma: 221–266.

Dalrymple, R.W., Zaitlin, B.A. and Boyd, R. (1992) Estuarine facies models: conceptual basis and stratigraphic implications. *Journal of Sedimentary Petrology,* **62**: 1130–1146.

Daniels, R.C. (1992) Sea level rise on the South Carolina coast: two case studies for 2100. *Journal of Coastal Research,* **8**: 56–70.

Davidson-Arnott, R.G.D. and Law, M.N. (1990) Seasonal patterns and controls on sediment supply to coastal foredunes, Long Point, Lake Erie. In: K.F. Nordstrom, N. Psuty and R.W.G. Carter (eds) *Coastal dunes: form and process.* John Wiley & Sons, Chichester, pp. 177–200.

Davies, J.L. (1957) The importance of cut and fill in the development of sand beach ridges. *Australian Journal of Science,* **20**: 105–111.

Davies, J.L. (1974) The coastal sediment compartment. *Australian Geographical Studies,* **12**: 139–151.

Davies, J.L. (1980) *Geographical variation in coastal development* (2nd edition). Longman, London.

Davies, P.J. (1983) Reef growth. In: D.J. Barnes (ed.) *Perspectives on Coral Reefs.* Manuka, Clouston, pp. 69–106.

Davis, J.H. (1940) The ecology and geological rôle of mangroves in Florida. *Publications of the Carnegie Institution,* **524**: 303–412.

Davis, R.A. (1989) Texture, composition and provenance of beach sands, Victoria, Australia. *Journal of Coastal Research,* **5**: 37–47.

Davis, W.M. (1896) The outline of Cape Cod. *Proceedings, American Academy Arts and Sciences,* **31**: 303–332.

Davis, W.M. (1928) *The Coral Reef Problem.* American Geographical Society Special Publication **9**.

Day, J.W., Scarton, F., Rismondo, A. and Are, D. (1998) Rapid deterioration of a salt marsh in Venice Lagoon, Italy. *Journal of Coastal Research,* **14**: 583–590.

De Martonne, E. (1909) *Traité de Géographie Physique.* Paris.

Dean, R.G. (1991) Equilibrium beach profiles: characteristics and applications *Journal of Coastal Research,* **7**: 53–84.

Delft Hydraulics Laboratory (1987) *Manual on Artificial Beach Nourishment.* Centre for Civil Engineering Research, Codes and Specifications, Rijkwaterstaat, Netherlands.

Dietz, R.S. (1963) Wave-base, marine profile of equilibrium, and wave-built terraces: a critical appraisal. *Bulletin, Geological Society of America,* **74**: 971–990.

Dionne, J.C. (1972) Caractéristiques des schorres des régions froides. *Zeitschrift fur Geomorphologie,* **13**: 131–162.

Dolan, R. (1971) Coastal landforms: crescentic and rhythmic. *Bulletin Geological Society of America,* **82**: 177–180.

Dolan, R. and Davis, R.E. (1992) An intensity scale for Atlantic coast northeast storms, *Journal of Coastal Research,* **8**: 840–853.

Donn, W.L., Farrand, W.R. and Ewing, J.R. (1962) Pleistocene ice volumes and sea level lowering. *Journal of Geology,* **70**: 206–214.

Donoghue, J.P. and White, N.M. (1995) Late Holocene sea level change and delta migration, Apalachicola River region, Northwest Florida. *Journal of Coastal Research,* **11**: 651–663.

Dubois, G. (1924) Recherches sur les terrains quaternaires du Nord de la France. *Memoirs du Société Géologique du Nord,* **8**: 1–356.

Dubois, R.N. (1977) Predicting beach erosion as a function of rising water level. *Journal of Geology,* **85**: 470–476.

Dubois, R.N. (1992) A re-evaluation of Bruun's Rule and supporting evidence. *Journal of Coastal Research,* **8**: 618–628.

Dugdale, R. (1981) Coastal processes. In: A. Goudie (ed.) *Geomorphological Techniques,* Allen and Unwin, London, pp. 247–265.

Dunbar, G.S. (1956) *Geographical history of the Carolina banks.* Coastal Studies Institute, Louisiana State University, Technical Report **8**.

Ehlers, J. (1988) *The morphodynamics of the Wadden Sea.* Balkema, Rotterdam.

Eitner, V. (1996) Geomorphological response of the Frisian barrier islands to sea level rise. *Geomorphology,* **15**: 57–65.

Ellison, J.C. and Stoddart, D.R. (1991) Mangrove ecosystem collapse during predicted sea-level rise: Holocene analogues and implications. *Journal of Coastal Research,* **7**: 151–165.

Emery, K.O. (1960) *The sea off Southern California.* John Wiley & Sons, New York.

Emery, K.O. (1961) Submerged marine terraces and their sediments. *Zeitschrift für Geomorphologie,* Supplementband **3**: 17–29.

Emery, K.O. (1969) *A coastal pond studied by oceanographic methods.* Elsevier, New York.

Emery, K.O. and Aubrey, D.G. (1991) *Sea levels, land levels and tide gauges.* Springer-Verlag, New York.

Emery, K.O. and Kuhn, G.G. (1982) Sea cliffs: their processes, profiles and classification. *Bulletin, Geological Society of America,* **93**: 644–654.

Emery, K.O. and Stevenson, R.E. (1957) Estuaries and lagoons, *Memoirs, Geological Survey, America*, **67**: 673–750.

Empsall, B. (1989) Workington ironworks reclamation. *Proceedings, Conference Institute of Engineers, Bournemouth*: 279–292.

Evans, G. (1965) Intertidal flat sediments and their environment of deposition in The Wash. *Quarterly Journal, Geological Society of London*, **121**: 209–241.

Evans, G. and Bush, P.R. (1969) Some oceanographical observations on a Persian Gulf Lagoon. In: A. Castañares and F.B. Phleger (eds) *Lagunas Costeras, un Simposio*, National University of Mexico, pp. 155–170.

Evans, O.F. (1942) The origin of spits, bars and related structures. *Journal of Geology*, **51**: 846–863.

Everard, C.E. (1962) Mining and shoreline evolution near St. Austell, Cornwall. *Transactions of the Royal Geological Society of Cornwall*, **19**: 199–219.

Everts, C.H. (1985) Sea-level rise effects on shoreline position. *Journal of Waterway, Port, Ocean and Coastal Engineering*, **111**: 985–999.

Fabbri, P. (1985) Coastal variations in the Po delta since 2500 BC *Zeitschrift für Geomorphologie*, Supplementband **57**: 155–167.

Fairbridge, R.W. (1961) Eustatic changes in sea level. *Physics and Chemistry of the Earth*, **4**: 99–185.

Fairbridge, R.W. (1967) Coral reefs of the Australian region. In: J.N. Jennings and J.A. Mabbutt (eds) *Landform studies from Australia and New Guinea*, pp. 386–417.

Fairbridge, R.W. (1980) The estuary: its definition and geodynamic cycle. In: E. Olausson and I. Cato (eds) *Chemistry and Biogeochemistry of Estuaries*, John Wiley & Sons, New York, pp. 1–36.

Fairbridge, R.W. (1983) Isostasy and eustasy. In: D. Smith and A.G. Dawson (eds) *Shorelines and Isostasy*. Academic Press, London, pp. 3–26.

Fairbridge R.W. and Krebs, O.A. (1962) Sea level and the Southern Oscillation. *Geophysical Journal*, **6**: 532–545.

Fairbridge, R.W. and Teichert, C. (1953) Soil horizons and marine bands in the coastal limestones of Western Australia. *Journal Royal Society of New South Wales*, **86**: 68–87.

Fisher, J.J. (1980) Shoreline erosion, Rhode Island and North Carolina coasts. In: M.L. Schwartz and J.J. Fisher (eds) *Proceedings of the Per Bruun Symposium*, University of Rhode Island, pp. 32–54.

Fisher, R.L. (1955) Cuspate spits of St. Lawrence Island, Alaska, *Journal of Geology*, **63**: 133–142.

Fisk, H.N. (1958) Padre Island and the Laguna Madre flats, coastal south Texas. *Proceedings Second Coastal Geography Conference*: 103–151.

Fisk, H.N. and McFarlan, E. (1955) Late Quaternary deltaic deposits of the Mississippi River. *Geological Society of America*, Special Publication **24**: 279–302.

Fleming, C.A. (1965) Two-storied cliffs at the Auckland Islands. *Transactions, Royal Society of New Zealand (Geology)* **3**: 171–174.

Fonseca, M.S. (1996) The role of seagrasses in nearshore sedimentary processes: a review. In: Nordstrom K.F. and Roman C.T. (eds) *Estuarine Shores*, John Wiley & Sons, Chichester, pp. 261–286.

Fotheringam, D.G. and Goodwins, D.R. (1990) Monitoring the Adelaide beach system. *Proceedings 1990 Workshop on Coastal Zone Management*, Yeppoon, Queensland, pp. 118–132.

Fox, W.T. (1985) Modeling coastal environments. In: R.A. Davis (ed.) *Coastal sedimentary environments*, Springer-Verlag, New York, pp. 665–705.

Fox W.T., Haney, R.L. and Curran, H.A. (1995) Penouille Spit, evolution of a complex spit, Gaspé, Quebec, Canada. *Journal of Coastal Research*, **11**: 478–493.

French, J.R. and Spencer, T. (1993) Dynamics of sedimentation in a tide-dominated backbarrier salt marsh, Norfolk, U.K. *Marine Geology*, **110**: 315–331.

French J.R. and Stoddart, D.R. (1992) Hydrodynamics of salt marsh creek systems: implications for marsh morphological development and material exchange. *Earth Surface Processes and Landforms*, **17**: 235–252.

French, P.W. (1996) Implications of a salt marsh chronology for the Severn estuary based on independent lines of dating evidence. *Marine Geology*, **135**: 115–125.

French, P.W. (1997) *Coastal and estuarine management*. Routledge, London.

Frey. R.W. and Basan, P.B. (1985) Coastal salt marshes. In: R.A. Davis (ed.) *Coastal sedimentary environments*, Springer-Verlag, New York, pp. 225–301.

Fuenzalida, H., Cooke, R., Paskoff, R., Segerstrom, K. and Weischet, W. (1965) High Stands of Quaternary Sea Level along the Chilean Coast. *Geological Society of America, Special Paper* **84**: 473–496.

Galvin, C.J. (1972) Waves breaking in shallow water. In: R. Meyer (ed.) *Waves on Beaches*. Academic Press, London, pp. 413–455.

Gares, P.A. (1990) Eolian processes and dune changes at developed and undeveloped sites, Island Beach, New Jersey. In: K.F. Nordstrom, N. Psuty and R.W.G. Carter (eds) *Coastal dunes: form and process*. John Wiley & Sons, Chichester, pp. 361–380.

Gell, R.A. (1978) Shelly beaches on the Victorian coast. *Proceedings, Royal Society of Victoria*, **90**: 257–269.

Gibbs, R.J. (1970) Circulation in the Amazon estuary and adjacent Atlantic Ocean. *Journal of Marine Research*, **28**: 113–123.

Gierloff-Emden, H.G. (1961) Nehrungen and Lagunen, *Petermanns Geographischen Mitteilungen*, **105**: 81–92 and 161–176.

Gilbert, G.K. (1890) *Lake Bonneville*. U.S. Geological Survey Monograph **1**.

Gill, E.D. (1973) Rate and mode of retrogradation on rocky coasts in Victoria, Australia. *Boreas*, **2**: 143–171.

Gill, E.D. and Lang, J.G. (1983) Micro-erosion meter measurements of rock wear on the Otway coast of south-east Australia. *Marine Geology*, **42**: 141–156.

Gimingham, C.H., Ritchie, W., Willetts, B.B. and Willis, A.J. (eds) (1989) Coastal sand dunes. *Proceedings, Royal Society of Edinburgh*, **B96**: 1–313.

Goldsmith, P. and Hieber, S. (1991) Space techniques support the monitoring of sea level. In: R. Frassetto (ed.) *Impacts of Sea Level Rise on Cities and Regions*, Marsilio, Venice, pp. 220–226.

Goldsmith V. (1985) Coastal dunes. In: R.A. Davis (ed.) *Coastal Sedimentary Environments*. Springer-Verlag, New York, pp. 303–378.

Goldsmith, V. (1989) Coastal sand dunes as geomorphic systems. *Proceedings, Royal Society of Edinburgh*, **B96**: 3–15.

Gowlland-Lewis, M., Fulton, I. and Audas, D. (1996) *Darwin Coastal Erosion Survey*. Resources Conservation Branch, Northern Territory, Australia, Technical Report **56**.

Greensmith, J.T. and Tucker, E.V. (1966) Morphology and evolution of inshore shell ridges and mud-mounds on modern intertidal flats near Bradwell, Essex. *Proceedings of the Geologists' Association*, **77**: 329–346.

Gresswell, R.K. (1953) *Sandy Shores in South Lancashire*. Liverpool.

Griggs, G.B. and Trenhaile, A.S. (1994) Coastal cliffs and platforms. In: R.W.G. Carter and C.D. Woodroffe (eds) *Coastal Evolution. Late Quaternary shoreline morphodynamics*, Cambridge University Press, Cambridge, pp. 425–450.

Guilcher, A. (1958) *Coastal and Submarine Morphology*. Methuen, London.

Guilcher, A. (1965) Drumlin and spit structures in the Kenmare River, south-west Ireland. *Irish Geography*, **5**: 7–19.

Guilcher, A. (1967) Origin of sediments in estuaries. In: G.H. Lauff (ed.) *Estuaries*. American Association for the Advancement of Science, Scientific Publication **83**: 149–157.

Guilcher, A. (1979) Marshes and estuaries in different latitudes. *Interdisciplinary Science Reviews* **4**: 158–168.

Guilcher, A. (1981) Shoreline changes in salt marshes and mangrove swamps (mangals) within the past century. In: E.C.F. Bird and K. Koike (eds) *Coastal Dynamics and Scientific Sites*, Komazawa University, Tokyo, pp. 31–53.

Guilcher, A. (1985a) Retreating cliffs in the humid tropics: an example from Paraiba, north-eastern Brazil. *Zeitschrift für Geomorphologie*, Supplementband **57**: 95–103.

Guilcher, A. (1985b) Angola In: E.C.F. Bird and M.L. Schwartz (eds) *The World's Coastline*, Van Nostrand Reinhold, New York, pp. 639–643.

Guilcher, A. (1988) *Coral Reef Geomorphology*. John Wiley & Sons, Chichester.

Guilcher, A. and Nicolas, J.P. (1954) Observations sur la langue de Barbarie et les bras du Sénégal aux environs de St. Louis. *Bulletin d'Information, Comité d'Océanographie et d'Etude des Côtes*, **6**: 227–242.

Guilcher, A., Vallantin, P., Angrand, J.P. and Galloy, P. (1957) Les cordons littoraux de la Rade de Brest. *Bulletin d'Information, Comité d'Océanographie et d'Etude des Côtes*, **9**: 21–54.

Guillen, J. and Palanques, A. (1997) A shoreface zonation in the Ebro delta based on grain size distribution. *Journal of Coastal Research*, **13**: 867–878.

Gulliver, F.P. (1899) Shoreline topography. *Proceedings, American Academy of Arts and Sciences*, **34**: 151–258.

Guza, R.T. and Inman, D.L. (1975) Edge waves and beach cusps. *Journal of Geophysical Research*, **80**: 2997–3012.

Hails, J.R. (1982) Humate. In: M.L. Schwartz (ed.) *Encyclopedia of Beaches and Coastal Environments*. Hutchinson Ross, Stroudsburg, Pennsylvania, pp. 466–467.

Hansom, J. (1983) Shore platform development in the South Shetland Islands, Antarctica. *Marine Geology*, **53**: 211–229.

Hansom, J. (1988) *Coasts*. Cambridge University Press, Cambridge.

Hardisty, J. (1990) *Beaches: form and process*. Unwin Hyman, London.

Hardisty, J. (1994) Beach and nearshore sediment transport. In: K. Pye (ed.) *Sediment transport and depositional processes*. Blackwell, Oxford, pp. 219–255.

Harris, R.L. (1955) *Restudy of Test Shore Nourishment by Offshore Deposition of Sand, Long Branch, New Jersey*. Beach Erosion Board, Technical Memorandum **62**.

Hart, B.S. (1995) Delta front estuaries. In: G.M.E. Perillo (ed.) *Geomorphology and Sedimentology of Estuaries*, Elsevier, Amsterdam, pp. 207–220.

Healy, T.R. (1968) Bioerosion on shore platforms in the Waitemata formation, Auckland, *Earth Sciences Journal*, **2**: 26–37.

Healy, T.R. (1977) Progradation of the entrance to Tauranga Harbour, Bay of Plenty. *New Zealand Geographer*, **30**: 90–91.

Healy, T.R. and Kirk, R.M. (1982) Coasts. In: J.M. Soons and M.J. Selby (eds) *Landforms of New Zealand*, Longman Paul, Auckland, pp. 81–104.

Hegg, B., Eliot, I. and Hsu, J. (1996) Sheltered sandy beaches of southwestern Australia. *Journal of Coastal Research*, **12**: 748–760.

Herd, D.G., Yourd, T.L., Hansjurgen, C., Person, W.J. and Mendoza, C. (1981) The great Tumaco, Colombia, earthquake of 12 December 1979. *Science*, **211**: 441–445.

Hesp, P.A. (1984) Foredune formation in southeast Australia. In: B.G. Thom (ed.) *Coastal Geomorphology in Australia*, Academic Press, Sydney, pp. 69–97.

Hesp, P.A. (1988) Surfzone, beach and foredune interactions on the Australian south east coast. *Journal of Coastal Research*, Special Issue **3**: 15–25.

Hesp, P.A. and Hilton, M.J. (1996) Nearshore-surfzone system limits and the impacts of sand extraction. *Journal of Coastal Research*, **12**: 726–747.

Hesp, P.A. and Thom, B.G. (1990) Geomorphology and evolution of active transgressive dunefields. In: K.F. Nordstrom, N. Psuty and R.W.G. Carter (eds) *Coastal dunes: form and process*. John Wiley & Sons, Chichester, pp. 253–288.

Heydorn, A.E.F. and Flemming, B.W. (1985) South Africa. In: E.C.F. Bird and M.L. Schwartz (eds) *The World's Coastline*. Van Nostrand Reinhold, New York, pp. 653–661.

Heydorn, A.E.F. and Tinley, K.L. (1980) *Estuaries of the Cape, Part I – Synopsis of the Cape Coast*. National Research Institute, Oceanology, Stellenbosch, South Africa.

Hicks, D.M. and Hume, T.M. (1996) Morphology and size of ebb-tidal deltas at natural inlets on open sea and pocket-bay coasts, North Island, New Zealand. *Journal of Coastal Research*, **12**: 47–63.

Hill, P.R., Barnes, P.W., Héquette, A. and Ruz, M.H. (1994) Arctic coastal plains shorelines. In: R.W.G. Carter and C.D. Woodroffe (eds) *Coastal Evolution. Late Quaternary shoreline morphodynamics*, Cambridge University Press, Cambridge, pp. 341–372.

Hills, E.S. (1949) Shore platforms. *Geological Magazine*, **86**: 137–152.

Hills, E.S. (1971) A study of cliffy coastal profiles based on examples in Victoria, Australia. *Zeitschrift für Geomorphologie*, **15**: 137–180.

Hinschberger, F. (1985) Ivory Coast. In: E.C.F. Bird and M.L. Schwartz (eds), *The World's Coastline*, Van Nostrand Reinhold, New York, pp. 585–589.

Hinton, A.C. (1998) Tidal changes. *Progress in Physical Geography*, **22**: 282–294.

Hodgkin, E.P. (1964) Rate of erosion of intertidal limestone. *Zeitschrift für Geomorphologie*, **8**: 385–392.

Hodgkin, E.P., Birch, P.B., Black, R.E., and Humphries, R.B. (1981) *The Peel-Harvey estuarine system study (1976–1980)*. Department of Conservation & Environment, Western Australia, Report No. **9**.

Hoekstra, P. (1993) Late Holocene development of a tide induced elongate delta, the Solo delta, East Java. *Sedimentary Geology*, **83**: 211–233.

Holman, R.A. and Bowen, A.J. (1982) Bars, bumps and holes: models for the generation of complex beach topography. *Journal of Geophysical Research*, **87**: 457–468.

Hopley, D. (1971) The origin and significance of North Queensland island-spits. *Zeitschrift für Geomorphologie*, **15**: 371–389.

Hopley, D. (1982) *Geomorphology of the Great Barrier Reef*. Wiley Interscience, New York.

Hopley, D. (1994) Continental shelf reef systems. In: R.W.G. Carter and C.D. Woodroffe (eds) *Coastal Evolution. Late Quaternary shoreline morphodynamics*, Cambridge University Press, Cambridge, pp. 303–340.

Hopley, D. and Kinsey, D. W. (1988) The effects of a rapid short-term sea-level rise on the Great Barrier Reef. In: G.I. Pearman (ed.) *Greenhouse: Planning for Climatic Changes, C.S.I.R.O. Division of Atmospheric Research*, Melbourne, Australia, pp. 189–201.

Horikawa, K. (ed.) (1988) *Nearshore dynamics and coastal processes: theory, measurement and predictive models*. University of Tokyo Press.

Horn, D.P. (1993) Sediment dynamics on a macrotidal beach, Isle of Man, U.K. *Journal of Coastal Research*, **9**: 189–208.

Hotten, R.D. (1988) Sand mining on Mission Beach, San Diego, California. *Shore and Beach*, **56**: 18–21.

Houghton, J.T., Jenkins, G.J. and Ephraums, J.J. (eds) (1990). *Scientific Assessment of Climate Change*. Cambridge University Press, Cambridge.

Hutchinson, J.N., Bromehead, E.N. and Lupini, J.F. (1980) Additional observations on the Folkestone Warren landslides. *Quarterly Journal of Engineering Geology*, **11**: 1–31.

Hutchinson, J.N., Brunsden, D. and Lee, E.M. (1991) The geomorphology of the landslide complex at Ventnor, Isle of Wight. In: R.J. Chandler (ed.) *Slope stability engineering*. Telford, pp. 213–218.

Ibe, A.C. (1988) Nigeria. In: Walker H.J. (ed.) *Artificial structures and shorelines*. Kluwer, Dordrecht, pp. 287–294.

Ibe, A.C., Awosika, L.F., Ibe, C.E. and Inegbedion, L.E. (1991) Monitoring of the beach nourishment project at Bar Beach, Victoria Island, Lagos, Nigeria. *Proceedings Coastal Zone '91*, pp. 534–552.

Ignatov, Y., Kaplin, P.A., Lukyanova, S.A. and Soloveiva, G.D. (1993) Evolution of Caspian Sea coasts under conditions of sea level rise. *Journal of Coastal Research*, **9**: 104–111.

Ingle, J.C. (1966) *The Movement of Beach Sand*. Elsevier, Amsterdam.

Inman, D.L. and Nordstrom, C.E. (1971) On the tectonic and morphologic classification of coasts. *Journal of Geology*, **79**: 1–21.

Isla, F.I. (1995) Coastal lagoons. In: G.M.E. Perillo (ed.) *Geomorphology and Sedimentology of Estuaries*, Elsevier, Amsterdam, pp. 241–272.

Jackson, J.M. (1985) Uruguay. In: E.C.F. Bird and M.L. Schwartz (eds) *The World's Coastline*, Van Nostrand Reinhold, New York, pp. 77–84.

Jackson, N.L. and Nordstrom, K.F. (1992) Site specific controls on wind and wave processes and beach mobility on estuarine beaches. *Journal of Coastal Research*, **8**: 88–98.

Jacobsen, E.E. and Schwartz, M.L. (1981) The use of geomorphic indicators to determine the direction of net shore-drift. *Shore and Beach*, **49**: 38–43.

Jacobson, H.A. (1988) Historical development of the salt marsh at Wells, Maine. *Earth Surface Processes and Landforms*, **13**: 475–486.

James, P. (1995) *The Sunken Kingdom*. Jonathan Cape, London.

Jennings, J.N. (1957) On the orientation of parabolic or U-dunes. *Geographical Journal*, **123**: 474–480.

Jennings, J.N. (1959) The coastal geomorphology of King Island, Bass Strait, *Records, Queen Victoria Museum, Launceston*, **11**: 1–39.

Jennings, J.N. (1963) Some geomorphological problems of the Nullarbor Plain. *Transactions, Royal Society of South Australia*, **97**: 41–62.

Jennings, J.N. (1964) The question of coastal dunes in humid tropical climates. *Zeitschrift für Geomorphologie*, **8**: 150–154.

Jennings, J.N. (1965) Further discussion on factors affecting coastal dune formation in the tropics. *Australian Journal of Science*, **28**: 166–167.

Jennings, J.N. (1967) Cliff-top dunes. *Australian Geographical Studies*, **5**: 40–49.

Jennings, J.N. and Bird, E.C.F. (1967) Regional geomorphological characteristics of some Australian estuaries. In: G.H. Lauff (ed.) *Estuaries,* American Association for the Advancement of Science, pp. 121–128.

Johnson, D.W. (1919) *Shore processes and shoreline development.* John Wiley & Sons, New York.

Johnson, J.W. (1956) Dynamics of nearshore sediment movement. *Bulletin, American Association of Petroleum Geologists*, **40**: 2211–2232.

Jolliffe, I.P. (1961) The use of tracers to study beach movements and the measurement of longshore drift by a fluorescent technique. *Revue de Géomorphologie Dynamique*, **12**: 81–95.

Jolliffe, I.P. (1964) An experiment designed to compare the relative rates of movement of beach pebbles. *Proceedings of the Geologists' Association*, **75**: 67–86.

Jones, B. and Hunter, I.G. (1992) Very large boulders on the coast of Grand Cayman: the effects of giant waves on rocky coastlines. *Journal of Coastal Research*, **8**: 763–774.

Jones, J.R., Cameron, B., and Fisher, J.J. (1993) Analysis of cliff retreat and shoreline erosion, Thompson Island, Massachusetts, USA. *Journal of Coastal Research*, **9**: 87–96.

Jones, O.A. and Endean, R. (eds) (1977) *Biology and Geology of Coral Reefs.* Academic Press, New York.

Kaplin, P.A. and Selivanov, A.O. (1995) Recent coastal evolution of the Caspian Sea as a model for coastal responses to the possible acceleration of global sea level rise. *Marine Geology*, **124**: 161–175.

Kaye, C.A. (1959) Shoreline features and Quaternary shoreline changes. Puerto Rico. *US Geological Survey Professional Paper 317–B*: 49–140.

Kaye, C.A. (1973) *Map showing changes on shoreline of Martha's Vineyard, Massachusetts, during the past 200 years.* US Geological Survey Miscellaneous Field Studies, Map MF-534.

Kearney, M.S. and Stevenson, J.C. (1991) Island land loss and marsh vertical accretion rate evidence for historical changes in Chesapeake Bay. *Journal of Coastal Research*, **7**: 403–416.

Kelletat, D. and Zimmerman, L. (1991) *Verbreitung und Formtypen rezenter und subrezenter organischer Gesteinsbildungen an der Küsten Kretas.* Essener Geographische Arbeiten, 23.

Kench, P.S. (1998) Physical controls on development of lagoon sand deposits and lagoon infilling in an Indian Ocean atoll. *Journal of Coastal Research*, **14**: 1014–1024.

Kidson, C. (1963) The growth of sand and shingle spits across estuaries. *Zeitschrift für Geomorphologie*, **7**: 1–22.

Kidson, C. (1971) The Quaternary history of the coasts of south-west England, with special reference to the Bristol Channel coast. In: K.J. Gregory and W. Ravenhill (eds) *Exeter Essays in Geography*, University of Exeter, pp. 1–22.

Kidson, C. (1986) Sea level changes in the Holocene. In: O. Van de Plassche (ed.) *Sea Level Research: A Manual for the Collection and Evaluation of Data.* Geo Books, Norwich, pp. 27–64.

King, C.A.M. (1972) *Beaches and Coasts.* 2nd edition. Arnold, London.

King, C.A.M. and McCullach, M.J. (1971) A simulation model of a complex recurved spit. *Journal of Geology*, **79**: 22–36.

King, L.C. (1962) *The Morphology of the Earth.* Oliver & Boyd, Edinburgh.

Kirk, R.M. (1977) Rates and forms of erosion on intertidal platforms at Kaikoura Peninsula, South Island, New Zealand. *New Zealand Journal of Geology and Geophysics*, **20**: 571–613.

Kirk, R.M. (1980) Mixed sand and gravel beaches: morphology, process and sediments. *Progress in Physical Geography*, **4**: 189–210.

Kjerfve, B. (ed.) (1993) *Coastal Lagoon Processes.* Elsevier, Amsterdam.

Klein, G. de V. (1985) Intertidal flats and intertidal sand bodies. In: R.A. Davis (ed.) *Coastal sedimentary environments*, Springer-Verlag, New York, pp. 187–224.

Koike, K. (1985) Japan. In: E.C.F. Bird and M.L. Schwartz (eds) *The World's Coastline*, Van Nostrand Reinhold, New York, pp. 843–855.

Komar, P.D. (1976) *Beach processes and sedimentation.* Prentice–Hall, New Jersey.

Komar, P.D. (1983) *Handbook of coastal processes and erosion.* CRC Press, Boca Raton, Florida.

Komar, P.D. (1986) The 1982–83 El Niño and erosion on the coast of Oregon. *Shore and Beach*, **64**: 3–12.

Kraft, J.C., Aschenbrenner, S.E. and Rapp, G. (1988) Paleogeographic reconstruction of coastal Aegean archaeological sites. *Science*, **195**: 941–947.

Kraus, N.C. and Pilkey, O.H. (eds) (1988) The Effects of Seawalls on the Beach. *Journal of Coastal Research*, Special Issue **4**.

Krumbein, W.C. (1963) *The Analysis of Observational Data from Natural Beaches.* Beach Erosion Board, Technical Memorandum **130**.

Kulm, L.M. and Byrne, J.V. (1967) Sediments of Yaquima Bay, Oregon. In: G.H. Lauff (ed.) *Estuaries.* American Association for the Advancement of Science, Scientific Publication **83**: 226–238.

Lakhan, V.C. and Pepper, D.A. (1997) Relationship between concavity and convexity of a coast and erosion and accretion patterns. *Journal of Coastal Research*, **13**: 226–232.

Landon, R.E. (1930) An analysis of pebble abrasion and transportation. *Journal of Geology*, **38**: 437–446.

Landsberg, S.Y. (1956) The orientation of dunes in Britain and Denmark in relation to the wind. *Geographical Journal*, **122**: 176–189.

Lasserre, P. (1979) Coastal lagoons. *Nature and Resources*, **15**: 2–21.

Lauff, G.H. (ed.) (1967) *Estuaries*. American Association for the Advancement of Science, Scientific Publication **83**.

Le Bourdiec, P. (1958) Aspects de la morphogenèse plio-quaternaire en basse Côte d'Ivoire. *Revue de Géomorphologie Dynamique*, **9**: 33–42.

Leatherman, S.P. (1990) Modelling shore response to sea-level rise on sedimentary coasts. *Progress in Physical Geography*, **1**: 447–464.

Leontiev, O.K. and Veliev, K.A. (1990) *Transgressive Coasts*. Proceedings, International Symposium, Baku, USSR, September 1990: 1–71.

Lewis, J.R. (1964) *The Ecology of Rocky Shores*. English Universities Press, London.

Lewis, W.V. (1931) Effect of wave incidence on the configuration of a shingle beach. *Geographical Journal*, **78**: 131–148.

Lewis, W.V. (1932) The formation of Dungeness foreland. *Geographical Journal*, **80**: 309–324.

Lewis, W.V. (1938) The evolution of shoreline curves. *Proceedings of the Geologists' Association*, **49**: 107–127.

Lewis, W.V. and Balchin, W.G.V. (1940) Past sea levels at Dungeness. *Geographical Journal*, **96**: 258–285.

Liu Cangzi and Walker, H.J. (1989) Sedimentary characteristics of cheniers and the formation of the chenier plain of East China. *Journal of Coastal Research*, **5**: 353–368.

Lotfy, M.F. and Frihy, O.E. (1993) Sediment balance in the nearshore zone of the Nile delta coast, Egypt. *Journal of Coastal Research*, **9**: 654–662.

Ly, C.K. (1980) The role of the Akosombo Dam on the Volta River in causing coastal erosion in central and eastern Ghana. *Marine Geology*, **37**: 323–332.

Mahinda Silva, A.T. (1986) Ecological and socio-economic aspects of environmental changes in two mangrove-fringed lagoon systems in southern Sri Lanka. In: P. Kunstadter, E.C.F. Bird and S. Sabhasri (eds) *Man in the Mangroves*, United Nations University, Tokyo: pp. 79–86.

Mandelbrot, B. (1967) How long is the coast of Britain? *Science*, **155**: 636–638.

Marker, M.E. (1967) The Dee estuary: its progressive silting and salt marsh development. *Transactions, Institute of British Geographers*, **41**: 65–71.

Martin, L. and Dominguez, J.M.L. (1993) Geological history of coastal lagoons. In: Kjerfve B. (ed.) *Coastal Lagoon Processes*, Elsevier, Amsterdam, pp. 41–68.

Masselink, G. and Pattiaratchi, C. (1998) Morphodynamic impact of sea breeze activity on a beach with beach cusp morphology. *Journal of Coastal Research*, **14**: 393–406.

Masselink, G. and Short, A.D. (1993) The effect of tide range on beach morphodynamics and morphology: a conceptual beach model. *Journal of Coastal Research*, **9**: 785–800.

Mathers, S. and Zalasiewicz, J. (1999) Holocene sedimentary architecture of the Red River delta, Vietnam. *Journal of Coastal Research*, **15**: 314–325.

May, V.J. (1972) Earth cliffs. In: R.S.K. Barnes (ed.) *The Coastline*, John Wiley & Sons, New York, pp. 215–235.

May, V.J. and Heeps, C. (1985) The nature and rates of change on chalk coastlines. *Zeitschrift für Geomorphologie*, Supplementband **57**: 81–94.

McCann, S.B. (1964) The raised beaches of north-east Islay and western Jura, Argyll. *Transactions, Institute of British Geographers*, **35**: 1–10.

McCave, I.N. (1978) Grain size trends and transport along beaches: example from East Anglia. *Marine Geology*, **28**: M43–M57.

McCave, I.N. and Geiser, A.C. (1979) Megaripples, ridges and runnels on intertidal flats of The Wash, England. *Sedimentology*, **26**: 353–369.

McGill, J.T. (1958) Map of coastal landforms of the world. *Geographical Review*, **48**: 402–405.

McKenzie, P. (1958) The development of beach sand ridges. *Australian Journal of Science*, **20**: 213–214.

McLean, R.F. (1978) Recent coastal progradation in New Zealand. In: J.L. Davies and M.A.J. Williams (eds) *Landscape Evolution in Australasia*, Australian National University, Canberra, pp. 168–196.

McLean, R.F. and Woodroffe, C.D. (1994) Coral atolls. In: R.W.G. Carter and C.D. Woodroffe (eds) *Coastal Evolution. Late Quaternary shoreline morphodynamics*, Cambridge University Press, Cambridge, pp. 267–302.

Michel, D. and Howa, H.L. (1999) Short-term morphodynamic response of a ridge and runnel system on a mesotidal sandy beach. *Journal of Coastal Research*, **15**: 428–437.

Milliman, J.D. and Haq, B.U. (1996) *Sea-level Rise and Coastal Subsidence: causes, consequences and strategies*. Kluwer, Dordrecht.

Møller, J.T. (1963) Accumulation and abrasion in a tidal area. *Geografisk Tidsskrift*, **62**: 56–79.

Møller, J.T. (1985) Denmark. In: E.C.F. Bird and M.L. Schwartz (eds) *The World's Coastline*, Van Nostrand Reinhold, New York, pp. 325–383.

Molnia, B.F. (1985) Processes on a glacier-dominated coast, Alaska. *Zeitschrift für Geomorphologie*, Supplementband **57**: 141–153.

Morang A., Larson, R. and Gorman, L. (1997) Monitoring the coastal environment, part I: waves and currents. *Journal of Coastal Research*, **13**: 111–133.

Morgan, J.P. (1967) Ephemeral estuaries of the deltaic environment. In: G.H. Lauff (ed.) *Estuaries*. American Association for the Advancement of Science, Scientific Publication **83**: 115–120.

Morgan, J.P. (1970) Depositional processes and products in the deltaic environment. In: J.P. Morgan (ed.) *Deltaic Sedimentation Modern and Ancient*. Society of Economic Palaeontologists and Mineralogists, Special Publication **15**, pp. 31–47.

Mörner, N.A. (1976) Eustacy and geoid changes. *Journal of Geology*, **84**: 123–151.

Morton, J.K. (1957) Sand dune formation on a tropical shore. *Journal of Ecology*, **45**: 495–497.

Morton, R.A., Leach, M.P., Paine, J.G. and Cardoza, M.A. (1993) Monitoring beach changes using GPS techniques. *Journal of Coastal Research*, **9**: 702–720.

Mottershead, D.N. (1989) Rates and patterns of bedrock denudation by coastal salt spray weathering: a seven year record. *Earth Surface Processes and Landforms*, **14**: 383–398.

Mottershead, D.N. (1998) Coastal weathering of greenschist in dated structures, south Devon, *Quarterly Journal of Engineering Geology*, **31**: 343–346.

Neumann, A.C. and Macintyre, I. (1985) Reef response to a sea level rise: keep-up, catch-up or give-up. *Proceedings, 5th International Coral Reef Congress*, **3**: 105–110.

Nichols, M.M. and Biggs, R.B. (1985) Estuaries. In: R.A. Davis (ed.) *Coastal sedimentary environments*, Springer-Verlag, New York, pp. 77–186.

Nichols, R.L. (1961) Characteristics of beaches formed in polar climates. *American Journal of Science*, **259**: 694–708.

Nielsen, N. (1979) Ice-foot processes: observations of erosion on a rocky coast, Disko, West Greenland. *Zeitschrift für Geomorphologie*, **23**: 321–331.

Nordstrom, K.F. (1992) *Estuarine beaches*. Elsevier, London.

Nordstrom, K.F. (1994) Developed coasts. In: R.W.G. Carter and C.D. Woodroffe (eds) *Coastal Evolution. Late Quaternary shoreline morphodynamics*, Cambridge University Press, Cambridge, pp. 477–509.

Nordstrom, K.F. and Roman, C.T. (eds) (1996) *Estuarine Shores*. John Wiley & Sons, Chichester.

Nordstrom, K.F., Psuty, N and Carter, R.W.G. (eds) (1990) *Coastal Dunes: form and process*. John Wiley & Sons, Chichester.

Norrman, J.O. (1980) Coastal erosion and slope development in Surtsey Island, Iceland. *Zeitschrift für Geomorphologie*, Supplementband **34**: 20–38.

Nossin, J.J. (1965) The geomorphic history of the Padang delta. *Journal of Tropical Geography*, **20**: 54–64.

Oaks, R.A. and Coch, N.K. (1963) Pleistocene levels, southeastern Virginia. Science, **140**: 979–984.

O'Brien, M.P. and Dean, R.G. (1972) Hydraulic and sedimentary stability of tidal inlets. *Proceedings 13th Coastal Engineering Conference*, American Society of Civil Engineers, New York: 761–780.

Off, T. (1963) Rhythmic linear sand bodies caused by tidal currents. *Bulletin, American Association of Petroleum Geologists*, **47**: 324–341.

Olson, J.S. (1958) Lake Michigan dune development. *Journal of Geology*, **66**: 254–263, 345–351 and 473–483.

Orford, J.D., Carter, R.W.G. and Jennings, S.C. (1996) Control domains and morphological phases in gravel-dominated coastal barriers of Nova Scotia. *Journal of Coastal Research*, **12**: 589–604.

Orme, A.R. (1985) California. In: E.C.F. Bird and M.L. Schwartz (eds) *The World's Coastline*, Van Nostrand Reinhold, New York, pp. 27–36.

Orme, A.R. (1990) The instability of Holocene coastal dunes: the case of the Morro dunes, California. In: K.F. Nordstrom, N. Psuty and R.W.G. Carter (eds) *Coastal dunes: form and process*. John Wiley & Sons, Chichester, pp. 315–336.

Orviku, K., Bird, E.C.F. and Schwartz, M.L. (1995) The provenance of beaches on the Estonian islands of Hiiumaa, Saaremaa and Muhu, *Journal of Coastal Research*, **11**: 96–106.

Ottmann, F. (1965) *Introduction a la Géologie Marine et Littorale*. Masson, Paris.

Parker, R. (1978) *Men of Dunwich*. Collins, London.

Parry, M., Magalhaes, A.R. and Nguyen Huu Ninh (eds) (1991) *The Potential Socio-Economic Effects of Climatic Change*. United Nations Environment Programme. Nairobi.

Paskoff, R. (1985) Malta. In: Bird, E.C.F. and Schwartz, M.L. (eds) *The World's Coastline*, Van Nostrand Reinhold, New York, pp. 431–437.

Paskoff, R. (1994) *Les Littoraux. Impact des aménagements sur leur évolution*. 2nd edition, Masson, Paris.

Paskoff, R. and Petiot, R. (1990) Coastal progradation as a by-product of human activity: an example from Chañaral Bay, Atacama Desert, Chile. *Journal of Coastal Research*, Special Issue **6**: 91–102.

Peres, J.M. (1968) Trottoir. In: R.W. Fairbridge (ed.) *Encyclopedia of Geomorphology*, Reinhold, New York, pp. 1173–1174.

Perillo, G.M.E. (ed) (1995) *Geomorphology and sedimentology of estuaries*. Elsevier, Amsterdam.

Pernetta, J.C. and Osborne, P.L. (1988) Deltaic floodplains: the Fly River and the mangroves of the Gulf of Papua. In: *Potential Impacts of Greenhouse Gas Generation on Climatic Change and Projected Sea Level Rise on Pacific Islands*. United Nations Environment Programme, Split, pp. 94–111.

Pethick, J.S. (1980) Salt marsh initiation during the Holocene transgression: the example of the North Norfolk marshes, England. *Journal of Biogeography*, **7**: 1–9.

Pethick, J. (1981) Long-term accretion rates on tidal salt marshes. *Journal of Sedimentary Petrology*, **51**: 571–577.

Pethick, J. (1984) *An Introduction to Coastal Geomorphology*. Edward Arnold, London.

Pethick, J.S. (1992) Salt marsh geomorphology In: J.R.L. Allen and K. Pye (eds) *Salt marshes*. Cambridge University Press, Cambridge, pp. 41–62.

Pethick, J.S. (1994) Estuaries and wetlands: function and form. In: R.A. Falconer and P. Goodwin (eds) *Wetland Management*. Institute of Civil Engineers, Telford: 75–87.

Pethick, J.S. (1996) The geomorphology of mudflats. In: K.F. Nordstrom and C.T. Roman (eds), *Estuarine Shores*. John Wiley & Sons, Chichester, pp. 185–211.

Pethick, J.S. (1998) Coastal management and sea level rise. In: S.Lane, K. Richards and J. Chandler (eds) *Landform monitoring, modelling and analysis*, John Wiley & Sons, Chichester, pp. 403–419.

Philip, A.L. (1990) Ice-pushed boulders on the shores of Gotland, Sweden. *Journal of Coastal Research*, **6**: 661–676.

Phillips, J.D. (1986) Coastal submergence and marsh fringe erosion. *Journal of Coastal Research*, **2**: 427–436.

Phleger, F.B. (1969) Some general features of coastal lagoons. In: A Castañares and F.B. Phleger (eds)

Lagunas Costeras, un Simposio, National University of Mexico, pp. 5–26.

Phleger, F.B. (1981) A review of some general features of coastal lagoons. *Coastal Lagoon Research, Past, Present and Future*, UNESCO Technical Papers in Marine Science, **33**: 7–14.

Piccazzo, M., Firpo, M., Corradi, N. and Campi, F. (1992) Coastline changes on beaches affected by defence works: a case study in Albissola (Western Liguria, Italy). *Bolletino di Oceanologica Teorica ed Applicata*, **10**: 197–210.

Picknett, R.G., Bray, L.G. and Stebber, R.D. (1976) The chemistry of cave waters. In: T.D. Ford and C.H.D. Cullingford (eds) *The Science of Speleology*. Academic Press, New York, pp. 213–266.

Pierce, J.W. (1969) Sediment budget along a barrier island chain. *Sedimentary Geology*, **3**: 5–16.

Pilkey, O.H. and Clayton, T.D. (1989) Summary of beach replenishment experience on US East Coast barrier islands. *Journal of Coastal Research*, **5**: 147–159.

Pilkey, O.H., Young, R.S., Riggs, S.R., Sam Smith, A.W., Wu, H. and Pilkey, W.D. (1993) The concept of shoreface profile of equilibrium: a critical review. *Journal of Coastal Research*, **9**: 255–278.

Pirazzoli, P.A. (1983) Flooding ('aqua alta') in Venice (Italy): a worsening phenomenon. In: E.C.F. Bird and P. Fabbri (eds) *Coastal Problems of the Mediterranean Sea*, University of Bologna, pp. 23–31.

Pirazzoli, P.A. (1986) Secular trends of relative sea-level changes indicated by tide-gauge records. *Journal of Coastal Research*, Special Issue **1**: 1–26.

Pirazzoli, P.A. (1989) Recent sea-level changes in the North Atlantic. In: D.B. Scott, P.A. Pirazzoli and C.A. Honig (eds) *Late Quaternary Sea-level Correlations and Applications*, Kluwer, Dordrecht, pp. 153–167.

Pirazzoli, P.A. (1991) *World Atlas of Holocene Sea-level Changes*. Elsevier Oceanography Series, **58**.

Pirazzoli, P.A. (1996) *Sea-level Changes: the last 20 000 years*. John Wiley & Sons, Chichester.

Prager, E.J. and Halley, R.B. (1999) The influence of seagrass on shell layers and Florida Bay mudbanks. *Journal of Coastal Research*, **15**: 1151–1162.

Prêcheur, C. (1960) *Le Littoral de la Manche: de Ste Andresse à Ault. Etude morphologique*. Norois special volume.

Price, W.A. (1947) Equilibrium of form and forces in tidal basins of the coast of Texas and Louisiana, *Bulletin of the American Association of Petroleum Geologists*, **31**: 1619–1663.

Pringle, A. (1981) Beach development and coastal erosion in Holderness, North Humberside. In: J. Neale and J. Flenley (eds) *The Quaternary in Britain*, pp. 194–205.

Pritchard, D.W. (1967) Observations of circulation in coastal plain estuaries. In: G.H. Lauff (ed.) *Estuaries*. American Association for the Advancement of Science, Scientific Publication **83**: 37–44.

Psuty, N.P. (1965) Beach–ridge development in Tabasco, Mexico. *Annals of the Association of American Geographers*, **55**: 112–124.

Psuty, N.P. (1985) Surinam. In: E.C.F. Bird and M.L. Schwartz (eds) *The World's Coastline*, Van Nostrand Reinhold, New York, pp. 99–101.

Psuty, N.P. (1986) Impacts of impending sea level rise scenarios: the New Jersey barrier island responses. *Bulletin of the New Jersey Academy of Science*, **31**: 29–36.

Psuty, N.P. (ed.) (1988) Dune–beach interaction. *Journal of Coastal Research*, Special Issue **3**.

Psuty, N.P. (1990) Foredune mobility and stability, Fire Island, New York. In: K.F. Nordstrom, N. Psuty and R.W.G. Carter (eds) *Coastal dunes: form and process*. John Wiley & Sons, Chichester, pp. 159–176.

Psuty, N.P. and Moreira, M.E. (1990) Nourishment of a cliffed coastline, Praia da Rocha, the Algarve, Portugal. *Journal of Coastal Research*, Special Issue **6**: 21–32.

Pugh, D.T. (1987a) *Tides, Surges and Sea Level. A Handbook for Engineers and Scientists*. John Wiley & Sons, Chichester.

Pugh, D.T. (1987b) The Global Sea Level Observing System. *Hydrographic Journal*, **45**: 5–8.

Pugh, D.T. (1991) Cities on water: sea level measurements and planning. In: R. Frassetto (ed.) *Impacts of Sea Level Rise on Cities and Regions*, Marsilio, Venice, pp. 217–219.

Pye, K. (1980) Beach salcrete and aeolian transport: evidence from North Queensland, *Journal of Sedimentary Petrology*, **50**: 257–261.

Pye, K. (1983) Formation and history of Queensland coastal dunes. *Zeitschrift für Geomorphologie*, Supplementband **45**: 175–204.

Pye, K. (1996) Evolution of the shoreline of the Dee estuary, United Kingdom. In: K.F. Nordstrom and C.T. Roman (eds) (1996) *Estuarine shores*. John Wiley & Sons, Chichester, pp. 15–37.

Ranwell, D.S. (1972) *Ecology of Salt Marshes and Sand Dunes*. London.

Redfield, A.C. (1972) Development of New England salt marshes. *Ecological Monographs*, **42**: 201–237.

Reed, D.J. (1990) The impact of sea level rise on coastal salt marshes. *Progress in Physical Geography*, **14**: 456–481.

Reed, D.J. (1995) The response of coastal marshes to sea level rise: survival or submergence. *Earth Surface Processes and Landforms*, **20**: 39–48.

Revelle, R. and Emery, K.O. (1957) Chemical erosion of beach rock and exposed reef rock. *US Geological Survey Professional Paper* **260-T**: 699–709.

Rey, V., Davies, A.G. and Belzons, M (1995) On the formation of bars by the action of waves on an erodible bed: a laboratory study. *Journal of Coastal Research*, **11**: 1195–1203.

Rink, W.J. (1999) Quartz luminescence as a light-sensitive indicator of sediment transport in coastal processes. *Journal of Coastal Research*, **15**: 148–154.

Ristaniemi, O., Eronen, M., Glückert, G. and Rantala, P. (1997) Holocene and recent shoreline changes on the rapidly uplifting coast of Western Finland. *Journal of Coastal Research*, **13**: 397–406.

Ritchie, W. (1977) Machair development and chronology in the Uists and adjacent islands. *Proceedings of the Royal Society of Edinburgh*, **77B**: 107–122.

Ritchie, W. and Mather, A.S. (1984) *The Beaches of Scotland*. Scottish Countryside Commission, Perth.

Ritchie, W. and Penland, S. (1988) Rapid dune changes associated with overwash processes on the deltaic coast of south Louisiana. *Marine Geology*, **81**: 97–112.

Robertson-Rintoul, M.J. (1990) A quantitative analysis of the near-surface wind flow pattern over coastal parabolic dunes. In: K.F. Nordstrom, N. Psuty and R.W.G. Carter (eds) *Coastal dunes: form and process*, John Wiley & Sons, Chichester, pp. 57–78.

Robinson, A.H.W. (1955) The harbour entrances of Poole, Christchurch, and Pagham, *Geographical Journal*, **121**: 33–50.

Robinson, A.H.W. (1960) Ebb-flood channel systems in sandy bays and estuaries. *Geography*, **45**: 183–199.

Robinson, A.H.W. (1961) The hydrography of Start Bay and its relationship to beach changes at Hallsands. *Geographical Journal*, **121**: 63–77.

Robinson, A.H.W. (1966) Residual currents in relation to shoreline evolution of the East Anglian coast. *Marine Geology*, **4**: 57–84.

Robinson, L.A. (1976) The micro-erosion meter technique in a littoral environment. *Marine Geology*, **22**: M51–M58.

Robinson, L.A. (1977) Marine erosive processes at the cliff foot. *Marine Geology*, **23**: 257–271.

Robinson, D.A. and Jerwood, L.C. (1987) Sub-aerial weathering of chalk shore platforms during harsh winters in southeast England. *Marine Geology*, **77**: 1–14.

Rooney, J.J. and Smith, S.V. (1999) Watershed land use and bay sedimentation. *Journal of Coastal Research*, **15**: 478–485.

Roy, P.S. (1984) New South Wales estuaries: their origin and evolution. In: B.G. Thom (ed.) *Coastal Geomorphology in Australia*, Academic Press, Sydney, pp. 99–121.

Roy, P.S., Cowell, P.J., Ferland, M.A. and Thom, B.G. (1994) Wave-dominated coasts. In: R.W.G. Carter and C.D. Woodroffe (eds) *Coastal Evolution. Late Quaternary shoreline morphodynamics*, Cambridge University Press, Cambridge, pp. 121–186.

Russell, R.J. (1967). *River plains and sea coasts*. University of California Press, Berkeley.

Russell, R.J. and Russell, R.D. (1966) *Australian tidal flats*. Louisiana State University, Coastal Studies **13**.

Rutin, J. (1992) Geomorphic activity of rabbits on a coastal sand dune, De Blink dunes, the Netherlands. *Earth Surface Processes and Landforms*, **17**: 85–94.

Ruz, M.H. (1987) Recent evolution of the southeast barrier coast of Ireland. *Journal of Coastal Research*, **5**: 523–540.

Ruz, M.H., Allard, M., Michaud, Y. and Héquette. A. (1998) Sedimentology and evolution of subarctic tidal flats along a rapidly emerging coast, Eastern Hudson Bay, Canada. *Journal of Coastal Research*, **14**: 1242–1254.

Sacchi, C.F. (1979) The coastal lagoons of Italy. In: R.L. Jeffries and A.J. Davy (eds), *Ecological processes in coastal environments*, pp. 593–601.

Saenger, P., Hegerl, E.J. and Davie, J.D.S. (eds) (1983) Global status of mangrove ecosystems. *The Environmentalist*, **3**: 1–88.

Safriel, U.N. (1975) The role of vermetid gastropods in the formation of Mediterranean and Atlantic reefs. *Oceanologia*, **20**: 85–101.

Sanderson, P.G. and Eliot, I. (1996) Shoreline salients, cuspate forelands and tombolos on the coast of Western Australia. *Journal of Coastal Research*, **12**: 761–773.

Sanderson, P.G., Eliot, I. and Fuller, M. (1998) Historical development of a foredune plain at Desperate Bay, Western Australia. *Journal of Coastal Research*, **14**: 1187–1201.

Sanlaville, P. (1985) Arabian Gulf coasts. In: E.C.F. Bird and M.L. Schwartz (eds) *The World's Coastline*, Van Nostrand Reinhold, New York, pp. 729–733.

Savigear, R.A.G. (1952) Some observations on slope development in South Wales. *Transactions, Institute of British Geographers*, **18**: 31–52.

Savigear, R.A.G. (1962) Some observations on slope development in North Devon and North Cornwall. *Transactions, Institute of British Geographers*, **31**: 23–42.

Sayles, R.W. (1931) Bermuda during the ice age. *Proceedings, American Academy, Arts and Science*, **66**: 381–467.

Schofield, J.C. (1975) Sea level fluctuations cause periodic postglacial progradation, South Kaipara barrier. *New Zealand Journal of Geology and Geophysics*, **18**: 295–316.

Scholl, D.W. (1968) Mangrove swamps: geology and sedimentology. In: R.W. Fairbridge (ed.) *Encyclopedia of Geomorphology*, Reinhold, New York, pp. 683–688.

Schou, A. (1945) *Det Marine Forland*. Copenhagen.

Schou, A. (1952) Direction determining influence of the wind on shoreline simplification and coastal dunes. *Proceedings, 17th Conference, International Geographical Union*, Washington: 370–373.

Schwartz, M.L. (1965) Laboratory study of sea level rise as a cause of erosion. *Journal of Geology*, **73**: 568–574.

Schwartz, M.L. (1967) The Bruun theory of sea level rise as a cause of shore erosion. *Journal of Geology*, **75**: 76–92.

Schwartz, M.L. (1971) The multiple causation of barrier islands. *Journal of Geology*, **79**: 91–93.

Schwartz, M.L. (ed.) (1972) *Spits and bars*. Dowden, Hutchinson and Ross, Stroudsburg, Pennsylvania.

Schwartz, M.L. (ed.) (1973) *Barrier Islands*. Dowden, Hutchinson and Ross, Stroudsburg, Pennsylvania.

Schwartz, M.L. and Bird, E.C.F. (eds) (1990) Artificial Beaches. *Journal of Coastal Research*, Special Issue **6**.

Schwartz, M.L., Fabbri, P., and Scott Wallace, R. (1987) Geomorphology of Dungeness Spit, Washington, USA. *Journal of Coastal Research*, **3**: 451–455.

Schwartz, M.L., Granö, O. and Pyökäri, M. (1989) Spits and tombolos in the southwest archipelago of Finland. *Journal of Coastal Research*, **5**: 443–452.

Schwartz, M.L., Mahala, J. and Bronson, H.S. (1985) Net shore-drift along the Pacific coast of Washington State. *Shore and Beach*, **53**: 21–25.

Schwartz, M.L. and Terich, T.A. (1985) Washington. In: E.C.F. Bird and M.L. Schwartz (eds) *The World's Coastline*, Van Nostrand Reinhold, New York, pp. 17–22.

SCOR Working Group 89 (1991) The response of beaches to sea-level changes: a review of predictive models. *Journal of Coastal Research*, **7**: 895–921.

Sestini, G. (1992) Implications of climatic change for the Po delta and Venice lagoon. In: L. Jeftic, J.F. Milliman and G. Sestini (eds) *Climatic changes and the Mediterranean*, Arnold, London, pp. 428–494.

Shackleton, N.J. and Opdyke, N.D. (1973) Oxygen isotope and paleomagnetic stratigraphy of Equatorial Pacific core V-28-238. *Quaternary Research*, **3**: 39–55.

Shennan, I. and Sproxton, I. (1991) Possible impacts of sea level rise: a case study from the Tees estuary. In: J.C. Doornkamp (ed.) *The Greenhouse Effect and rising sea levels in the U.K.* M1 Press, Nottingham: pp. 109–133.

Shepard, F.P. (1973) *Submarine Geology* (3rd edition). Harper and Row, New York.

Shepard, F.P. (1976) Coastal classification and changing coastlines. *Geoscience and Man*, **14**: 53–64.

Shepard, F.P. and Wanless, H.R. (1971) *Our Changing Coastlines*. McGraw-Hill, New York.

Sherman, D.J. and Hotta, S. (1990) Aeolian sediment transport: theory and measurement. In: K.F. Nordstrom, N. Psuty and R.W.G. Carter (eds) *Coastal dunes: form and process*, John Wiley & Sons, Chichester, pp. 17–37.

Short, A.D. (1991) Macro-meso tidal beach morphodynamics – an overview. *Journal of Coastal Research*, **7**: 417–436.

Short, A.D. (1992) Beach systems of the central Netherlands coast: processes, morphology and structural impacts in a storm driven multi-bar system. *Marine Geology*, **107**: 103–137.

Short, A.D. (ed.) (1993) Beach and surf zone morphodynamics. *Journal of Coastal Research*, Special Issue **15**.

Short, A.D. and Hesp, P. (1982) Wave, beach and dune interactions in southeastern Australia. *Marine Geology*, **48**: 259–284.

Shuisky, Y.D. (1994) An experience of studying artificial ground terraces as a means of coastal protection. *Ocean and Coastal Management*, **22**: 127–139.

Silvester, R. (1974) *Coastal Engineering*, Elsevier, Amsterdam.

Smith, T.H. and Iredale, T. (1924) Evidence of a negative movement of the strand line of 400 feet in New South Wales. *Proceedings, Royal Society of New South Wales*, **58**: 157–68.

Snead, R.E. (1982) *Coastal Landforms and Surface Features*. Stroudsburg, Pennsylvania.

Snedaker, S.C. (1982) Mangrove species zonation: why? In: D.N. Sen and K.S. Rajpurohit (eds) *Tasks for Vegetation Science*, Junk, The Hague, pp. 111–125.

So, C.L. (1965) Coastal platforms of the Isle of Thanet. *Transactions, Institute of British Geographers*, **37**: 147–152.

So, C.L. (1982) Wind-induced movements of beach sand at Portsea, Victoria, *Proceedings, Royal Society of Victoria*, **94**: 61–68.

Spencer, T (1988) Coastal biogeomorphology In: H.A. Viles (ed.) *Biogeomorphology*, Blackwell, Oxford, pp. 255–318.

Spencer, T (1995) Potentialities, uncertainties and complexities in the response of coral reefs to future sea-level rise. *Earth Surface Processes and Landforms*, **20**: 49–64.

Sprigg, R.C. (1959) Stranded sea beaches and associated sand accumulations of the upper south-east coast of South Australia. *Transactions, Royal Society of South Australia*, **82**: 183–193.

Stanley, D.J. and Warne, A.G. (1998) Nile delta in its destruction phase. *Journal of Coastal Research*, **14**: 794–823.

Stapor, F.W., Mathews, T.D. and Lindfors-Kearns, F.E. (1991) Barrier island progradation and Holocene sea-level history in southwest Florida. *Journal of Coastal Research*, **7**: 815–838.

Stearns, H.T. (1974) Submerged shorelines and shelves in the Hawaiian islands. *Bulletin, Geological Society of America*, **85**: 795–804.

Steers, J.A. (1937) The coral islands and associated features of the Great Barrier Reefs. *Geographical Journal*, **89**: 1–28 and 118–146.

Steers, J.A. (1953) *The Sea Coast*. Collins, London.

Steers, J.A. (ed.) (1960) *Scolt Head Island*. Heffer, Cambridge.

Steers, J.A. (1964) *The Coastline of England and Wales*, Cambridge University Press, Cambridge, (2nd edition).

Steers, J.A. (1973) *The Coastline of Scotland*. Cambridge University Press, Cambridge.

Steers, J.A., Stoddart, D.R., Bayliss-Smith, T.P., Spencer, T. and Durbridge, P.M. (1979) The storm surge of January 1978 on the east coast of England. *Geographical Journal*, **145**: 192–205.

Stephenson, W.J. (1997) Improving the traversing micro-erosion meter. *Journal of Coastal Research*, **13**: 236–241.

Stephenson, W.J. and Kirk, R.M. (1998) Rates and patterns of erosion on inter-tidal shore platforms, Kaikoura Peninsula, South Island, New Zealand. *Earth Surface Processes and Landforms*, **34**: 1071–1085.

Sterr, H., Schwarzer, K. and Kliewe, H. (1998) Holocene evolution and morphodynamics of the German Baltic Sea coast – recent research advances. In: D. Kelletat (ed.) *German Geographical Coastal Research: the last decade*. Institute for Scientific Co-operation, Tübingen, Germany, pp. 107–134.

Stevenson, J.C. and Kearney, M.S. (1996) Shoreline dynamics on the windward and leeward shores of a large temperate estuary. In: K.F. Nordstrom and C.T. Roman (eds) *Estuarine shores*. John Wiley & Sons, Chichester, pp. 233–259.

Stoddart, D.R. (1965) Re-survey of hurricane effects on the British Honduras reefs and cays. *Nature*, **207**: 589–592.

Stoddart, D.R. (1990) Coral reefs and islands and predicted sea-level rise, *Progress in Physical Geography*, **14**: 521–536.

Stoddart, D.R., McLean, R.F., Scoffin, T.P. and Gibbs, P.E. (1978) Forty-five years of change on low wooded islands, Great Barrier Reef. *Philosophical Transactions, Royal Society of London*, **B-284**: 63–80.

Stoddart, D.R. and Scoffin, T.P. (1979) *Micro-atolls: review of form, origin and terminology*. Atoll Research Bulletin, **224.**

Stumpf, R.P. (1983) The process of sedimentation on the surface of a salt marsh. *Estuarine, Coastal and Shelf Science*, **17**: 495–508.

Suanez, S. and Provansal, M. (1998) Large scale evolution of the littoral of the Rhône delta (southeast France). *Journal of Coastal Research*, **14**: 493–501.

Suess, E. (1906) *The Face of the Earth*, Clarendon Press, Oxford.

Sunamura, T. (1992) *Geomorphology of Rocky Coasts*. John Wiley & Sons, Chichester.

Swan, B. (1979) Sand dunes in the humid tropics: Sri Lanka. *Zeitschrift für Geomorphologie*, **23**: 152–171.

Syvitski, J.P.M. and Shaw, J. (1995) Sedimentology and geomorphology of fiords. In: G.M.E. Perillo (ed.) *Geomorphology and Sedimentology of Estuaries*, Elsevier, Amsterdam, pp. 113–178.

Tait, J.F. and Griggs, G.B. (1990) Beach response to the presence of a sea wall: a comparison of field observations. *Shore and Beach*, **58**: 11–28.

Tanner, W.F. (1982) Ripple marks In: Schwartz M.L. (ed.) *The Encylopaedia of beaches and coastal environments*, Hutchinson Ross, Stroudsburg, Pennsylvania, pp. 695–698.

Tanner, W.F. (1995) Origin of beach ridges and swales. *Marine Geology*, **129**: 141–161.

Tanner, W.F. and Stapor, F. (1972) Accelerating crisis in beach erosion. *International Geography*, **2**: 1020–1021.

Taylor, M. and Stone, G.W. (1996) Beach ridges: a review. *Journal of Coastal Research*, **12**: 612–621.

Teh Tiong Sa (1985) Pensinsular Malaysia. In: E.C.F. Bird and M.L. Schwartz (eds) *The World's Coastline*, Van Nostrand Reinhold, New York, pp. 789–795.

Teh Tiong Sa (1992) The permatang series of Peninsular Malaysia: an overview. In: H.D. Tjia and Sharifah Mastura Syed Abdullah (eds), *The Coastal Zone of Peninsular Malaysia*. Penerbit Universiti Kebangsaan Malaysia, Bangi, pp. 42–62.

Terich, T.A. and Komar, P (1974) Bayocean spit, Oregon: history of development and erosional destruction. *Shore and Beach*, **42**: 3–10.

Thom, B.G. (1965) Late Quaternary coastal morphology of the Port Stephens–Myall Lakes area, *Proceedings, Royal Society of New South Wales*, **98**: 23–36.

Thom, B.G. (1967) Mangrove ecology and deltaic geomorphology: Tabasco, Mexico. *Journal of Ecology*, **55**: 301–343.

Thom, B.G. (1984) Sand barriers of eastern Australia: Gippsland – a case study. In: B.G. Thom (ed.) *Coastal Geomorphology in Australia*. Academic Press, Sydney, pp. 233–261.

Thom, B.G. and Wright, L.D. (1983) The geomorphology of the Purari delta. In: T. Petr (ed.) *The Purari: tropical environment of a high rainfall river basin*, Junk, Amsterdam, pp. 47–65.

Tietze, W. (1962) Ein beitrag zum geomorphologischen problem der strandflate. *Petermanns Geographichen Mitteilungen*, **106**: 1–20.

Titus, J.G. (ed.) (1988) *Greenhouse Effect, Sea Level Rise and Coastal Wetlands*. US Environment Protection Agency, Washington, D.C.

Tooley, M.J. and Jelgersma, S. (1992) *Impacts of sea-level rise on European coastal lowlands*. Institute of British Geographers, Special Publication 27. Blackwell, Oxford.

Trenhaile, A.S. (1978) The shore platforms of Gaspé, Quebec. *Annals of the Association of American Geographers*, **68**: 95–114.

Trenhaile, A.S. (1987) *The Geomorphology of Rock Coasts*. Clarendon Press, Oxford.

Trenhaile, A.S. (1999) The width of shore platforms in Britain, Canada and Japan. *Journal of Coastal Research*, **15**: 355–364.

Trenhaile, A.S., Pérez Alberti, A., Martinez Cortizas, A, Costa Casais, M. and Blanco Chao, R. (1999) Rock coast inheritance: an example from Galicia, northwest Spain. *Earth Surface Processes and Landforms*, **24**: 605–621.

Tricart, J. (1956) Aspects morphologiques du delta du Sénégal. *Revue de Géomorphologie Dynamique*, **7**: 65–85.

Tricart, J. (1959) Problèmes géomorphologiques du littoral oriental du Brésil. *Cahiers Océanographiques*, **11**: 276–308.

Tricart, J. (1962) Observations de géomorphologie littorale à Mamba Point, Liberia. *Erdkunde*, **16**: 49–57.

Turner, I.L. and Leatherman, S.P. (1997) Beach dewatering as a soft engineering solution to coastal erosion – a history and critical review. *Journal of Coastal Research*, **13**: 1050–1063.

US Army Corps of Engineers (1984) *Shore Protection Manual* (2 volumes), 4th edition.

Usoro, E.J. (1985) Nigeria. In: E.C.F. Bird and M.L. Schwartz (eds) *The World's Coastline*, Van Nostrand Reinhold, New York, pp. 607–613.

Valentin, H. (1952) Die Küsten der Erde. *Petermanns Geographische Mitteilungen*, **246.**

Van der Muelen, F. (1990) European dunes: consequences of climatic changes and sea level rise. *Catena Supplement*, **18**: 51–60.

Van Straaten, L.M.J.U. (1953) Megaripples in the Dutch Wadden Sea and in the Basin of Arcachon. *Geologie en Mijnbouw*, **15**: 1–15.

Van Straaten, L.M.J.U. (1959) Littoral and submarine morphology of the Rhône delta. *Proceedings, Second Coastal Geography Conference*: 233–264.

Van Straaten, L.M.J.U. (1965) Coastal barrier deposits in south and north Holland. *Mededelingen van de Geologische Stichting*, **17**: 41–75.

Vanney, J.R., Menanteau, L. and Zazo, C. (1979) Physiographie et évolution des dunes de Basse-Andalousie. In:

A. Guicher (ed.) *Les Côtes Atlantiques de l'Europe*, University of Western Brittany, Brest, pp. 277–285.

Vaughan, T. (1910) The geologic work of mangroves in southern Florida. *Smithsonian Miscellaneous Collection*, **52**: 461–464.

Verger, F. (1968) *Marais et wadden du littoral Francais*. Biscaye Frères, Bordeaux.

Vermeer, D.E. (1985) Mauritania. In: E.C.F. Bird and M.L. Schwartz (eds) *The World's Coastline*, Van Nostrand Reinhold, New York, pp. 549–553.

Viles, H. and Spencer, T. (1995) *Coastal Problems*. Edward Arnold, London.

Wadsworth, A.H. (1966) Historical deltation of the Colorado River, Texas. In: M.L. Shirley (ed.) *Deltas in their Geologic Framework*, Houston Geological Society, Texas, pp. 99–105.

Walker, D.J. and Jessup, A. (1992) Analysis of the dynamic aspects of the River Murray mouth, South Australia. *Journal of Coastal Research*, **8**: 71–76.

Walker, H.J. (1988) *Artificial structures and shorelines*. Kluwer, Dordrecht.

Walker, H.J. (1998) Arctic deltas. *Journal of Coastal Research*, **14**: 718–739.

Wanless, H.R. (1982) Sea level is rising – so what? *Journal of Sedimentary Petrology*, **52**: 1051–1054.

Warrick R.A., Barrow E.M., and Wigley T.M.L. (1993) *Climate and sea level change*. Cambridge University Press, Cambridge.

Watson, J.D. (1928) Mangrove forests of the Malay Peninsula. *Malay Forest Records*, **6**: 1–275.

Watson, R.L. (1971) Origin of shell beaches, Padre Island, Texas. *Journal of Sedimentary Petrology*, **41**: 1105–1111.

Webb, J.E. (1958) The ecology of Lagos Lagoon. *Philosophical Transactions Royal Society*, London, **241**: 307–318.

Wells, J.T. (1995) Tide-dominated estuaries and tidal rivers. In: G.M.E. Perillo (ed.) *Geomorphology and Sedimentology of Estuaries*, Elsevier, Amsterdam, pp. 179–205.

Wells, J.T., Adams, C.E., Park, Y.A. and Frankenberg, E.W. (1990) Morphology, sedimentology and tidal channel processes on a high-tide-range mudflat, west coast of South Korea. *Marine Geology*, **95**: 111–130.

Wentworth, J.K. (1939) Marine bench-forming processes: solution benching. *Journal of Geomorphology*, **2**: 3–25.

Wigley, T.M.L. and Raper, S.C.B (1993) Thermal expansion of sea water associated with global warming. *Nature*, **330**: 127–131.

Wilcock, P.R., Miller, D.S. and Shea, R.H. (1998) Frequency of effective wave activity and the recession of coastal bluffs: Calvert Cliffs, Maryland. *Journal of Coastal Research*, **14**: 256–268.

Williams, A.T. and Davies, P. (1980) Man as a geological agent: the sea cliffs of Llantwit Major, Wales. *Zeitschrift für Geomorphologie*, Supplementband **34**: 129–141.

Williams, M.A.J., Dunkerley, D.L., De Deckker, P., Kershaw, A.P. and Stokes, T. (1991) *Quaternary environments*. Arnold, London.

Williams, W.W. (1956) An east coast survey: some recent changes in the coast of East Anglia. *Geographical Journal*, **122**: 317–334.

Wilson, G. (1952) The influence of rock structures on coastline and cliff development around Tintagel, North Cornwall. *Proceedings of the Geologists' Association*, **63**: 20–48.

Wiseman, W.J., Coleman, J.M., Gregory, A., Hsu, S.A., Short, A.D., Suhayda, J.N., Walters, C.D. and Wright, L.D. (1973) Alaskan Arctic coastal processes and morphology: wave climate and the role of subaqueous profile. *Science*, **176**: 282–289.

Wiseman, W.J., Chappell, J., Thom, B.G., Bradshaw, M.P. and Cowell, P. (1979) Morphodynamics of reflective and dissipative beach and inshore systems, south-eastern Australia. *Marine Geology*, **32**: 105–140.

Wolanski, E. and Chappell, J. (1996) The response of tropical Australian estuaries to a sea level rise. *Journal of Marine Systems*, **7**: 267–279.

Wolanski, E. and Gibbs, R.J. (1995) Flocculation of suspended sediment in the Fly River estuary, Papua New Guinea. *Journal of Coastal Research*, **11**: 754–762.

Woodroffe, C.D. (1993) Late Quaternary evolution of coastal and riverine plains of south-east Asia and northern Australia – an overview. *Sedimentary Geology*, **83**: 163–175.

Woodroffe, C.D. (1995) Response of tide-dominated mangrove shorelines in northern Australia to anticipated sea level rise. *Earth Surface Processes and Landforms*, **20**: 65–85.

Woodroffe, C.D. (1996) Late Quaternary infill of macrotidal estuaries in northern Australia. In: K.F. Nordstrom and C.T. Roman (eds) *Estuarine shores*. John Wiley & Sons, Chichester, pp. 89–114.

Woodworth, P.L. (1987) Trends in UK mean sea level. *Marine Geodesy*, **11**: 57–87.

Woodworth, P.L. (1991) The Permanent Service for mean sea level and the Global Sea Level Observing System. *Journal of Coastal Research*, **7**: 699–719.

Wright, L.D. (1985) River deltas. In: R.A. Davis (ed.) *Coastal sedimentary environments*, Springer-Verlag, New York, pp. 1–76.

Wright, L.D. and Coleman, J.M. (1973) Variations in morphology of major river deltas as functions of ocean wave and river discharge regimes. *Bulletin American Association of Petroleum Geologists*, **57**: 370–398.

Wright, L.D., Coleman, J.M. and Thom, B.G. (1973) Processes of channel development in a high-tide-range environment: Cambridge Gulf – Ord River delta. *Journal of Geology*, **81**: 15–41.

Wright, L.D. and Short, A.D. (1984) Morphodynamic variability of surf zones and beaches: a synthesis. *Marine Geology*, **56**: 93–118.

Wright, L.D., Thom, B.G. and Higgins, R. (1980) Sediment transport and deposition at wave-dominated river mouths. Examples from Australia and Papua

New Guinea. *Estuarine and Coastal Marine Science*, **11**: 263–277.

Wright, L.W. (1967) Some characteristics of the shore platforms of the English Channel and the northern part of North Island, New Zealand. *Zeitschrift für Geomorphologie*, **11**: 36–46.

Wright, L.W. (1970) Variation in the level of the cliff/shore platform junction along the south coast of Great Britain. *Marine Geology*, **9**: 347–353.

Yang, S. (1999) Tidal wetland sedimentation in the Yangtze delta. *Journal of Coastal Research*, **15**. 1091–1099.

Yap, H.T. (1989) Implications of expected climatic changes on natural coastal ecosystems. In: Chou Loke Ming (ed.) *Implications of Climatic Change in the East Asian Seas Region*, University of Singapore, pp. 128–148.

Yasso, W.E. (1965) Fluorescent tracer particle determination of the size–velocity relation for foreshore sediment transport, Sandy Hook, New Jersey. *Journal of Sedimentary Petrology*, **35**: 989–993.

Zenkovich, V.P. (1959) On the genesis of cuspate spits along lagoon shores. *Journal of Geology*, **76**: 169–177.

Zenkovich, V.P. (1967) *Processes of Coastal Development* (trans. O.G. Fry; ed. J.A. Steers), Oliver & Boyd, Edinburgh.

Zenkovich, V.P. (1973) Geomorphological problems of protecting the Caucasian Black Sea coast. *Geographical Journal*, **139**: 460–466.

Zhuang, W.Y. and Chappell, J. (1991) Effects of seagrass beds on tidal flat sedimentation. In: D.G. Smith et al. *Clastic tidal sedimentology*. Canadian Society of Petroleum Geologists Memoir **16**, pp. 291–300.

Index